Solid Waste Management

This is an authorized facsimile
of the original book, printed
by xerography on acid-free paper.

UNIVERSITY MICROFILMS INTERNATIONAL
Ann Arbor, Michigan, U.S.A.
London, England
1981

```
2014903
Books on Demand
TD          Hagerty, D. Joseph.
788            Solid waste management / D. Joseph
.H33       Hagerty, Joseph L. Pavoni, and John E.
1981       Heer, Jr. -- New York : Van Nostrand
           Reinhold, c1973.
              xiii, 302 p. : ill. -- (Van Nostrand
           Reinhold environmental engineering
           series)
              Includes bibliographical references
           and index.
              Photocopy. Ann Arbor, Mich. :
           University Microfilms International,
           1981.
              1. Refuse and refuse disposal--United
           States.  I. Pavoni, Joseph L.
           II. Heer, Joh    n E.  III. Title
           IV. Series

UnM       03 SEP 81      7723677     EEUBs2
```

Solid Waste Management

D. Joseph Hagerty
Associate Professor of Civil Engineering
University of Louisville

Joseph L. Pavoni
Associate Professor of Civil and Environmental Engineering
University of Louisville

and

John E. Heer, Jr.
Professor and Chairman, Civil Engineering
University of Louisville

VAN NOSTRAND REINHOLD ENVIRONMENTAL ENGINEERING SERIES

 VAN NOSTRAND REINHOLD COMPANY
New York Cincinnati Toronto London Melbourne

Van Nostrand Reinhold Company Regional Offices:
New York   Cincinnati   Chicago   Millbrae   Dallas

Van Nostrand Reinhold Company International Offices:
London   Toronto   Melbourne

Copyright © 1973 by Litton Educational Publishing, Inc.

Library of Congress Catalog Card Number: 73-10281
ISBN: 0-442-23026-5

All rights reserved. No part of this work covered by the copyright hereon may be reproduced or used in any form or by any means—graphic, electronic, or mechanical, including photocopying, recording, taping, or information storage and retrieval systems—without permission of the publisher.

Manufactured in the United States of America

Published by Van Nostrand Reinhold Company
450 West 33rd Street, New York, N.Y. 10001

Published simultaneously in Canada by Van Nostrand Reinhold Ltd.

15 14 13 12 11 10 9 8 7 6 5 4 3

Library of Congress Cataloging in Publication Data
Hagerty, D   J
  Solid waste management.

  (Van Nostrand Reinhold environmental engineering series)
  Includes bibliographical references.
  1. Refuse and refuse disposal—United States.
I. Pavoni, Joseph L., joint author. II. Heer, John E., joint author. III. Title.
TD788.H33   363.6   73-10281
ISBN 0-442-23026-5

To
Pat, Judy and Ag

Van Nostrand Reinhold Environmental Engineering Series

THE VAN NOSTRAND REINHOLD ENVIRONMENTAL ENGINEERING SERIES is dedicated to the presentation of current and vital information relative to the engineering aspects of controlling man's physical environment. Systems and subsystems available to exercise control of both the indoor and outdoor environment continue to become more sophisticated and to involve a number of engineering disciplines. The aim of the series is to provide books which, though often concerned with the life cycle—design, installation, and operation and maintenance—of a specific system or subsystem, are complementary when viewed in their relationship to the total environment.

Books in the Van Nostrand Reinhold Environmental Engineering Series include ones concerned with the engineering of mechanical systems designed (1) to control the environment within structures, including those in which manufacturing processes are carried out, (2) to control the exterior environment through control of waste products expelled by inhabitants of structures and from manufacturing processes. The series will include books on heating, air conditioning and ventilation, control of air and water pollution, control of the acoustic environment, sanitary engineering and waste disposal, illumination, and piping systems for transporting media of all kinds.

Van Nostrand Reinhold Environmental Engineering Series

ADVANCED WASTEWATER TREATMENT, by Russell L. Culp and Gordon L. Culp

ARCHITECTURAL INTERIOR SYSTEMS—Lighting, Air Conditioning, Acoustics, John E. Flynn and Arthur W. Segil

SOLID WASTE MANAGEMENT, by D. Joseph Hagerty, Joseph L. Pavoni and John E. Heer, Jr.

THERMAL INSULATION, by John F. Malloy

HANDBOOK OF SOLID WASTE DISPOSAL: Materials and Energy Recovery, by Joseph L. Pavoni, John E. Heer, Jr., and D. Joseph Hagerty

AIR POLLUTION AND INDUSTRY, edited by Richard D. Ross

INDUSTRIAL WASTE DISPOSAL, edited by Richard D. Ross

MICROBIAL CONTAMINATION CONTROL FACILITIES, by Robert S. Runkle and G. Briggs Phillips

SOUND, NOISE, AND VIBRATION CONTROL, by Lyle F. Yerges

Preface

This book is intended as a presentation of general management considerations in the field of solid waste collection and disposal. As such it has been written for the practitioner who is unfamiliar with the problems of solid waste management, as well as for the undergraduate engineering student pursuing study in environmentally oriented curricula. Emphasis is placed upon the practical aspects of problem solving with discussions of: sources and characteristics of present-day solid wastes; current collection methods, systems, and equipment; available disposal techniques and facilities; recycling and resource conservation; economics of solid waste collection and disposal; new developments in collection and disposal technology; prediction of future trends in solid waste generation and characteristics; and the role of public relations in solid waste management.

The treatment of material in this text is general but comprehensive and not cursory. Sufficient numbers of references are listed at the ends of the chapters for individuals interested in detailed study of particular subjects. Content and style are suited to technologically oriented readers; however, the non-scientist or non-engineer should find the material comprehensible and straightforward.

The authors wish to express their appreciation for the help and encouragement furnished by their wives and their students. Also, the patient services of Misses LeAnne Whitney and Angela Reed, who typed this manuscript several times, must be graciously acknowledged.

D. Joseph Hagerty
Joseph L. Pavoni
John E. Heer, Jr.

Contents

PREFACE — IX

1 INTRODUCTION — 1

Background to the problem, mention of legislative actions including Solid Waste Disposal Act and Resource Recovery Act

2 BASIC DATA ON SOLID WASTES IN THE UNITED STATES TODAY — 3

Survey and discussion of generation and characteristics of solid wastes in present circumstances; review of community collection methods and disposal systems; evaluation of adequacy of collection and disposal facilities; financing and economics of collection and disposal

3 VOLUME REDUCTION — 23

Inherent advantages of volume reduction prior to or during collection are presented. Discussion of volume reduction methods at the source, including residential composting, incineration (private residence), garbage grinders, and other methods. Evaluation of volume reduction at a central location

4 ACCESSORY OPERATIONS — 30

Description and discussion of size reduction and densification operations accessory to collection and disposal techniques. Presentation is made in the following order: sanitary landfill accessory operations; auxiliary operations to composting; and incineration accessory operations.

5 COLLECTION SYSTEMS 44

Discussion of management systems used in waste collection with evaluation of advantages and disadvantages of public agency, municipality-contractor and individual-contractor arrangements. Also description of factors in design of collection systems. Special collection problems associated with agricultural wastes also mentioned

6 REUSE AND RECYCLING 55

Introduction to general aspects of reuse of wasted materials, with general methods of waste alleviation. Attention focused on packaging wastes as an area of much potential improvement. Detailed discussion of recycling of paper, plastic, glass, metals (steel, aluminum, etc.). Includes description of associated problems of waste separation for recycling

7 SOLID WASTE DISPOSAL MICROBIOLOGY 100

Prelude to discussion of solid waste disposal by composting and landfill. Also health problems associated with organisms and degradation in wastes. Description of microorganisms in soil and their influence on waste disposal

8 DISPOSAL METHODS—COMPOSTING 115

Background and history of composting as a disposal method. Description of the unit operations associated with the composting method. Discussion of potential for composting as a disposal method. Presentation of public relations and marketing problems associated with composting

9 INCINERATION 132

General introduction to incineration as a disposal method. Technical and economic aspects of incinerator operation. Location and design considerations for physical plant. Components and unit operations for waste incinerators. Description of furnaces, grates, scales, and accessory apparatus. Description of operating problems, waste heat recovery, instrumentation and control devices, emission control

10 SANITARY LANDFILL 178

Landfilling as a disposal method. Description of unit operations associated with landfill. Evaluation of economic factors including financing for capital outlays, operating costs, and resale potential for landfill sites. Discussion of landfilling techniques, operational restrictions, and equipment used in fill construction. Site selection and land-use planning also presented

11 RECENT DEVELOPMENTS 223

General discussion of relatively recent technological advancements in solid waste collection and disposal. Included in this chapter are high-temperature incineration; pyrolysis of selected materials; disposal of hazardous wastes; and other topics

12 LEGAL ASPECTS OF SOLID WASTE MANAGEMENT 263

Presentation of general background of environmental legislation pertinent to solid wastes management. Discussion of Solid Wastes Act of 1965 and Resource Recovery Act of 1970. Presentation of case history illustrating legal and political difficulties in solid wastes disposal.

13 EFFECTIVE SOLID WASTE MANAGEMENT 274

Synthesis of preceding chapters into a summarized statement of the problem of waste management. Discussion of the necessity for public education and persuasion concerning all aspects of waste management, but particularly, managed solutions to collection and disposal problems

APPENDIX: RESOURCE RECOVERY ACT OF 1970   281

INDEX   293

1
Introduction

Environmental pollution has been receiving more and more attention in recent years. Water and air pollution have been examined thoroughly, significant control measurements have been proposed, and a comprehensive volume of legislation has been passed for pollution control and prevention. The pollution of the land surfaces of the United States, which has been called by some the "third pollution," consists essentially of disposal of that which is termed solid waste. This third pollution has been neglected until recent years and only since the mid-sixties, with the passage of the Solid Waste Disposal Act (1965), has concerted action been taken to control and prevent land pollution.

Although rather late in starting, the movement to control and prevent pollution of the land has accelerated rapidly within the last few years, and today the effort for management of national solid wastes has reached a rather significant stature. The general public is no longer apathetic to the issue of solid waste collection and disposal. Considerable amounts

2 SOLID WASTE MANAGEMENT

Fig. 1-1 The ever-increasing mountain of solid wastes—a management problem requiring an urgent solution (*courtesy* RayGo, Inc.).

of money are being spent in investigations of the problem and in the planning of solutions. This volume proposes to focus attention on some of the major issues in the problem of solid waste collection and disposal, and to direct interested persons to possible solutions to these same problems. The general intent herein is to provide an overview and as far as possible an in-depth presentation of the material. Sufficient numbers of references are included so that further intensive investigations of any phase of the problem may be made by interested individuals.

2

Basic data on solid wastes in the United States today

2-1. GENERAL

In order to put the problem of solid waste management in perspective, it is necessary to gather and present some basic data in regard to the amounts of solid wastes which are generated in the United States today and the characteristics of these generated wastes. These data should be considered necessary for a variety of reasons. First, any future system for collection and disposal of this refuse will have to be geared to the total amount and quantity of the materials produced and also the system selected will be dependent upon the characteristics of the generated refuse. In particular, the disposal method, whether it be incineration, composting, or sanitary landfill, may depend in large measure on the physical and chemical characteristics of the refuse generated. Second, the overall management of the developed system for collection and disposal of the refuse will rely quite heavily on preliminary data concerning the quantities and character of that refuse. For example, selection of collection and transfer equipment, selection of routes for such equipment, choice of auxilary equipment such as compactors, and the overall economics of the procedure used for collection and disposal can be economically determined only on the basis of adequate basic data.

While it appears that data on the categories and amounts of solid wastes are quite necessary and desirable from the point of view of optimal planning and design, a survey of the engineering literature reveals that very little comprehensive data are available at the present time. The majority of the published data have come from governmental sources such as the Office of Solid Waste Management Programs. The lack of comprehensive data can be attributed to several factors: 1) the measurement and categorization of solid wastes is inherently difficult since the material is heterogeneous and appears in seasonally variable quantities; 2) there has been no development of a standardized method of gathering and presenting this data; and 3) in many cases, the need for such data has not been recognized by persons or agencies which have the ability or opportunity to collect it, and the information, therefore, has remained lacking.

In connection with much of the published data, the information is of only partial benefit to a designer or planner, since the national averages presented are applicable only very broadly to the entire nation; that is, to adequately design a solid waste management system for a particular locality comprehensive information about the solid waste in *that* locality must be obtained. Since solid wastes reflect the life-style of the generating populace, the character and amounts of various components in the solid waste stream will vary from locale to locale in the United States. For example, there is little general agreement even in the definition of some rather common terms which are applied to solid wastes. "Refuse," which consists of all putrescible and non-putrescible solid wastes except for body wastes, is often confused with "garbage" and "rubbish." "Garbage" simply refers to the wasted or rejected food constituents which have been produced during the preparation, cooking, or storage of meat, fruit, vegetables, etc. "Rubbish," on the other hand, refers to the non-putrescible solid waste constituents and includes such items as paper, tin cans, glass, wood, etc. Certainly a first step in the gathering of data should be the standardization of just these same common terms.

In discussing solid waste, generally and traditionally certain categories of wastes are recognized. For example, solid wastes are classified as domestic, commercial, industrial, due to construction and demolition, agricultural, institutional, and miscellaneous. Many times domestic and commercial wastes are considered together as so-called urban wastes. Included in this category are the garbage materials which result from food preparation both in the home and in restaurants, and also the rubbish which is produced in residences and commercial establishments. Generally the garbage consists of rapidly decomposable materials while the rubbish is either slowly decomposable or non-degradable. The collected densities of this sort of material will vary from approximately 250 to about 650 lb/cu yd depending upon the relative proportions of rubbish and garbage in the collected sample. For example, if the material is proponderately garbage, the collected density and moisture content will be quite high. Densities will generally be in the range of 600 to 650 lb/cu yd. The non-putrescible materials in rubbish, on the other hand, are generally much bulkier and less compact and may have densities as low as 250 pounds per cubic yard. Taken together domestic and commercial refuse in the United States today amounts to an average of approximately 5.3 pounds per person per day.* Frequently included in the general category of metropolitan or municipal solid wastes are other materials which are collected and must be disposed of as a result of urban activities. For example, sewage treatment plant residues, street sweepings, deciduous leaf residue, etc. are types of municipal miscellaneous solid wastes. If these wastes are added, the total amount is about 7-8 pounds per person per day.

A second major category of wastes is industrial, the refuse produced by industrial processes. Generally the character of the refuse produced in any manufacturing or processing operation will depend very much on the mechanics of that particular manufacturing operation. No comprehensive statements can be made about the type, amounts, or the character of wastes produced by industry. Obviously, the wastes produced by steel manufacturers will differ considerably from those produced in the chemical industry. However, several very general statements can be made. Usually the wastes produced in any food-processing operation will closely resemble the garbage produced in residential areas. In addition, the waste materials from the paper and plastics industries are similar to the paper and plastic packaging materials found in domestic rubbish. On the other hand, the metal-processing industry will obviously generate metallic wastes, but in addition will also create large quantities of slags, processing chemicals, and other residues, many of which are produced in air-pollution-control and water-pollution-control activities. The wastes produced by chemical industries and other more specialized industries will in general depend upon the particular end product of the manufacturing process.

A very significant proportion of the solid wastes produced each year in this country are derived from agricultural operations. Included in this category are both crop residues and animal wastes, such as manure, urine, and bedding wastes from confined feeding operations for livestock. Whereas domestic and commercial wastes amount to between 5 and 6 pounds per person per day in the United States, agricultural wastes amount to between 50 and 60 lb/cap/day. Much of this waste was formerly recycled into the land by such practices as plowing stalks, leaves, and other crop residues back into the soil. Manures also

―――――――――――
*Elements of Solid Wastes Management, EPA, HEW, 1970.

6 SOLID WASTE MANAGEMENT

TABLE 2-1 Physical Characteristics of Municipal Refuse: Typical 100-lb Sample, Municipal Refuse
(USPHS Data)

Item	Wet Wt. (lb)	Dry Wt. (lb)
Paper	48.0	35.0
Garbage	16.0	8.0
Leaves and grass	9.0	5.0
Wood	2.0	1.5
Synthetics	2.0	2.0
Cloth	1.0	0.5
Noncombustibles		
Glass	6.0	6.0
Metals	8.0	8.0
Ashes, stone, dust, etc.	8.0	6.0

Data from Nine Separate Studies, 300 Samples—USPHS, APWA, Universities

Chemical Characteristics

Measure	Minimum	Maximum	Average	
% Moisture	20	60	38	
% Carbon	8	35	24	WET
% Nitrogen	0.2	3.0	1.0	BASIS
Btu/lb	3000	6000	4500	
Ash %	4	9	6.5	
Carbon %	20	50	40	DRY
Nitrogen %	0.3	5	1.0	BASIS
Btu/lb	6000	10000	7700	

were disposed of by recycling them into agricultural lands. Increased use of feeder lot operations and the increasing loss of agricultural land to urbanization has produced a significant problem of waste disposal in the agricultural industry today.

Other more specialized categories are construction and demolition wastes and institutional solid wastes. Construction and demolition wastes are quite heterogeneous in character and may be composed of discarded building materials which are wasted during the construction of new structures and facilities, or they may consist of the wood, steel, bricks, concrete, and other construction materials which are wasted during the demolition of existing structures and facilities. Generally this sort of waste material is in large part non-degradable and, except for the wood waste, will exhibit very little decomposition with time. Since it does exhibit little decomposition, it is quite often useful as solid fill. Institutional wastes include those materials produced in hospitals, schools, nursing homes, prisons, and other large facilities for great numbers of persons. In general these wastes are similar to domestic and commercial types but contain slightly larger amounts of paper and cloth. However, in such facilities as hospitals and nursing homes, contaminated wastes may pose a special disposal problem. Table 2-1 lists physical and chemical characteristics of typical samples of municipal refuse. Obviously refuse in other categories such as agricultural or industrial solid wastes will differ significantly as to their characteristics from those listed in the table.

2-2. RESULTS OF 1968 NATIONAL SURVEY

2-2.1 Background

There is little reliable data available on the subject of generated quantities and characteristics of solid wastes. However, *some* data have been obtained.

In July 1968, a survey was completed by the United States Public Health Service in which 30 states participated. These data represent 6,259 communities, or a population of 92.5 million persons (46 percent of the population of the United States).

The population sample consisted of about 75 percent urban residents with the remainder rural. This corresponds to the national distribution of 73 percent urban, 27 percent rural inhabitants, as of the 1966 population.

Information was obtained either by direct observation or by personal interview. Data on land disposal and disposal facilities is very reliable, except for some of the quantity information.

2-2.2 Collection and Disposal Systems

Some salient points from this survey may be presented. For instance, 41 percent of all communities operate their own collection systems (64 percent of urban communities and 24 percent of rural ones). On a population basis, 64 percent of the total population lives in communities which operate their own collection systems (77 percent urban, 22 percent rural; see Fig. 2-1).

For collection and disposal, these figures become (approximately) 64 percent and 70 percent (see Fig. 2-2). In other words about 70 percent of the national population resides in communities which actually operate a disposal site of some kind.

Who actually does the collecting? For so-called "household" wastes, about 56 percent is collected by public collectors, 32 percent by private collectors, and

8   SOLID WASTE MANAGEMENT

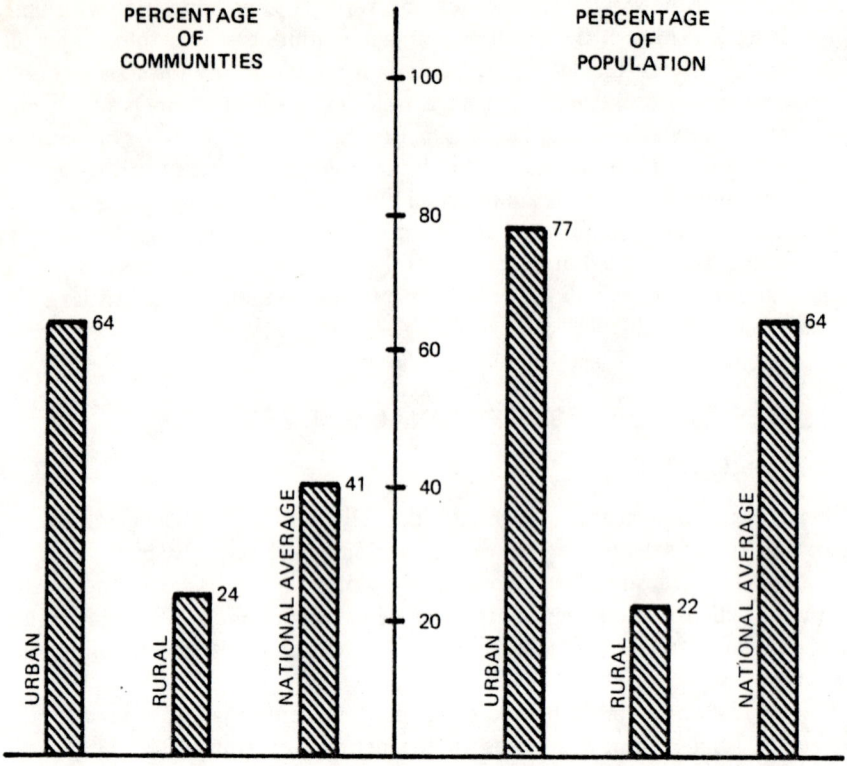

Fig. 2-1   Community operation of collection systems (1968).*

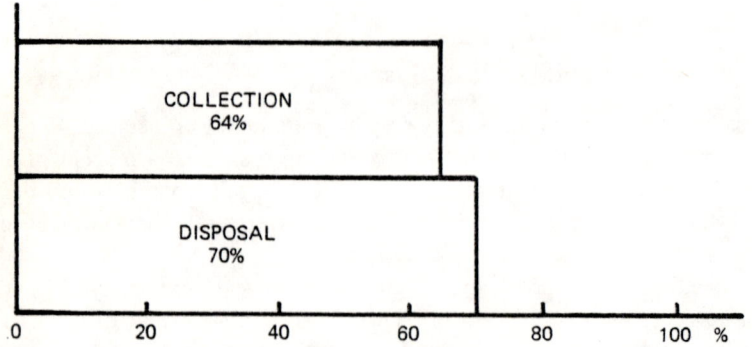

Fig. 2-2   Percentage of population in communities operating collection and disposal systems (1968).

*Source of Figs. 2-1–2-4: Ref. 2-1.

BASIC DATA AND SOLID WASTES 9

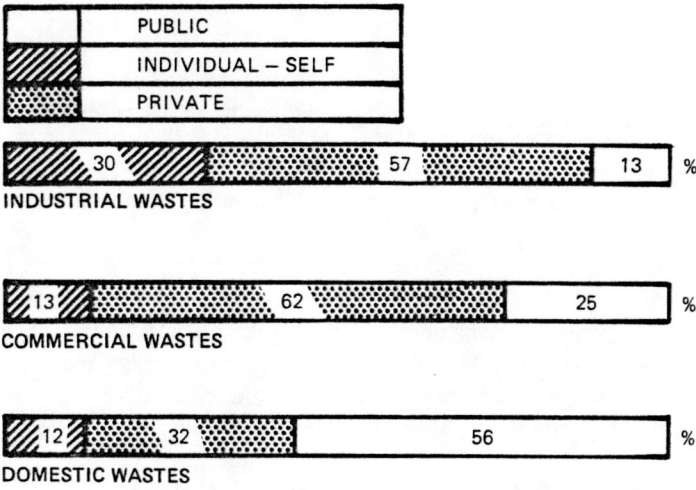

Fig. 2-3 Solid waste collection in the U.S. (1968).

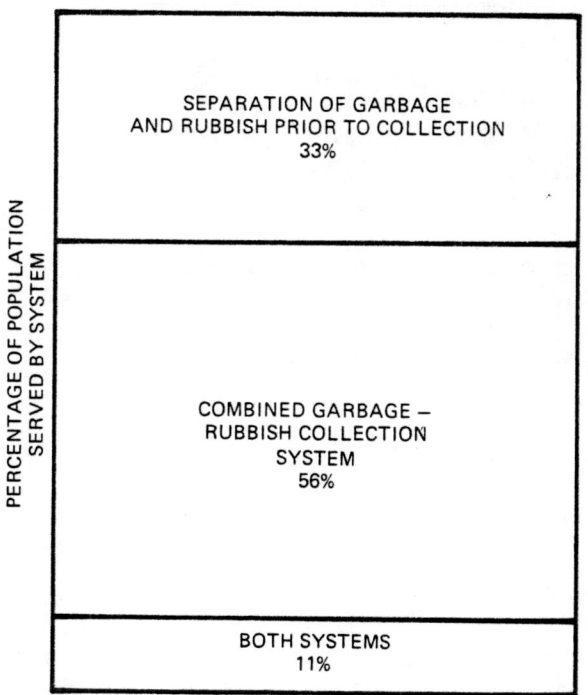

Fig. 2-4 Collection systems (by types) in the U.S. (1968).

about 12 percent must be collected by the individual homeowner. This is shown in Fig. 2-3.

Only about 25 percent of commercial wastes is collected by public collectors, while 62 percent is gathered by private haulers, and about 13 percent is handled by the commercial establishment itself.

Regarding industrial wastes, 13 percent is collected by public agency, 57 percent by private haulers, and 30 percent is handled by the industrial facility itself.

The collection systems employed in this operation are comprised of both those wherein all refuse is collected together, and those systems where garbage is separated from the remainder of the disposable materials (see Fig. 2-4). About a third of the nation separates its refuse into garbage and a non-putrescible remainder. Approximately 11 percent of the population is served in communities which have both "separate" and "don't separate" systems. The remaining 56 percent lives in communities where the combined system (no separation) is utilized. In general, rural areas are served preponderantly by combined systems and have very few of the systems employing both "separate" and "don't separate" techniques.

Frequency of collection varies tremendously with locality and collection system. For "don't separate" systems (see Fig. 2-5):
1. 48 percent have weekly pickup
2. 32 percent have twice weekly pickup
3. 20 percent have some other frequency

For systems wherein the *garbage* must be separated from the other refuse:
1. 61 percent have weekly pickup
2. 29 percent have twice weekly pickup
3. 3 percent have some other frequency
4. 7 percent have no collection of the separated garbage

For systems wherein *rubbish* is separated from the other refuse:
1. 66 percent have weekly pickup
2. 17 percent have twice weekly pickup
3. 9 percent have another frequency
4. 8 percent have no collection of the separated rubbish

2-2.3 Labor

A tremendous number of workers are required to accomplish the massive collection assignment described in the previous pages. On the average, one solid waste collector or truck driver is employed for every 590 persons in the population; in other words, there are more than 330,000 persons employed in the total nationwide collection effort. A slight majority of these workers are public employees.

BASIC DATA AND SOLID WASTES   11

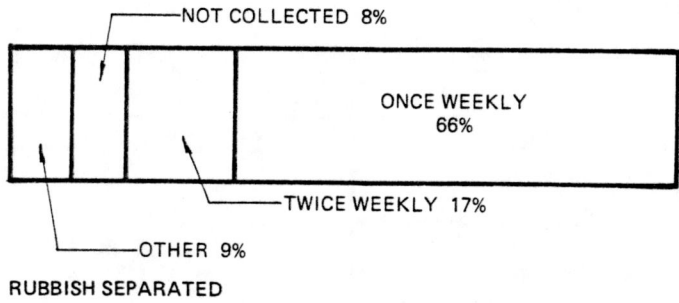

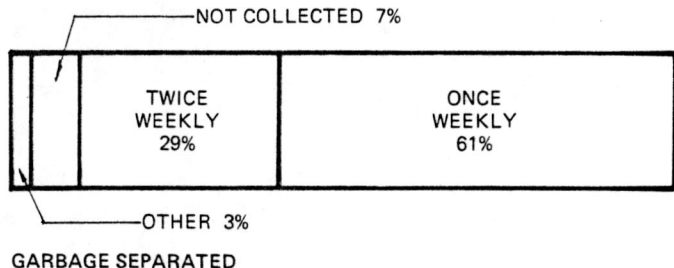

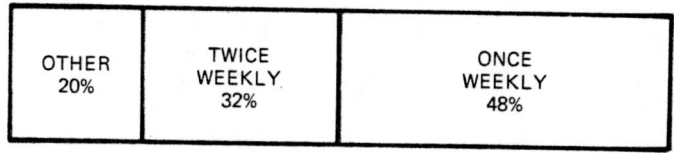

Fig. 2-5 Frequency of collection (1968).

About 53 percent of all solid waste collectors are employed by public collection agencies, while the remaining 47 percent are employed privately. Figure 2-6 shows graphically the distribution of workers according to job assignments. Of the number of public employees:

1. 75 percent collect household and/or commercial solid wastes
2. 5 percent collect industrial solid wastes
3. 20 percent are engaged in street cleaning

12 SOLID WASTE MANAGEMENT

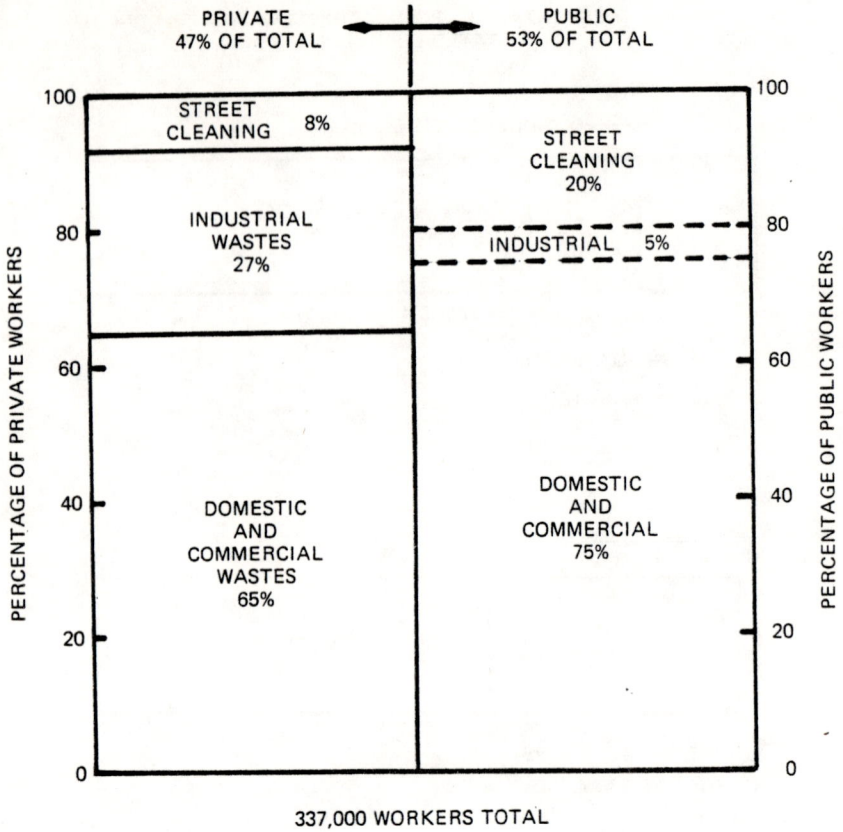

Fig. 2-6 Employment in solid waste collection (1968).

For private-sector employees:
1. 65 percent collect household and commercial solid wastes
2. 27 percent collect industrial solid wastes
3. 8 percent clean streets

2-2.4 Equipment

The large labor force previously described makes use of an equally large number of machines in their collection efforts. For example, more than 93,000 collection compactor trucks are used in the national collection operation; this is equivalent to one such truck for every 2,100 or so persons in the country. About half (53 percent) of these trucks are privately owned, and the remainder are public property. Private owners use compactor trucks in somewhat dif-

BASIC DATA AND SOLID WASTES   13

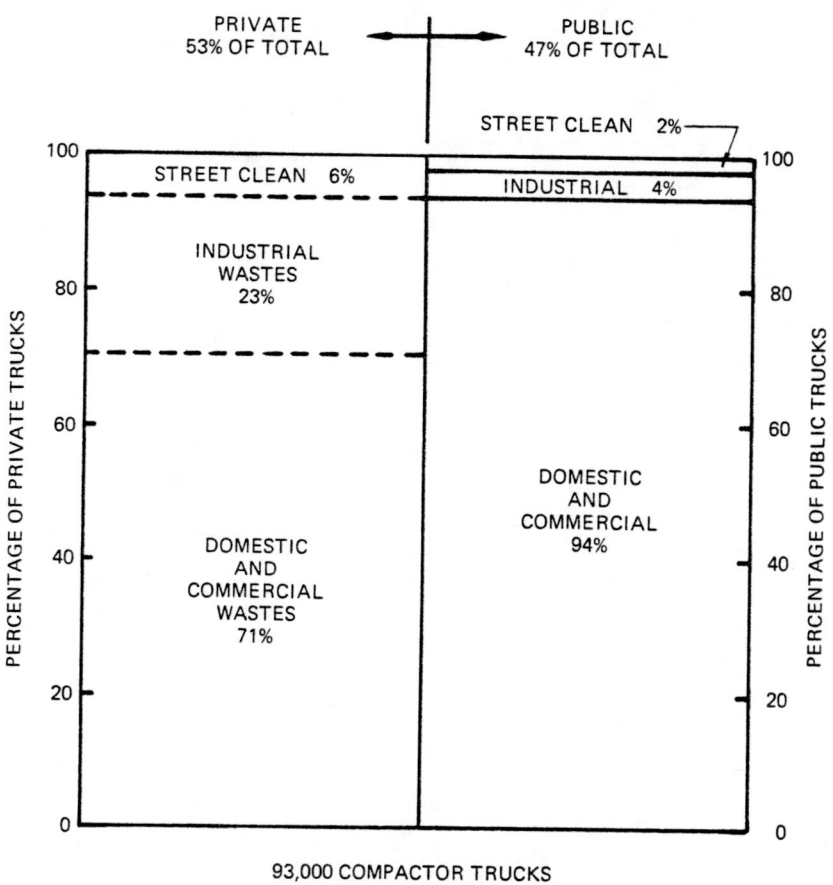

Fig. 2-7 Compactor truck use in the U.S. (1968).

ferent ways than do public collection agencies, as shown in the tabulation below (see also Fig. 2-7):

Activity	Number of Trucks in Use, %	
	Private Sector	Public Sector
domestic-commercial collection	71	94
industrial collection	23	4
street sweepings collection	6	2

Other collection vehicles total more than 179,000 nationally. These vehicles include non-compactor trucks, sweepers, vans, and other assorted items of equipment. Approximately 80 percent of all such vehicles are privately owned and are used primarily in industrial collection, as shown in Fig. 2-8.

14 SOLID WASTE MANAGEMENT

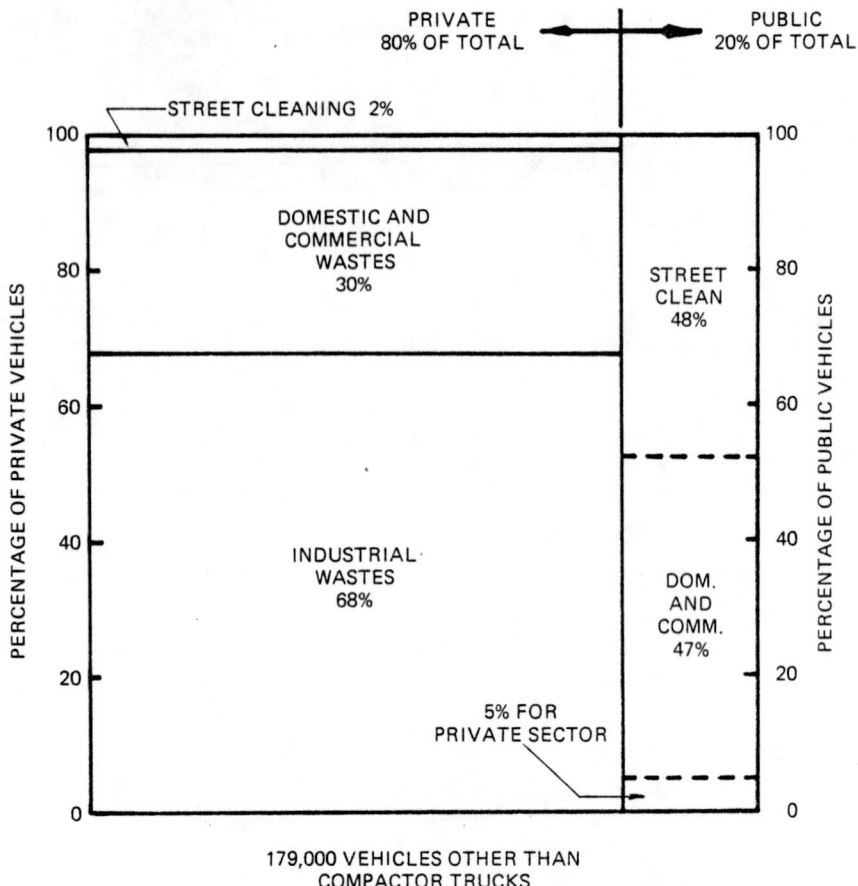

Fig. 2-8 Vehicle use in solid waste collection (1968).

2-2.5 Quantities

It is apparent that a considerable army of workers equipped with numerous machines is engaged in collecting the nation's solid wastes. It could be said facetiously that these workers "have their hands full." A veritable mountain of waste materials is collected each day in the United States. Table 2-2 shows survey results for determinations of quantities of wastes collected.

The national average figure of 5.32 pounds per person per day (pcd) consists of about 3 pcd from household sources, 1 pcd from commercial sources, about 0.6 pcd from industrial sources, and about 0.2 pcd from construction/demolition operations.

TABLE 2-2 Solid Wastes Collected Daily

Item	National Averages (lb/cap/day) Urban	Rural	Total
Domestic	1.26	0.72	1.14
Commercial	0.46	0.11	0.38
Combined	2.63	2.60	2.63
Industrial	0.65	0.37	0.59
Demolition-construction	0.23	0.02	0.18
Street sweepings	0.11	0.03	0.09
Miscellaneous	0.38	0.08	0.31
TOTALS	5.72	3.93	5.32

Source: Ref. 2-1.

These figures are approximate, and since they represent only that material *collected*, they do not reflect all wastes generated. More realistic waste generation estimates based on collected plus uncollected refuse would probably be

domestic	4 pcd
municipal	1 pcd
commercial	2-3 pcd
industrial	2-3 pcd
Total	8-10 pcd

This figure of 8-10 pcd does not include amounts of solid wastes generated in industry that are collected and disposed of in-house, nor are agricultural and mineral industry wastes included.

2-2.6 Financial Aspects

Solid wastes management costs are significant from any point of view. The average community budget allots $1.42/cap/year for disposal activities (about 15 percent for capital investments), as shown in Table 2-3. For actual operation by communities of disposal facilities, about $2.17/cap/year is spent (one-third for capital expenditures). For collection activities communities spend $5.60/cap/year for once-a-week collection (national average) and $6.82/cap/year for twice-a-week collection.

To establish perspective for this financial situation an example may be given:

In a typical community of 400,000 people about $900,000 is spent each year for waste disposal and about $2.5 million is spent for collection during the same period

16　SOLID WASTE MANAGEMENT

TABLE 2-3　Budgeted Community Expenditures

Activity	Dollars/cap/year Excluding Capital	Capital	Total
Disposal			
national average	1.17	0.25	1.42
community-operated disposal facility	1.46	0.71	2.17
Collection			
national average	4.86	0.53	5.39
areas with collection weekly	4.85	0.75	5.60
areas with collection twice weekly	5.67	1.15	6.82

In general, approximately 80 cents of every waste management dollar is spent for collection. A more detailed breakdown of average community expenditures is given in Table 2-3.

2-2.7　Adequacy of Disposal Operations

In the aforementioned 1968 study of solid waste practices, a survey of land disposal sites was made. More than 6,000 sites were investigated (see Fig. 2-9).

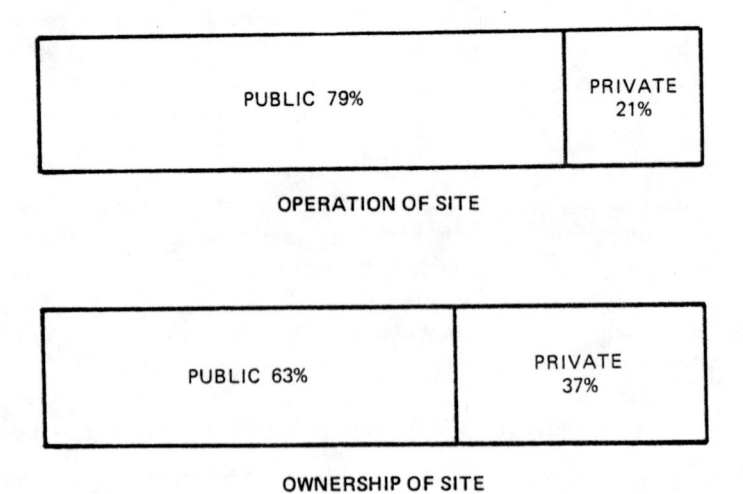

```
| PUBLIC 79%              | PRIVATE 21% |
```
OPERATION OF SITE

```
| PUBLIC 63%        | PRIVATE 37%    |
```
OWNERSHIP OF SITE

Fig. 2-9　Land disposal sites information (1968).

BASIC DATA AND SOLID WASTES   17

```
┌─────────────────────────────────────────────┐
│  PASS  │                                    │
│  14%   │        86% INADEQUATE              │
│        │                                    │
└─────────────────────────────────────────────┘
```
PRELIMINARY EVALUATION

```
┌─────────────────────────────────────────────┐
│ │                                           │
│ │          94% INADEQUATE                   │
│↑│                                           │
└─────────────────────────────────────────────┘
```
REVISED EVALUATION

──── ONLY 6% OF ALL SITES SURVEYED
SATISFIED ALL THREE CRITERIA:
- DAILY COVER
- NO OPEN BURNING
- NO WATER POLLUTION

Fig. 2-10 Adequacy of land disposal sites (1968).

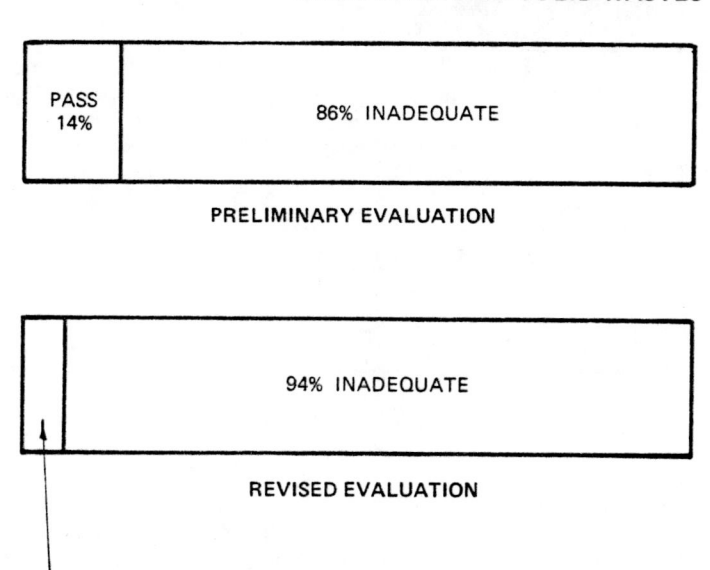

Of these sites:
1. 79 percent were publicly operated
2. 63 percent were publicly owned
3. 14 percent were judged by the interviewers to be "sanitary landfills" (see Fig. 2-10)

However, if the term "sanitary" connotes 1) daily cover; 2) no open burning; and 3) no water pollution problems; then only 6 percent of the 6,000 sites qualified. Only 14 percent had daily cover; 41 percent had *no* cover. Only about 25 percent had an acceptable appearance, and about 75 percent had some open burning. Figure 2-11 shows survey results.

The characteristics of the acceptable sites included: 1) capacity, about 27,000 tons/year; and 2) cost, about $1.05/ton for operating and amortizing capital. For the inadequate sites surveyed, similar data were 1) capacity, about 11,000 tons/year; and 2) cost, about $0.96/ton. To make all inadequate sites "sanitary landfills," an additional capital outlay of $244 million would be required, and an added $81 million/year in operating cost would result.

In addition to landfills, other methods of disposal were also surveyed. Survey results are given in the following tabulation:

18 SOLID WASTE MANAGEMENT

Facility	Public Ownership	Sightly Appearance	Average Input (tons/day)
Incinerators	96%	80%	188[a]
Transfer stations	76%	68%	375
Conical burners	59%	33%	41

[a] For incinerators built after 1950, average capacities range from 230 T/D to 400 T/D.

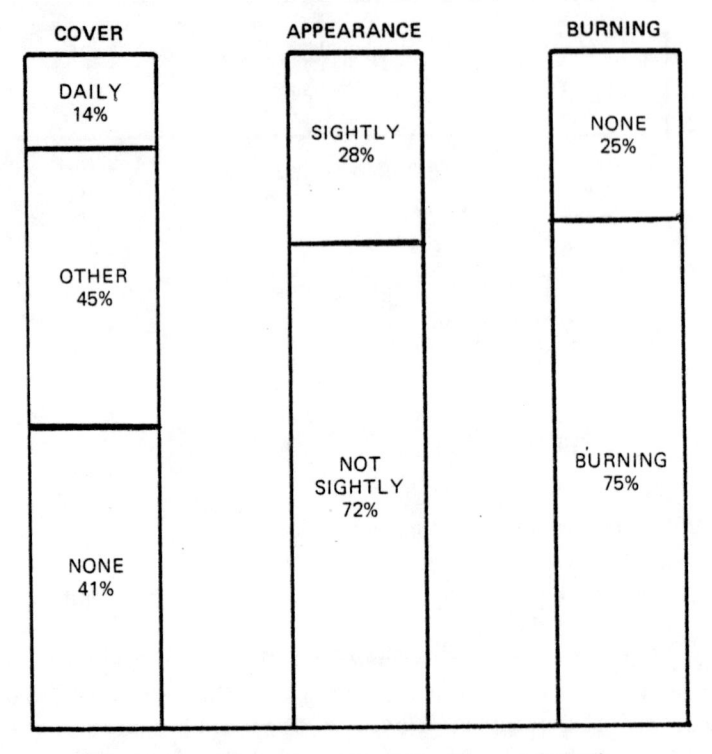

```
     COVER              APPEARANCE              BURNING

    DAILY
     14%
                        SIGHTLY                  NONE
                          28%                     25%

    OTHER
     45%


                          NOT                   BURNING
                        SIGHTLY                    75%
                          72%
    NONE
     41%
```

Fig. 2-11 Character of land site disposal (1968).

The capacities and appearance figures listed above represent survey results based on the study of 142 incinerators, 43 transfer stations, and 23 conical burners, as shown in Fig. 2-12.

Average costs for these disposal facilities were determined as:

Item	Costs Ton of Daily Capacity Operating cost/ton	Capital costs/ton
Incinerators	$4.50	$7,100
Transfer stations	1.10	1,100
Conical burners	1.60	1,700

BASIC DATA AND SOLID WASTES

1 = INCINERATORS, BURNERS
2 = TRANSFER STATIONS
3 = CONICAL BURNERS

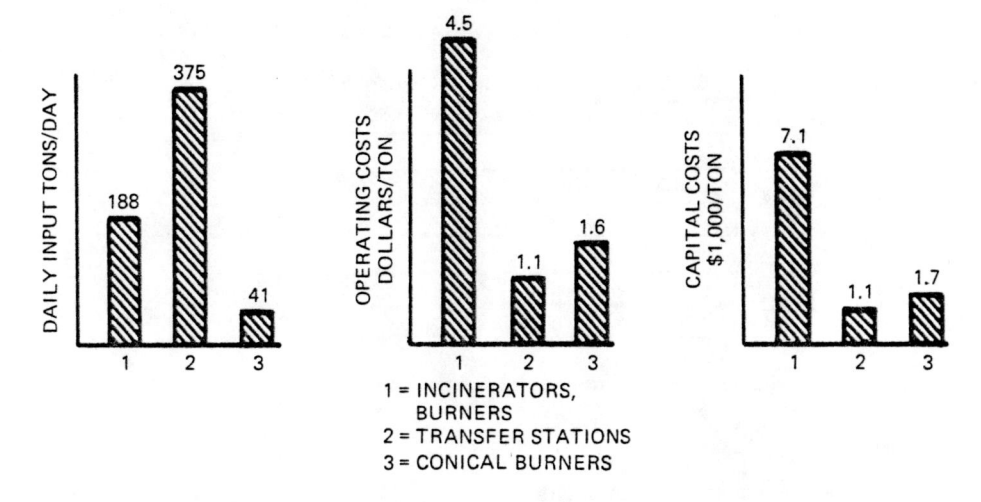

Fig. 2-12 Characteristics of disposal facilities (1968).

Such general cost data can become more meaningful if it is modified to include a measure of the adequacy of the disposal operation; that is, a low-cost operation may be quite undesirable if savings are achieved due to inadequate disposal. Various parameters of performance for incinerators may be considered: volume reduction; compliance with air quality standards; sightliness; etc.

In the 1968 survey incinerators were grouped into three classes on the basis of "performance" criteria:

1. class 1 incinerators
   constructed during or after 1950
   attain at least 75 percent volume reduction
   have 2 or more furnaces
   have operational and adequate air pollution control devices

TABLE 2-4 Incinerator Performance

Average[a] Incinerator Age	Weight Reduction %	Volume Reduction %	Costs ($/ton in)	Costs ($/ton reduced)
1962	79	85	3.27	4.06
1958	70	82	4.05	5.79
1945	65	76	5.37	8.26

[a] "1962" indicates that if a sample of incinerators is selected so that their average date of construction is 1962, weight reduction in those incinerators averages 79%, etc.

Source: Ref. 2-1.

20 SOLID WASTE MANAGEMENT

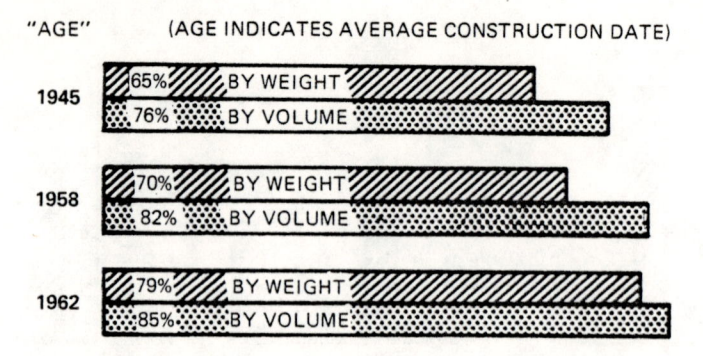

"AGE"   (AGE INDICATES AVERAGE CONSTRUCTION DATE)

1945 — 65% BY WEIGHT / 76% BY VOLUME

1958 — 70% BY WEIGHT / 82% BY VOLUME

1962 — 79% BY WEIGHT / 85% BY VOLUME

Fig. 2-13 Reduction of solid wastes by incineration (1968).

2. class 2 incinerators are identical with Class 1 except that they have no air pollution control devices
3. class 3 includes all other incinerators

The newer incinerators in Classes 1 and 2 were found to be more efficient disposal devices. Improvements in recent years in incinerator design, construction, and operation are reflected in the cost figures shown in Table 2-4 and in Figs. 2-13 and 2-14.

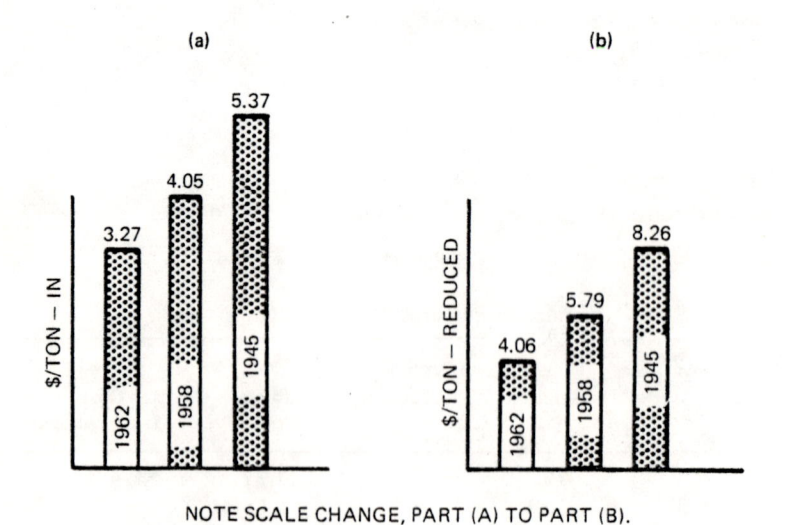

(a) $/TON – IN: 1962: 3.27; 1958: 4.05; 1945: 5.37

(b) $/TON – REDUCED: 1962: 4.06; 1958: 5.79; 1945: 8.26

NOTE SCALE CHANGE, PART (A) TO PART (B).

Fig. 2-14 Incinerator operating costs vs. "age" in the U.S. (1968).

(a)

(b)

Fig. 2-15 (a) A packer-type truck for municipal refuse; (b) packer truck in position for unloading (*courtesy* New Way Manufacturing Co.).

2-3. SUMMARY

It is pertinent to recall that:
1) at present about 12 percent of the total population receives no collection service
2) another 11 percent of the population receives only partial service
3) 14 percent of the populace is served by "separate" collection systems, but their *disposal* systems are of the combined type
4) of the approximately 12,000 land disposal sites in use, 94 percent are unacceptable and are potential sources of disease, pollutants, and land devaluation
5) of the 300 incinerators in the country, about 70 percent have inadequate air pollution control equipment
6) by 1980 the solid wastes generated in the United States will have increased 50–60 percent because of population growth and increased use of packaging, convenience goods, and disposable items

REFERENCES

2-1. *1968 National Survey of Community Solid Wastes Practices*, an Interim Report, EPA, USPHS, Cincinnati, 1968.

3

Volume reduction

3-1. VOLUME REDUCTION AT THE SOURCE

A review of the data presented in the preceding section indicates that a considerable portion of all the expenses incurred in solid wastes management result from collection. Therefore any operation which will reduce the time or labor, or any other cost facet of solid waste collection, will produce a significant savings in the overall expenses of management. One such operation for reducing the amount collected and transported is volume reduction. Since a reduction in the volume of materials (or essentially any compaction or densification) will result in less collection time and fewer trips by the collection vehicles to the disposal site, it therefore results in less expense. Additional incentives for the practice of volume reduction, particularly at the source of the wastes, include both economic and health reasons. The economic reasons for volume reduction at the source include lower handling costs for less refuse (attendant lower labor costs), as mentioned previously. Additionally, since less work is done by collectors, there is less chance for human error in the collection operation and consequently a better, more economical operation results. Finally, on-site volume diminution also yields lower costs through reductions in required storage area and, in addition, may lead to collection of wastes in areas which were not formerly served by collection agencies. Many times, municipal collection agencies will not accept refuse from outlying apartment districts unless this has been collected throughout the apartment complex and reduced in volume in some manner.

The health-oriented reasons for on-site volume reduction include elimination of harborage and food sources for disease vectors such as mosquitoes, flies, rats, and roaches. In addition, safety hazards to people and dangers of fires are reduced, since less material must be temporarily stored. In general, because less refuse must be stored, the storage area will also be much cleaner and neater.

The methods of volume reduction that may be practiced at the source of generation of solid wastes in general depend for their success on the size of the waste generating facility. For example, for individual residences only certain means or mechanisms of volume reduction may be practiced. Formerly such practices as burning of rubbish and refuse in backyard incinerators and dumping of refuse on private property were tolerated. These practices are no longer considered acceptable means for reducing the amount of solid wastes to be collected. Some hope has been held out for the use of composting of refuse at the individual home. However, definite problems exist in any residential composting attempt. For example, the use of standard farm composting techniques in the individual home may be very difficult since biodegradation in a composting operation usually requires some preliminary preparation of the refuse such as shred-grinding. Additionally, open refuse piles will furnish some aforementioned harborage and food sources for disease vectors and generally will encourage the production of vermin. A second method of residential composting which has been proposed is the use of bottomless cans which are fitted with tightly closing lids and which are set into the ground with only the tops exposed. In these cans the food wastes from a residence decompose and volume reduction of from 70 to 80 percent occurs in the decomposable wastes. However, a certain amount of non-decomposable material is included in every charge into the can and the latter must therefore be cleaned out every six to eight months. The refuse or residue which is retrieved from the can at this time is then very difficult to handle and is quite odorous. Finally, the can itself will rust and decompose and will become very difficult to remove from its location in the ground. Therefore, little promise is held out for these methods of volume reduction at the individual homesite.

Another method of volume reduction in the individual home involves the use of garbage grinders. In the typical model the refuse which is fed into the grinder is reduced in volume about 10 percent in a manner very convenient for the

Fig. 3-1 Stationary compactor for volume reduction at an industrial site (*courtesy* New Way Manufacturing Co.).

homeowner. However, garbage grinders will not handle all materials in domestic refuse, and large bones or other bulky items must be disposed of in an alternate manner. In essence, the garbage grinder is really a modification procedure used to render garbage suitable for transportation in a water flow system to a disposal site. The use of the garbage grinder simply transfers a certain amount of the solid wastes produced in the home from disposal in a landfill or incinerator operation to disposal by a sewage treatment plant.

Finally, the use of home compaction units has been suggested as another means of on-site volume reduction. Home compaction units, if manufacturers' claims are to be believed, hold much promise for reduction in collection and transportation costs since significant densification will be possible with these units. However, a significant amount of experience with these units in individual homes has not been achieved and the overall balance of benefits and disadvantages for them cannot be assessed at this particular time. There is some doubt in the minds of many persons as to the desirability of placing a capital expenditure cost on the individual homeowner in the form of the purchase price for a home compaction unit. Additional research is needed in this area so that the overall effectiveness of the unit can be adequately assessed.

Whenever a number of families are grouped together in a multiple-unit housing facility, the larger amount of refuse at this particular location will suggest additional means of volume reduction for solid wastes at point of generation. For example, in the past, incinerators have been used in many apartment buildings and multiple housing units. Generally these incinerators can be classed in one or two different categories. Some incineration units consist of dehydrating hot plates on which the refuse is dumped; a gas flame or electric heating device not directly in contact with the refuse heats the plate, dries the refuse, and causes it to ignite. In other incinerators, refuse is simply stored in some type of container until it is ignited by the action of a preset timer which activates a burner supplied with an auxiliary fuel. In some of the timer-controlled incineration units an afterburner has been included so that the refuse is dried and partially burned in a primary chamber and the gases leaving that chamber are more completely combusted in a secondary chamber. In some apartment house systems the incinerator may be a flue-fed device. In this type of apparatus a single flue is used as the charging chute for incoming refuse and also as the effluent stack for the gases produced during burning of the deposited refuse. All of these types of incinerators have been quite popular throughout the country, since they function as very convenient means of refuse disposal for the person dwelling in the multiple-unit system. They are seemingly economical to the apartment dweller, for example, since the cost of operation of the incinerator is generally included in the apartment rent and nowhere is this cost directly passed on to the resident. Another means of economy is the reduction in the amount of refuse to be collected and disposed of by a municipality serving the area of the apartment complex. In contrast to these advantages, most domestic or apartment unit incinerators present severe difficulties with air pollution, incomplete combustion of waste, and safety hazards. The limited volume of the combustion chambers, the poor air supply to them, the low combustion temperatures, and the poor dispersion characteristics of flue-fed incinerators lead to incomplete combustion of the refuse and the discharge of incompletely oxidized materials to the atmosphere. Included in the air pollutants produced by this type of incinerator are flyash, smoke, nuisance gases (such as highly odorous aldehydes and esters), harmful gases (such as sulfur oxides and nitrogen oxides), and small pieces of charred paper and other particulate matter. The haphazard manner of charging and regulating combustion in this type of incinerator may also present significant safety hazards. Finally, incomplete combustion of the incoming refuse will usually generate a residue which may contain putrescible materials and incompletely burned combustible materials. All of the residue matter will require alternate disposal means such as a sanitary landfill. For these reasons, incinerator systems for volume reduction at the source in multiple housing units are considered unsuitable. In many areas of the country, local air pollution standards have effectively eliminated the use of apartment house incinerators as means of volume reduction. In very large concentrations of housing, institutions such as hospitals, or certain facilities such as office buildings, a very large production of solid waste may make it possible to use newer techniques of on-site volume reduction. For example, wet pulverization and pulping operations may prove feasible. Although the installation of such a system in an institution or in a high-rise office building will represent a significant capital outlay and will require special equipment and special maintenance, several distinct advantages for such an operation may make it the most desirable means of on-site volume reduction for a given situation. For example, volume reduction in a pulping operation may be as high as 80 percent. Moreover, modern pulping operations can include facilities for extraction of salvageable materials such as paper pulp, magnetic and nonmagnetic metals, and glass. Since a certain amount of processing and separation prior to pulverization is included in almost all modern pulping devices produced in this country, significant savings may be realized by the installation of a pulper in a particular locality. For example, restaurants using pulpers may save on silverware recovered in separation before pulping, silverware which otherwise would have vanished in the midst of the solid wastes generated there.

Finally, in situations where pulpers may be used, and where large amounts of solid waste are generated in small areas, compaction equipment may also prove attractive. Large pneumatic or hydraulic compactors have been developed for use in high-rise structures such as office buildings or apartment complexes. The

Fig. 3-2 A stationary baler receiving various types of refuse ranging from garbage to discarded refrigerators (*courtesy* American Solid Waste Systems).

28 SOLID WASTE MANAGEMENT

high compaction ratios (up to 75–80 percent) attainable in these devices produce a significant reduction in the volume of wastes which have to be collected and transported. Additionally, the essentially automatic operation of such a system, which includes refuse chutes and automatic compactors (and in some a conveyor or similar means for transfer of the compacted refuse to the collection trucks), reduces manpower requirements and handling costs. The operation is therefore generally neater, quieter, and simpler than with formerly used systems. However, some attendant disadvantages must also be recognized. Highly automated systems will be subject to clogging and plugging in chutes and disruption of the automatic transfer facilities, and will thus entail higher maintenance costs. In spite of these disadvantages, large automated pulping and compacting systems appear to be quite promising for the reduction of solid waste volume at the point where the wastes are generated.

3-2. VOLUME REDUCTION AT A CENTRAL LOCATION

In addition to the reduction in volume of wastes at the point of generation, in some localities and in some situations it appears quite advantageous to practice

Fig. 3-3 Baled cans ready for recycling; small bale weighs about 850 lb (*courtesy* American Solid Waste Systems).

volume reduction at a centralized location. In such a procedure, refuse is collected and brought to a focal point where grinding, baling, and/or compaction equipment is employed in a volume reduction operation. Such centralized volume reduction is especially advantageous when the solid waste must be transferred from the centralized volume reduction station over long distances to an ultimate disposal location. The general purpose and means of accomplishment of centralized volume reduction are the same as those for volume reduction at the source. However, some specialized equipment has been developed for the purpose of accomplishing this reduction. This equipment includes hammer mills, rasps, and other types of grinders; baling equipment which may be used in conjunction with compactors or used to contain uncompacted refuse; and various types of compaction equipment, such as presses, compactors, etc. The auxiliary equipment used in such operations is described in the next section.

REFERENCES

3-1. Day and Zimmerman, Engrs. and Archs., *Special Studies for Incinerators*, for the Government of the District of Columbia, Dept. of Sanitation, USPHS Publ. No. 1748, Washington, D.C., 1968.

3-2. *Elements of Solid Waste Management*, A Training Manual, EPA, 1970.

4
Accessory operations

4-1. GENERAL

In many solid waste processing operations, accessory procedures are required in addition to the primary one, such as incineration, composting, or landfill. Two general classifications of these accessory operations may be made. All may be classed either as densification operations or size reduction operations. In each of these, different equipment is used—equipment which differs in function, performance, and character. Densification equipment is used to increase the weight of material per unit volume. For example, a scrap baling press is a type of densification apparatus.

"Comminution" is the general term which is applied to size reduction processes. In size reduction the equipment commonly used includes rock crushers, flour mills, domestic garbage grinders, hammer mills, flail mills, etc.

There is a wide range of types and capacities of both densification and size reduction equipment currently available from a large number of manufacturers. Moreover, many of the existing types of equipment are easily varied and adapted to particular ways of handling almost any given type of solid waste.

Almost all types of solid waste disposal procedures include some sort of densification or size reduction operation. For example, sanitary landfilling operations employ a variety of machines to compact and reduce the bulk volume of the refuse. Size reduction equipment has become used on a limited scale in Europe and in several experimental operations in the United States. In addition, size reduction equipment has been used in conjunction with incineration operations to reduce bulky wastes to practical size. Finally, composting operations make extensive use of various types of size reduction equipment.

4-2. SANITARY LANDFILL ACCESSORY OPERATIONS

In most sanitary landfills, rubber-tired or crawler-mounted bulldozers have been used to work the filled material. These machines commonly produce a compaction ratio of about 3 to 1. Thus, the final volume of material in the completed fill occupies a third of the volume occupied by the original uncompacted refuse. Several machines have been designed specifically for sanitary landfill work; a 25-ton compactor bulldozer using lug or gear-tooth-like wheels has been developed for more efficient compaction of landfill refuse. Another type of landfill compaction device is a ballasted-drum-type compactor similar to a sheepsfoot roller; it is pulled behind a dozer or tractor and also vibrates to improve compaction. In addition to these types of compactors many other smaller devices have been developed for limited scale operations. These devices are described in more detail in Chapter 10.

4-2.1 Collection Compaction

In addition to compaction of the refuse at the landfill site, refuse in the United States has also been compacted at the point of collection since about 1950. At that time packer-type refuse trucks were introduced; these compactors normally carry 10-30 cubic yard bodies on standard truck chasses. In Europe, several specialized collection vehicles have been developed: one system employs two-stage compaction with a reciprocating ram followed by a swiveled compression plate that moves the refuse into the truck body; two other systems use helical screws which compact and advance the refuse into the truck; a fourth system consists of a truck body which is cylindrical in shape and is fitted with an internal worm drive—the drum rotates at about 4 rpm, crushing, compressing and pushing the refuse toward the front of the truck.

In addition to the indicated compactor systems, mention should also be made

32 SOLID WASTE MANAGEMENT

Fig. 4-1 Landfill compactor fitted with oversize blade for spreading refuse and metal feet on wheels for crushing and compacting wastes (*courtesy* RayGo, Inc.).

of transfer stations. The development of these has come about because landfills are becoming more and more remotely removed from collection areas. Thus it becomes economical to employ central collection stations where local collection trucks may dump their refuse into large-capacity transfer semitrailers for hauling to the landfill itself. At these stations the local trucks dump the refuse into a hopper where it is hydraulically compacted and pushed into the transfer trailer. In addition to the compaction of refuse, other experiments have been tried at transfer stations and other processes have been developed wherein the compacted materials are confined in a baling operation, so that the compacted refuse may be more easily transported and placed in a landfill. However, baling operations have not been completely successful. One unsuccessful operation in Los Angeles County, California, has been discussed in the engineering literature. In this particular operation, the compressors and balers that were used did not perform satisfactorily, and the bulldozers at the landfill site had difficulty in placing the finished bales. Much additional work needs to be done in this area.

4-2.2 Compaction Systems

Before leaving the subject of compaction it would be pertinent to discuss several items which have been introduced into the solid waste management field in

Fig. 4-2 Semitrailer waste carrier being unloaded at landfill site (*courtesy* Peerless Trailer and Truck Service, Inc.).

34   SOLID WASTE MANAGEMENT

recent years. Included in these are the Tezuka-Kosan compressor and various American pulverizer-compressor operations, such as the Mil-Pac compressor and the compressing system developed by Reclamations Systems, Inc.

The Tezuka refuse compression system consists basically of three major operations: (1) the preliminary compression system; (2) the main compression system, and (3) the additional equipment required to provide covering materials such as asphalt or cement to the finished bales. Various types of preliminary compressor systems are available, including one- or two-stroke precompression devices which generally exert a pressure on the refuse in the system of about 220 psi. The refuse leaves the precompression system and goes through a feeding hopper wherein further compression is attained; the maximum hopper pressure is about 425 psi. From the loading hopper the refuse goes into the main compression system, which includes the main press, a mobile compression chamber, an associated bale enclosure, and a push-up device to remove the bale from the compression chamber. The main ram and its pressure face are designed to perform four successive compression operations; the pressure face is divided into three sections. The dimensions of the latter range from 63 × 63 in. to about 71 × 71 in. In the first step of the compression the full amount of the available force is applied to all three pressure face sections so that a pressure of approximately 675 psi is exerted on the refuse. In the second step a central cylinder about 18 inches in diameter and about 18 inches in length is forced into the center of the confined refuse: a nominal pressure of about 5,000 psi is exerted on the refuse beneath the cylinder head. The pressure face of the center cylinder is only about one-sixteenth of the total area of the pressure face. In the third step twelve smaller cylinders distributed over the total original pressure face are pushed downward in either a simultaneous or individual pattern. Each of these twelve cylinders is about 6.8 in. in diameter and about 18 in. in length and exerts a pressure in excess of 5,000 psi on the refuse. The total pressure face area of these twelve cylinders is only about 11 percent of the initial pressure phase area. In essence then, a pressure of more than 5,000 psi is concentrated on small disconnected areas inside the bale. In the fourth and final compression step the remainder of the original compression face is moved downward until an average pressure of about 800 psi is applied to the refuse. After the compression process has been thus completed, a bale-enclosure closing device folds the enclosure used over the top of the bale; quite commonly chicken wire is used as enclosure or encasement for the bale. Operating time for the Tezuka system is about 10–15 minutes per 5-ton bale. This does not include any additional time needed for bale treatment; additional treatment includes encasement of the bale in concrete, dipping of the bale into asphalt, wrapping the bale in vinyl or other materials, and strapping the bale with metal bands or wires. The density achieved, according to Tezuka reports, ranges from about 60 to 120 lb/cu ft, with the variation resulting from variable composition of the compacted wastes. The

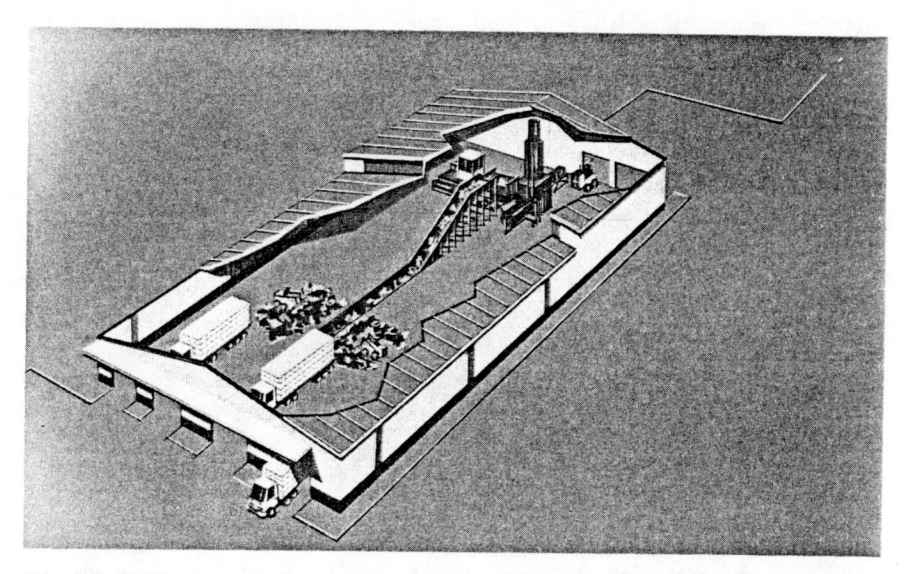

Fig. 4-3 Schematic view of baler station. Notice large floor area for dumping refuse on both sides of conveyor (*courtesy* American Solid Waste Systems).

physical stability of these large refuse bales is not established. The high densities of the refuse should be viewed in light of the fact that the water content of the bales may be as high as 50 percent. Although the manufacturer claims that aerobic bacteria activity is zero within the completed bale, certainly some partial degradation does take place. Fully enclosed, sealed, and moist bales could represent significant hazards, since anaerobic biodegradation could produce explosive methane gas. The volume reduction ratio claimed by Tezuka ranges from 5 to 1 to 7 to 1. In order to produce these large bales while using only a limited total pressure force, a complicated machine, complicated hydraulic switching arrangement, and large number of sequentially operating system elements must be employed. In addition to the complicated equipment involved in a Tezuka system, the investment cost must include a substantial transportation figure from Japan to the final destination. Typical costs for the equipment alone may be estimated at roughly $400,000 for 150-ton per day installation and at over $7 million for a 3,000-ton per day installation. Transportation costs and import duty are estimated to increase the cost by 25–30 percent. A significant plant site must also be allotted for the Tezuka equipment. A 750-ton per day installation requires about 720 square yards of space for the equipment alone, and a significant foundation must be placed beneath the press since it weighs approximately 400 tons. In addition to the 720 sq yds of space for the equipment itself, a further area approximately ten times that large should be allotted for associated operations. In addition to capital costs, the operating costs for a 750-ton per day installation are close to $2 per ton. These costs do not include maintenance expenditures, depreciation for the physical plant, and depreciation and operating costs for any additional materials handling equipment. Finally, delivery time for the press is about ten months; assembly and erection would require an additional two months.*

A much simpler system is the Mil-Pac vertical hammer mill which has been in use for approximately ten years in the United States. The primary use for this mill has been pulverizing coarse screenings from sewerage, and sugar bagasse, the wastes which are generated during the refinement of sugar cane. Increasing use of such hammer mills, however, is forecasted for the immediate future in solid wastes handling. In the hammer mill, refuse is ground, chopped, and shredded to a fraction of its original size by means of a top flow plate which includes two carbon-tipped breaker bars. The incoming material encounters the flow plate and is pulverized and must pass between the plate and the metal housing; a clearance of approximately three-quarters of an inch is allowed between the plate and the mill housing, where the refuse encounters eight tungsten-coated hammers. The choice of fine or coarse pulverization is accomplished by moving the breaker bars in or out toward the hammer tips. After pulverization, a lower set of sweep hammers blows the pulverized material into a mechanical screw compactor. The pulverized extruded material leaves the latter in the form of 9-in. logs. The advantages of this system are that in addition to volume reduction with less handling, lower hauling costs, and less final required disposal volume, the pulverized and compacted refuse is extruded in a dense, easily handled product which is virtually odorless and vermin free. The final log material is useful, however, only in landfill. It is not suitable for building. The final densities attained in this process are roughly 1600 to 1700 lb/cu yd for "average" refuse; final densities of 3,000 lb/cu yd have been attained after landfill.

The Reclamation Systems compactor and baler is another currently operational apparatus. The first high-pressure compaction and transfer terminal in the United States built by Reclamation Systems began operations in 1970, in East Cambridge, Massachusetts. At this location rubbish brought to the plant by refuse collection trucks is dumped onto one of two 120-ft-long, 8-ft-wide conveyor belts and transferred to a precharging conveyor which feeds the refuse to the compaction press. Approximately 1,000 hp drive the compacting press onto the dumped waste. A force of about 4,000,000 lb is expended in the compactor. A hydraulic ejection device pushes the completed 4 ft × 4 ft × 16 in. block out of the chamber. After compaction, blocks are piled three high and moved to a strapping unit to be bound with two metal bands. An automated monorail assembly then transfers the completed and finished bound bales to the trans-

---

*For a complete discussion of the Tezuka-Kosan Compaction System see Ref. 4-2.

Fig. 4-4 Completed baled fill. Solid waste bales left uncovered for several months during winter remained intact (*courtesy* American Solid Waste Systems).

portation facility for subsequent movement to a landfill site. The presses in this compactor operate on 2-minute cycles and can compact more than 2,000 tons of refuse per 16-hr workday. The compaction and transfer plant and the related baling equipment costs approximately $2 million. The system will handle all types of solid waste except building demolition materials, rubber tires, and tree stumps. In addition to the items previously presented about this system several miscellaneous data may be reported: the juices in the garbage which is put under high pressure (1800 psi) will go into a liquid waste disposal unit ($80,000) which meets city health standards; little difference in density of the finished bale has been noted, regardless of changes in moisture content; the final density of the material is approximately 70 lb/cu ft, or approximately 2,000 lb/cu yd.

4-2.3 Size Reduction Before Landfilling

In addition to compaction as a preparatory operation for landfilling, size reduction has also been practiced in Europe at several locations. However, such applications of size reduction are rather unusual in the United States. While it may appear that grinding municipal refuse is a needless operation, the system developed in France (Heil-Gondard) has several unique advantageous characteristics. It is a combined system which is rather unusual for a hammer mill-type grinder. In this hammer mill the items which cannot easily be ground up are automatically ejected; hammer rotation is such that ungrindable material is batted up into a recovery chute where it is collected. Thus, grinding and sorting are accomplished in one operation. An increase in density of from two and a half to three times is possible using the milling operation to increase the compactability of the material. Furthermore, a significant improvement in the refuse as a disposal item is achieved by the grinding and milling operation. The main advantage of the latter operation is that it renders the refuse much less of a potential home for disease-carrying vectors. The grinding operation itself destroys many of the insect eggs and larvae contained within domestic refuse, and little or no fly or other insect breeding is accomplished in the final pulverized ground-up material. Second, little harborage is furnished by the ground refuse for larger vectors such as mice and rats. In tests at Madison, Wisconsin, materials ground in a Gondard mill were found to be unsuitable for the feeding and harborage of rats. In essence, a rat feeding on this material would expend more energy in obtaining the edible portions of the refuse (which is intimately intermixed with inedible portions) than he would obtain by metabolism of the consumed edible portions. Second, the ground refuse has little harborage potential—burrowing animals are not able to find shelter in the refuse since it readily collapses on the animal. In an extensive series of tests conducted at Purdue University, a test group of Norway rats given as food only ground refuse from Madison, Wisconsin, all died within a short time. At first the test rats sought to eat the ground refuse, with little success. After a few days' time, the starving rats turned to cannibalism. Eventually, the entire rat population died in the midst of abundant milled refuse.

In addition to these types of equipment the Hydrasposal equipment developed by the Black Clawson Company (described in Chapter 6) typifies another type of compacting-grinding operation which may or may not be applicable to landfill uses. Furthermore, other equipment, such as brush chippers and paperboard box shredders, is being used. The brush chippers represent on-site solid waste size reduction and consist of hammer mills with knifelike hammers which reduce the size of particles of brush and tree trimmings about 80 percent. Paperboard box shredders offer just another method of reducing bulk volume at the site of the waste generation. All of these operations used in conjunction with landfills or incineration require a different series of accessory operations.

4-3. INCINERATION ACCESSORY OPERATIONS

As an auxiliary operation for incineration, densification is practically unheard of in the United States. In Europe, the only instance in which densification processes have been used in conjunction with incineration is that of some experimental work which has been done in England and Switzerland. In this work, the refuse coming to an incinerator was sorted to remove noncombustible materials, and the remaining combustible ones were ground for greater uniformity; the combustibles were dried to bring the moisture content below 10 percent; and finally they were extruded in the form of 2 to 3 in. diameter briquettes. The calorific value of the briquettes was found to be in the range of 8,000 to 10,000 Btus/lb. However, the system was generally unsuccessful since approximately 6 tons of normal domestic refuse produced only a ton of briquettes and the additional 5 tons of remaining material required other disposal means.

4-3.1. Size Reduction Prior to Incineration

On the other hand, size reduction operations have found significant employment as accessory operations for incineration. The present application of size reduction equipment to municipal incinerators has generally been limited to reduction of bulky or oversized items. Bulky wastes may be too large for direct charging, may burn too slowly, or may contain noncombustible portions that interfere with the mechanical operations of incinerator grates or residue discharge systems. Examples of such bulky wastes include construction and demolition refuse, furniture, mattresses, and tree stumps. At many European incinerator locations large oversized materials are segregated and moved by crane to specially designed multiple-type shears or impact crushers. After this shredding and size reduction operation, the once-bulky material is then mixed in a common receiving pit with the normal refuse.

In many instances, improved incinerator operation and savings on cost and size of other components have been claimed through the use of grinding mills or shredders. Certainly explosive hazards from containers of volatile liquids or materials under pressure can be eliminated through the use of preliminary grinding and shredding. Moreover, feed rates into an incinerator may be much more uniform and easier to control if the material is previously shredded. However, no clear-cut data on the successful use of size reduction prior to incineration have been documented to date, and additional research is needed in this area.

4-4. ACCESSORY OPERATIONS FOR COMPOSTING

Generally, composting operations do not require the use of densification equipment. The only densification in a composting operation is a very nominal one during pelletizing operations for the final composted product.

4-4.1 Size Reduction During Composting

The most extensive application of size reduction equipment in the enitre field of solid waste disposal has been as an auxiliary operation to composting. Some

40 SOLID WASTE MANAGEMENT

sort of size reduction operation is necessary in almost all of the many types of composting processes in use in the United States today. The size reduction achieved varies from operation to operation, depending upon the process itself and the intended use of the resultant end product. If the final product from a composting operation is to be useful and acceptable for agricultural uses, it must be free of large fragments of metal, glass, rock, or other undesirable and injurious components. Consequently, extensive screening, separation, and final reduction (in equipment such as hammer mills) are usually used in a composting operation, between the compost digesting operation and final preparation of the end product for sale. No general statements may be made concerning the use of size reduction equipment in composting plants because the type varies extensively from plant to plant. However, since very extensive use of such equipment is made in composting operations, some description of the particular types of mills or other apparatus may now be given. Six general categories of equipment are recognized: hammer mills, flail mills, vertical-axis rasps, drum rasps, roll crushers, and pulpers.

Hammermills, the broadest category of size reduction equipment, include all types of crushers, grinders, chippers, and shredders that employ pivoted or fixed hammers, or cutters. Hammermills usually have a simple horizontal or

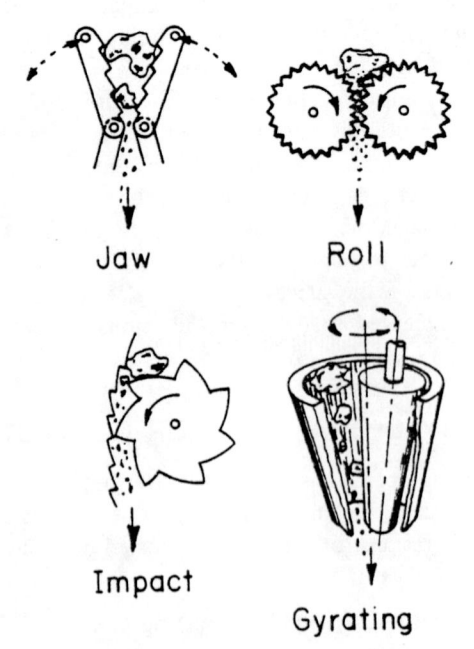

Fig. 4-5 Schematic of crushing apparatus (from *Recovery and Utilization of Municipal Solid Waste*, USPHS Publ. 1908).

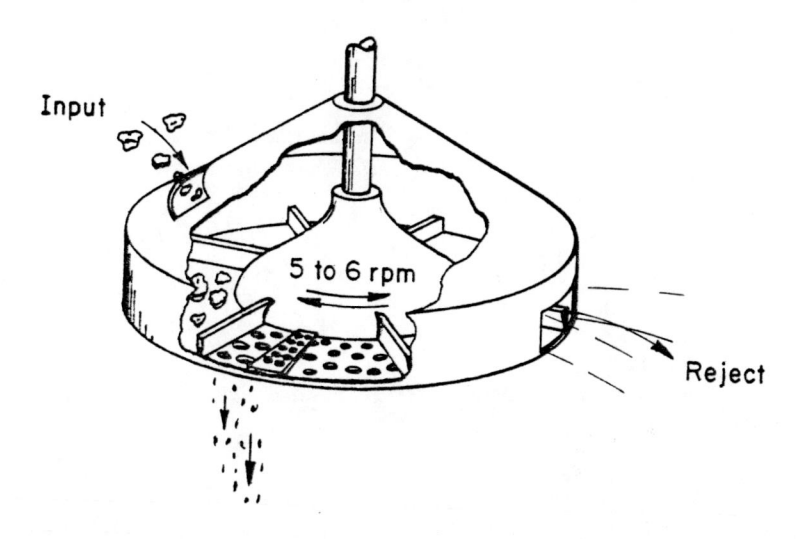

Fig. 4-6 Schematic of a rasp mill (from *Recovery and Utilization of Municipal Solid Waste*).

vertical rotor, but may have twin rotors. Most such mills are of the single-rotor, swing-hammer, high-speed type. Two common types are the Tollemache mill and the Heil-Gondard mill. The Tollemache mill is a vertical-shaft type which can be manufactured in any size; the standard size is a 150-hp unit capable of reducing 12 tons of refuse per hour of operation. The Tollemache mill is a mobile operation which features ballistic separation of ungrindable materials. Maintenance for this mill will amount to approximately 25 cents to 50 cents per ton of refuse processed.*

The Heil-Gondard hammermill has also been used in the United States, particularly at Madison, Wisconsin. Some cost figures for the Heil-Gondard mill at Madison are given in Table 4-1.

TABLE 4.1. Madison, Wisconsin, Pilot Plant Cost Figures $/ton

		Grate size	
Cost	2 in.	4 in.	6 in.
Labor	$2.98 (2.42)[a]	2.35 (2.04)	1.83 (1.50)
Depreciation	3.85 (3.17)	3.24 (2.86)	2.46 (2.04)
Hammer wear	0.34 (0.26)	0.32 (0.23)	0.28 (0.16)
Utility	0.92 (0.76)	0.63 (0.56)	0.70 (0.58)

[a]Costs shown in parentheses are adjusted for downtime excess.
Source: Office of Solid Waste Management Programs.

*See Ref. 4-1.

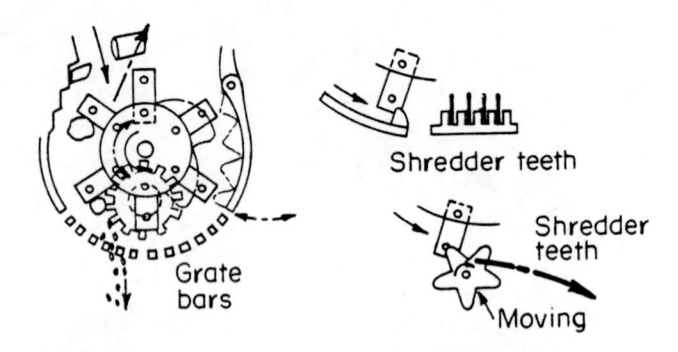

Fig. 4-7 Schematic of a hammermill (from *Recovery and Utilization of Municipal Solid Waste*).

Flail mills contain a horizontal, single-rotor unit with a studded shell and hammers attached to the rotor by means of small chains. Flail mills have been used primarily in composting operations on the incoming refuse and during a regrind procedure halfway through the digestion phase.

Vertical-axis rasps are machines which consist of a vertical axle carrying horizontal arms hinged to rotate upward. The axle rotates at about 15 to 25 rpm, moving the arms over the grinding floor of the unit. The floor consists of alternate plate sections containing perforations or welded extensions. The material to be ground is dropped into the unit, and the rotating arms move it into contact with the protruding rasping pins on the floor. Material sufficiently reduced in size drops through the holes in the alternate plate sections. Rasping units are commonly about 16 ft in diameter and are about 7 ft high. Generally capacities are in the range of 5 to 15 tons of refuse per hour; the capacity depends primarily on the size of the holes in the perforated bottom plates. In comparison to a hammermill, a rasping unit has a higher first cost and is somewhat larger; the advantages of the rasp are that it has reduced power requirements and requires much less maintenance.

A drum rasp is another type of unit also in use in the United States. As an example, a drum rasp has been developed under the trade name of "Pulverator." The Pulverator is a 6 ft-diameter, 16 ft-long drum covered on its inside wall with triangular steel plates. The axis of the drum is inclined to the horizontal. This unit has been used primarily to break up large masses of refuse, tear open bags, and mix incoming refuse before it is placed in another type mill.

Roller crushers have only limited application to composting operations. In the Netherlands a roll crusher has been used to reduce glass particles and other brittle materials in compost in a final product upgrading process.

Occasionally pulpers also have been used in conjunction with composting. At Altoona, Pennsylvania, a wet pulper consisting of a 6 ft-diameter steel bowl with a rotatable steel plate mounted in the bottom is used. The bowl of the pulper is partially filled with water, and the solid material to be composted is then added. The steel plate is rotated at about 650 rpm, and in 5 minutes, the input refuse is reduced to a slurry (about 5 percent solids). The slurry is subsequently discharged through a screen to a dewatering screw where the moisture content of the material is reduced to about 75 percent. Before the pulp goes into the digester, a second pressing operation further reduces the moisture content to about 50-60 percent.

4-5. SUMMARY

Many types of equipment have been discussed in this presentation of accessory equipment. Other specialized types are in extensive use in salvaging operations in the United States. A discussion of such equipment exceeds the scope of this presentation.

REFERENCES

4-1. Day and Zimmerman, Engrs. and Archs., *Special Studies for Incinerators*, previously cited in Chap. 3.

4-2. American Public Works Assoc. Research Foundation, "The Tezuka Refuse Compression System," A Preliminary Report to USPHS, Cincinnati, 1969.

4-3. Koentop, D. C., "Pulverizing and Compacting," *Proc.*, New Directions in Solid Wastes Processing Institute, Framingham, Mass., 1970, p. 146.

4-4. LaBarbara, J., "Solid Waste Compaction," *ibid.*, p. 152.

5

Collection systems

5-1. GENERAL

The type of collection system which is used in any particular area or in any given locale will depend in large part upon the allocation of responsibility for collection. In other words, the collection agency itself, in many cases, will be the governing factor in the design and planning of the collection system.

5-1.1 Collection Agencies—Pro and Con

Refuse collection may be performed by municipal agencies, by private companies under contract to a municipality, or by private companies individually engaged by homeowners. Each of these systems has certain inherent advantages and disadvantages. If refuse is collected by the public employees of a municipal agency and the operation is conducted under the direction of an official of the city, then certain advantages accrue: first of all, the official who is in charge of the collection agency has as his basic motivation the maintenance of sanitary conditions in the area wherein refuse is collected; additionally, the directors and employees in a municipal collection agency are aware that they are very much in the public eye and that public scrutiny is always directed at their work. Of course, certain disadvantages also exist for municipal collection of refuse. For example, since American political life is dominated by the spoils system, there may be a frequent turnover in the supervisory personnel associated with a municipal collection agency, and there is a possibility that operations of the agency may be vested in incompetent and untrained officials. Furthermore, since the major issue in relations between governmental officials and the voting public many times appears to be taxes and monies spent, the emphasis in operating a municipal collection system may be placed upon saving money and reducing costs, to the detriment of the collection operation. To avoid some of the pitfalls inherent in municipal agency operation, many municipalities throughout the country have contracted with private companies for refuse collection services. The reasoning behind such a contract procedure is that the collection of refuse will be conducted as a business venture, and political considerations will not affect the efficiency of the operation. Additionally, the burden of expenditures for equipment and other capital outlay are placed on the private company. Also, the cost to the city is definitely established before the operation so that, other than financial, the city's part in the operation is well defined—simply the enforcement of adequate collection standards. However, there are also certain disadvantages to this system: (1) the basic motivation of the private collector is profit, instead of the maintenance of sanitary conditions; (2) the contractual obligations of the private collector are definite and in general do not vary with changing conditions—any alteration in collection practices to fit changing patterns in waste generation, will have to be met with further contractual negotiations; (3) the entire system of contracting for collection services to a private concern is somewhat hazardous in that the private concern is exposed to the risk of non-renewal of contract after extremely high capital outlay, and the contracting municipality is exposed to the risk of no collection if the private company fails, or chooses not to renew the contract for any reason whatsoever.

The third system for refuse collection, that of a private contract between the individual householder or business and the collector, is found most frequently in suburban and rural areas. An obvious advantage of this arrangement is that individuals in these areas are thus furnished with collection service which may not have been provided by neighboring municipalities. However, certain distinct disadvantages are inherent in any system based upon individual contract between homeowner and private collector. These disadvantages include: possible uneconomical operation of collection, since this method is highly competitive, and overlapping of collection routes may occur; price-cutting and consequent lowering of service standards; simple failure to give adequate service by incompetent or untrustworthy smaller companies—a distinct possibility; a monopoly on services in a given area leading possibly to unjustified price schedules; and always the possibility of collusion and price-fixing by collectors.

5-1.2 Factors in Collection

After the responsibility for collection of refuse in any given area is established, the collecting agency or company must then consider certain factors in organizing the collection operation. The organization of the collection operation will include the establishment of methods of refuse pickup to be used, the types of equipment that are to be employed, and the assembling of the collecting crews. In planning and developing such a collection organization, certain factors must be considered. One group of factors will always have to be considered no matter who is doing the collection. These relatively unchanging factors include: (1) population distribution and density in the collection area; (2) the overall topography or configuration of the area; (3) the rainfall, average wind speeds, temperatures, and other climatic factors there; (4) the characteristics of the refuse produced in different subsections of the area; and (5) the land-use zoning regulations which are in effect throughout the area.* These factors must always be considered. In addition to these relatively permanent ones, another set of variable factors must also be considered. For example, in planning the collection system to be employed in any particular locality, the designer or planner must give some consideration to the disposal methods which are currently being used in that community. Furthermore, the available type of collection equipment, the customary collection frequencies in various areas, and the overlap and extent of responsibilities among municipal, contract, and private companies must be considered. Finally, the traditional or required labor practices in the given area must also be taken into account.

5-2. COLLECTION EQUIPMENT

It is pertinent to point out that the success of any collection system will depend in large part on the proper selection of equipment. The major reason for the overall importance of collection equipment is the need for mechanization of the collection operation. In the previous discussions it was pointed out that approximately 80 cents of the entire solid waste management dollar goes to pay for collection. Of this 80 cents about three-quarters customarily is paid to the labor force. In addition to being expensive, labor is also in this day and age increasingly difficult to manage. For these reasons, mechanization of the collec-

---

*Land-use zoning may restrict location of transfer stations or disposal facilities and may thus affect collection practices.

COLLECTION SYSTEMS

tion of solid waste is more and more advantageous in recent times. The two most common types of mechanized equipment for collection of solid waste in the United States today are first, trucks with compactor bodies which are manually loaded; and second, equipment which accomplishes mechanical loading of the compactor body.

The high cost of labor makes it imperative to use some type of compacting body on collection equipment in lieu of open-bodied or makeshift equipment. A conventional compactor-body truck as used in common practice in the United States has a capacity of from 16 to 24 cu yd and generally will be staffed by a crew of 2 or 3, a driver and 1 or 2 loaders. Since the vehicles are designed specifically for collection of refuse, some advantages are, in effect, built into the vehicle: a leakproof, covered body is installed to prevent corrosion and chemical attack by the contained refuse; the body is easily cleaned and therefore generally presents a respectable picture to the public; it is designed for convenience in manual loading and ease of mechanical unloading; appurtenances such as handholds, steps, mirrors, etc., make the designed collection truck safer for use than a conventional one pressed into service as a refuse collection vehicle.

5-2.1 Equipment Innovations

The increasing trend toward mechanization has created some innovative practices in particular areas of the country. For example, in some areas where

Fig. 5-1 Modern compactor truck for solid waste collection (*courtesy* The Heil Co.).

houses are situated at considerable distance from the street or roadway the refuse loaders are equipped with small motor scooters or pickup scooters so that economies are achieved by savings in walking time. Mechanization has been carried to the ultimate degree in a new experimental collection device developed and currently put into operation by the Gulf Oil Company, in which one man operates the entire truck; by means of a large overhead grab, he dumps a can, replaces the empty can, compacts the load into the truck body, and drives on to the next residence. The use of this one-man truck must still be considered experimental. A more practical and proven system involves the use of a mechanically loaded body. In these systems a detachable container or body is left in place at the refuse source and most of the labor associated with collecting the refuse is furnished by the customer who is generating it. Operation of these systems varies from locale to locale and also with manufacturer. In almost all cases the customer puts the refuse into the detachable container; the container may then be emptied into a collection truck and replaced at the storage site or it may be carried to a disposal site and emptied there. Likewise, the refuse may be compacted by means of appurtenances on the detachable body, so that compaction is achieved at the point of generation, or the refuse may be compacted within the collection vehicle during its trip to the disposal facility. The different types of detachable containers vary in size from 1 to 40 cu yd or more capacity. In some instances the containers have been equipped with wheels so that a so-called refuse train may be formed by attaching the containers to a single tractor which hauls them to the disposal site. Another variation includes the use of container stations for rural collection, as has been demonstrated in Chilton County, Alabama. Each of the systems described herein offers some distinct advantages and also possesses certain disadvantages in comparison to the others. It is apparent that a continuing trend toward a greater amount of mechanization is existent in the field of solid waste collection and handling. It also appears that some form of mechanized detachable-body truck might well be the predominant system of waste collection in the future. The exceptions may possibly be vacuum-cleaner-type trucks which can be used for collecting leaves, and refuse from litter baskets. Finally, some small portion of the collection in any locality will have to be done in open-bodied dump trucks; such trucks will be used for hauling bulky items such as wrecked cars, large appliances, discarded furniture, etc.

5-3. TRANSFER STATIONS

The apparent advantages of volume reduction and mechanization of collection equipment and collection practices have been combined to create a rather specialized facility for solid waste collection systems—the transfer station. The basic idea of a transfer station is that refuse is collected by a large number of small collector trucks and crews, and is brought to a central location where it is compacted to a high density. It is then hauled to an ultimate disposal site in larger trucks, which are especially suited to the long-haul operation. Transfer stations may appear to be a recent innovation, but they have actually been in use since the era of horse-drawn collection vehicles. The older transfer stations were made obsolete by the introduction of modern trucks.

5-3.1 Advantages of Transfer Stations

At the present time transfer stations are again appearing as a portion of the total solid waste collection system. The reasons behind the attractiveness of such a facility are that, first of all, disposal sites near the sources of solid waste generation are becoming harder to locate; second, the responsibilities for collection and disposal in many localities are separated and placed in the hands of different agencies or companies; and finally, the high cost of labor militates against any waste of time for the collection crew, during hauling of collected refuse to a disposal site. For a transfer station to function adequately, the system including the station must in essence be equal to or better than the system without it. In other words, the adequacy and reliability of the system with a station should exceed the capabilities of a system where the collection trucks haul the refuse directly to the ultimate disposal site. If the inclusion of a transfer station into the system creates a lowering of the standards of operation, a sanitary hazard, or confusion and loss of time on the part of collection personnel, then the composite system will not be as economical as the original system without the transfer station.

5-3.2 Requirements of a Transfer Station

To achieve this overall economy, a transfer station must first be located near the center of the weighted* collection area it is designed to serve. Additionally, it must be located and built in such a way that public objection to its construction is minimized. Finally, since the refuse is to be transferred at this location for long haul to a disposal site, the station should be conveniently located with respect to several different forms of transportation. The detailed design of a transfer station and its ancillary equipment will depend upon a number of factors including such items as: the amount and characteristics of the refuse to be processed; the equipment used to collect it; the type of ultimate disposal facility in use; and the general willingness on the part of the community to furnish and accept a site for construction of such a station. It may be designed so that it can operate with rail haul transport, with waterborne transport, or

---

*"Weighted" signifies consideration of a collection area with respect to both geographic form and population refuse distribution.

with over-the-road motortrucks. Obviously, the transfer station is also an ideal location for centralized volume reduction equipment.

5-3.3 Analysis for Transfer Stations

The decision to use or not use a transfer station in a refuse collection system must be based upon a comprehensive investigation of the suitability of transfer operations to the particular area under investigation. The factors which must be considered include the haul distance to the ultimate disposal site, the time of travel to the latter, and the overall efficiency of the particular type of transfer operation. A complete cost analysis of the system must be made including collection, transfer station operation, and long-haul cost versus the cost of hauling by the collection equipment alone. In order for the transfer station to be economically viable the total cost of collection, transfer, and disposal must obviously be less than the total cost of collection plus direct haul plus disposal. The economics of a typical transfer station system may be graphically illustrated as shown in Fig. 5-2. Typical costs for a transfer station for Los Angeles County, California (see Ref. 5-1), include the following:

Operation and maintenance	$2.49/ton
Amortization	0.19/ton
Total	2.68/ton

For a transfer station located in Orange County, California:*

Operation and maintenance	$0.72/ton
Cost of transfer haul	0.92/ton
Total	1.64/ton

5-4. OTHER TRANSPORT SYSTEMS

In addition to the use of transfer stations, several other innovative practices are being considered for the transport and collection of solid waste. Included in the methods being developed are the transfer of wastes by pneumatic or hydraulic means. Currently in the United States vast quantities of mineral commodities are transferred from point to point, sometimes over great distances, in pipelines. For example, in one pipeline over 100 miles long from a West Virginia coalfield to a consumer in the state of Ohio, more than 1 million tons of coal are transferred each year. In addition to transfer in water pipelines, other industrial transfer systems make use of pneumatic conveyance; light commodities such as alumina, cement, raw grains, and flour are currently conveyed pneumatically.

---

*Includes no collection costs; 26 miles round trip; four stations-average capacity 910 tons/day for each station (see Ref. 5-2).

COLLECTION SYSTEMS   51

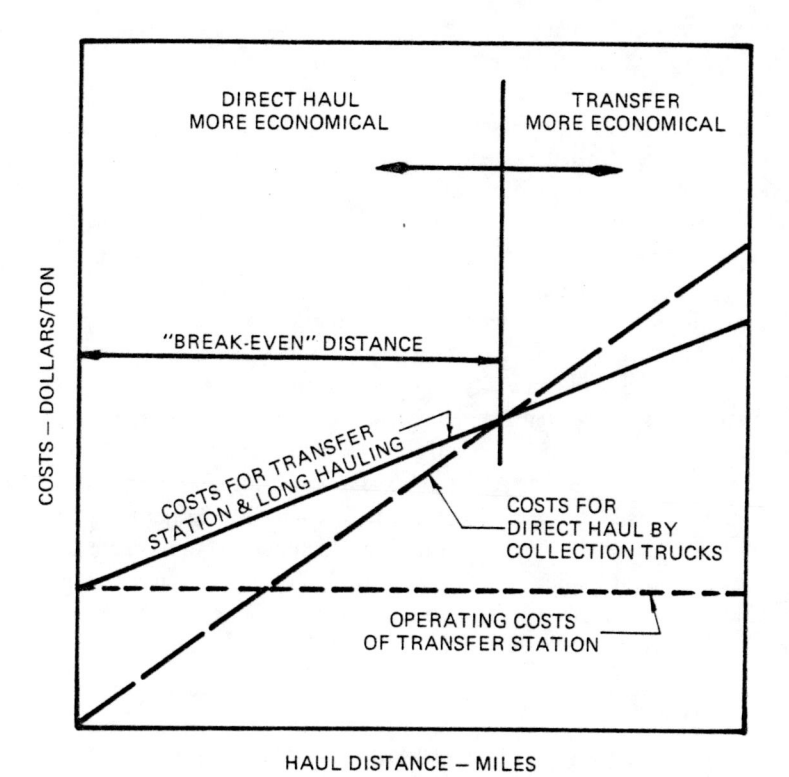

Fig. 5-2 Economic analysis—transfer stations.

Transport of solid waste by means of hydraulic systems has been considered for a number of years, but early considerations in the 1930s were centered on the gravity transport of solid waste in water pipelines. For example, in the Garchey system, the wastes are fed through a 5 in.-diameter opening in a kitchen sink to a 3-gal retention compartment. The compartment is periodically flushed and the wastes are transferred to a central holding tank. Excess water is drained from the latter and is wasted to a sewer. The collected refuse is then removed and placed, by means of suction transfer, into compactor trucks with a further removal of excess water again to sewer facilities. This system has many advantages in that almost all household wastes may be rapidly and quietly removed; it seems to be particularly well suited for such installations as multiple-unit housing where the use of on-site incinerators has been outlawed. However, several distinct disadvantages exist for this system: 1) economically, this system can only be installed in new construction and even there requires separate costly plumbing facilities; 2) large items of waste must be reduced in size so that they can enter the 5 in. diameter for flushing in the system; 3) the system is more

52 SOLID WASTE MANAGEMENT

complex than other means of collection and is subject to plugging, generation of odors, and special maintenance. Moreover, sewage treatment requirements are increased. Hydraulic transport of solid waste at the present time appears to offer more hope in rather specialized situations. For example, in a research study sponsored by the United States Public Health Service, Zandi (Ref. 5-3), at the University of Pennsylvania, determined that the waste from the center city area of Philadelphia could be effectively and economically collected and transferred by means of pipelines to a disposal facility. Further use of hydraulic transport in similar situations now appears feasible.

Pneumatic transport systems in contrast have several disadvantages that hydraulic systems do not, in that more complex control valves and mechanisms must be installed and generally high speed turbines are required. However, pneumatic transport systems are planned for installation in the Martin Luther King Hospital in Los Angeles, California. Existing pneumatic systems are presently being used in the Walt Disney World amusement park in Orlando, Florida, and Sundyborg, a suburb of Stockholm, Sweden. The Sundyborg system has been in use since late 1966 in an apartment complex where five thousand units are being served. Installation costs are quite high for pneumatic systems but the expected life of the system is quite long—up to 30 years in the case of the Sundyborg one. Specific details of the Sundyborg system are available elsewhere (see Ref. 5-4).

5-5. SPECIAL COLLECTION PROBLEMS—AGRICULTURAL WASTES

In recent years because of changes in the food processing industry, the amounts of agricultural wastes have grown alarmingly (Ref. 5-5). In addition to changes in food processing, several other circumstances have contributed to the problem of handling of agricultural wastes. For example, each year less and less good farmland is available in the United States as a result of "suburban sprawl." The technological and economic situation in the country today has forced farmers into specialization, with only a limited number of crops, and has created a greater dependence upon fertilizers and biocides. The dependence upon fertilizers has in turn led to a widespread reduction in the use of manure, other animal wastes, and plant residues as fertilizers and soil conditioners.

Agricultural wastes can be classified as either animal or plant residues. The primary generators of animal wastes include dairies, beef feedlots, hog confinement feeding lots, and poultry and egg production facilities. The predominant waste generated in all of these facilities is manure; to a lesser extent urine is produced. Additionally, contaminated bedding must be disposed of along with a certain number of dead animals. Because of the compact layout and plant of modern dairies and feedlots another problem has been created: many times insufficient amounts of land are available for the spreading of manure and subsequent reworking of it into the soil. If the floor of the feedlot or the dairy is hard-surfaced, a problem may result with the contamination or pollution of surface waters as a result of rainfall runoff from feedlots and wash water from dairy floors. If, on the other hand, the feedlot surfaces are porous and permeable a potential for pollution of groundwater resources exists. In addition to the problems of pollution of surface and subsurface waters, other problems arise in the harborage and nurturing of disease vectors such as flies and other insects and the generation of noisome odors. Adequate treatment of animal wastes from these feedlots, dairies, etc. includes adequate treatment of the process waters or rainfall runoff by means of anaerobic digestors, aerated lagoons, oxidation ditches, drying processes, or storage pits. Composting of animal wastes, particularly manure, has been attempted in some areas. While certain limited successes have been attained, little hope is held for widespread treatment of manure by composting.

A second major category of agricultural wastes consists of plant residues. Formerly, many of these were recycled or disposed of in other ways which are presently considered unacceptable. For example, formerly stalks, stems, and stubble were burned in such a way that air pollutants were released into the atmosphere. To some degree recycling was achieved by the spreading of plant residues over the land and plowing under the wasted material. Recycling in this manner is still the most desirable means for disposal of these wastes in that it essentially conserves this natural resource. However, in today's situation where multiple crop planting practices are prevalent, there is a reluctance on the part of farmers to forgo additional crops and income while working stalks, stubble, leaves, and culls back into the soil. Some plant residues are quite high in food value and therefore can be used as a food source for livestock. Fruit and vegetables particularly are high in energy and may be combined with bulkier vegetable wastes to from a livestock feed which is high in energy value and also furnishes the necessary roughage. Additionally, some hope is held out for the disposal of bulky plant residue items in conjunction with the disposal of manure. When the bulky portion of the plant remains are mixed with livestock manure and composted, the end result is that the residues decompose in the soil and also to some extent reduce the fly and odor problems which commonly occur in the composting of manure alone. In situations where none of the previously suggested disposal techniques appears feasible, plant residues should be buried in a landfilling operation in such a way that no problems with water pollution, disease vectors, or odors are created.

REFERENCES

5-1. American Public Works Assoc., *Refuse Collection Practice*, Proj. No. 101, Chicago, Public Admin. Service, 1966.

5-2. Orange County Road Dept., "The Orange County Refuse Disposal Program," Santa Ana, Calif., 1965.

5-3. Zandi, Iraj, "Pipeline Collection and Removal of Domestic Solid Wastes," Engineering Foundation Research Conference, Solid Waste Research and Development II, 1968.

5-4. *Elements of Solid Waste Management*, A Training Manual, Environmental Protection Agency, 1970.

5-5. Pennsylvania Conference on Agricultural Waste Management, *Proc.*, Pennsylvania Dept. of Agriculture, 1970.

6

Reuse and recycling

6-1. GENERAL ASPECTS OF RECYCLING

From previous chapters it is obvious that significant savings in costs for solid waste management can be achieved if reductions in collection requirements are made. Additionally, compositional analyses have shown that significant amounts of highly valuable materials such as steel, aluminum, and other metals form a significant part of the solid waste stream. Therefore, the removal from that stream and reuse of such materials would reduce volumes of wastes to be collected and would also yield significant salvage and resale income (see Fig. 6-1). Thus "recycling" seems to be a one-word solution to a large part of the current solid waste problem.

In discussing the reuse and recycling of materials which now appear in the solid waste stream several facts are obvious:
 1. the predominant material in recyclable solid waste is paper
 2. of the materials in the solid waste stream a large portion of the material is recoverable

56 SOLID WASTE MANAGEMENT

METALS RECYCLING

From discarded copper wire to copper ingots to new copper products.

PAPER RECYCLING

Graded and baled waste paper is converted into new paper and manufactured into books.

TEXTILE RECYCLING

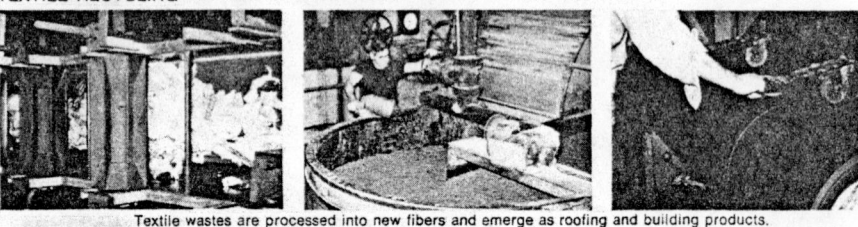

Textile wastes are processed into new fibers and emerge as roofing and building products.

Fig. 6-1 New raw materials and products from recycled solid waste (*courtesy* NASMI).

3. vast potential exists for expanding the current resource recovery rate (see Fig. 6-2)
4. the amount of solid wastes has increased phenomenally during the 1960s and 1970s and will continue to do so
5. there is no simple solution, no simple way to recycle and reuse solid waste materials

6-2. POTENTIAL FOR RECYCLING

Undoubtedly recycling is at present the most constructive approach to alleviation of the solid waste problem. Solid waste will always be produced as long as man inhabits the planet: however, it can be qualitatively controlled so that recycling would be much more feasible. Recycling and reuse are not magic actions to get rid of the problem. Recycling must be economical as part of the total system of industry in the United States. In other words, recycling means

REUSE AND RECYCLING

MAJOR MARKETS FOR RECYCLED MATERIALS

MATERIAL	MAJOR MARKETS	PERCENT OF TOTAL CONSUMPTION BY EACH MARKET
ALUMINUM	CASTING ALLOYS WROUGHT ALUMINUM PRODUCTS	71 24 95
COPPER	BRASS MILL PRODUCTS BRASS AND BRONZE FOUNDRY PRODUCTS WIRE AND WIRE PRODUCTS	47 25 20 92
LEAD	STORAGE BATTERY LEAD TETRAETHYL LEAD SOLDER	68 13 5 86
ZINC	GALVANIZING SLAB OXIDES AND CHEMICALS DUST	40 25 19 84
NICKEL	STAINLESS STEEL NONFERROUS ALLOYS	52 14 66
STAINLESS STEEL	STAINLESS STEEL ROLLED PRODUCTS EXPORTS	74 14 88
PRECIOUS METALS	JEWELRY PHOTO CHEMICALS CATALYSTS ELECTRICAL AND ELECTRONIC	NOT APPLICABLE
PAPER	PAPERBOARD CONSTRUCTION, PAPER AND BOARD (INCLUDING GYPSUM WALLBOARD)	71 17 88
TEXTILES	WIPERS PAPER EXPORTS PADDING AND BATTING ROOFING FLOCK AND FOLDER	16 14 13 11 7 7 68

RESOURCE RECOVERY RATE

MATERIAL	SHORT TONS AVAILABLE FOR RECYCLING	SHORT TONS RECYCLED	% RECYCLED
Aluminum	2,215,000	1,056,000	48
Copper	2,456,000	1,489,000	61
Lead	1,406,000	585,000	42
Zinc	1,271,000	182,000	14
Nickel	106,000	42,100	40
Steel	141,000,000	36,700,000	26
Stainless Steel	429,000	378,000	88
Precious Metals	105,000,000 (troy ounces)	79,000,000 (troy ounces)	75
Paper	46,800,000	11,400,000	19
Textiles	4,700,000	800,000	17

Fig. 6-2 Expansion of resource recovery rates appears plausible in view of current recovery rates and major secondary material markets.

not merely collecting materials for reuse, but also the development of a market for the recycled materials. At the present time in the United States, for example, 600 pounds per person per year of paper is used. Of this amount less than 20 percent is recycled. Three times as much raw pulp is used in paper manufacture as recycled pulp. This ratio is likely to be increased with the present market status. Since it is currently more profitable to cut down trees for virgin pulp than to use recycled paper, a demand must be created for recycled paper. Several factors cause the present "low demand" situation. For example, it is cheaper to transport some raw materials over the nation's railroads than it is to transport recycled materials involving the same basic substance. An example is raw paper pulp, which is cheaper to transport than recycled paper scrap.

Finally, it would seem that an entire area has been neglected. There has been practically no market research done on the effects on brand loyalties, buying habits, and economic profits to manufacturers of the "ecological restriction."

In other words, what will be the economic effects on the sales efforts of a given company if one or more of the products of that company are banned from sale for ecological reasons? There has been little or no research on the level of consumer concern over environmental pollution and his willingness to pay increased cost for pollution-free consumer items.

It is now pertinent to investigate some materials in detail for their recycling potential and their present impact on the recycling situation in the country.

6-2.1 Recycling Paper

A visit to any solid waste landfill will give the impression that an overwhelming percentage of all solid waste consists of paper. Basic data presented in Chapter 2 shows that this impression is a correct one. Approximately 50 percent (by weight) of all solid waste is paper; this corresponds to nearly 70 percent by bulk. What can be done about this large amount of paper in the solid waste stream? The obvious answer to the problem is recycling. It has been calculated that recycling one ton of paper will save seventeen trees. This type of simplistic statement typifies the approach of many people to the problem of recycling today. This answer is an unfortunately simple approach to a very complex problem. When discussing recycling of paper, it is pertinent to examine the limitations of any recycling process and also the economics specifically of paper recycling. First of all, recycling does not save trees in the strictest sense of the word *save*. Like any other vegetable product, trees grow, mature, and die. As we have done with cereal grains and other agricultural products, it seems practical to harvest trees at the peak of their growth and use them to their fullest potential before they pass the peak of maturity and die. Furthermore, a growing or young forest will produce more oxygen than will a mature or dying forest, so cropping of trees, in some senses, may actually produce more generated oxygen per acre than simply following a policy of noninterference in virgin forests. The practice of growing trees for harvest, including advanced techniques of forest management, has produced a situation in the United States today where more trees and more board feet of timber are grown each year than are cut. For example, in the southeastern United States an unmanaged acre of timberland has an average production of approximately one-third cord of wood per year; under proper management techniques the same acre will produce two and one-half cords of wood per year.

During the same time that forestry methods have improved tree yields, recycling practices have declined. During World War II about 35 percent of all our paper was recycled, but in 1970 only about 19 percent of all paper was. Why has recycling decreased? The answer lies in economics. For one thing, paper companies prefer virgin fiber as a more stable, uniform, and dependable supply source than recycled paper which fluctuates considerably in price and supply. Additionally, sources for virgin fiber have expanded considerably. For example, in 1970, about 25 percent of all paper was made from wastes which formerly would not have been utilized. At the present time sawdust, chips, and other mill wastes which formerly were burned are being used in more efficient pulping operations.

Other factors affecting the decrease of recycling and the increase in overall paper use are consumer preference and disposal practices. The self-service practices in modern food markets have reduced the price of food, increased the number of items available, and decreased the amount of time spent by homemakers in food preparation. Thus throwaway paper packaging has increased in popularity.

In retrospect, however, paper packaging has also reduced solid waste to some extent. For example, vast quantities of citrus juices are consumed in the United States each year. Because of packaging capabilities, fresh fruit can be processed so that more than half the original fruit is left behind in the growing area in the form of peels and pulp. These peels and pulp are often recycled into animal feed at the present time. As another example, frozen food packaging now keeps much of the waste portion of fresh foods (bones and stalks, for example) out of the urban waste stream. Much of this, however, must be disposed of as solid waste at the source of processing.

As mentioned previously, economics are today the primary factor in recycling of paper. Approximately 90 percent of the cost of wastepaper recycling is consumed in the collecting, sorting, and transporting of the waste material. Because of rising labor costs and little applicability for mechanization, collection costs for paper have increased rapidly in recent years. Consequently, wastepaper products have been replaced in material markets by virgin materials and plastics. At the present time recycled pulp is being used in only a limited number of products. In order for recycling of paper to become widespread, there must be a change in the public attitude toward recycled products, and development of new markets for recycled paper.

First, a realistic approach to the problem of recycling is necessary. Total recycling of all paper is not possible. Approximately one-eighth of all paper production is unrecoverable. Much of this paper is in permanent use such as in structural materials or books. Another significant portion is lost in residential fireplaces, or is disposed of as tissue products in sewage systems.

A second factor in the recycling concept is that paper, unlike metals, is not as good as new after it has been recycled. Each time paper is recycled the fibers become somewhat shorter and more frayed, with the consequence that the resulting recycled product is weaker than a similar product made from virgin fibers. If strength is not a factor in the use of the pulp materials, then the recycled fibers may be as suitable as the raw fibers, although color may still be a factor. However, in other products such as cartons or containers weaker fibers of the recycled product would necessitate thicker and wider gauges for boxes and other containers. The total weight of containers would therefore increase, thereby magnifying the overall solid waste problem.

Additional limitations on the use of recycled paper also exist. For example, reuse of recycled paper is prohibited by laws or by rulings of the federal Food and Drug Administration which require that paper products which come into direct contact with food contain no materials or substances which have been used previously in any sort of contaminatory use. Of course, these regulations do not apply to containers in which food is contained in an inner pouch or wrapping.

The most complex issue in the whole concept of recycling of paper is the socioeconomic problem. Many paper mills are presently located in rural areas and function as the predominant employer in those areas. They are major buyers of agricultural products (trees) and could not subsist, if they were converted to a recycling operation, on the wastepaper produced in these rural localities. The alternative is to move paper mills from their rural locations to the major sources of wastepaper, the cities. In such a move a "wastepaper" mill would probably be built for less money than would be required to rebuild a "virgin forest pulp" paper mill. There is some question also about the suitability of moving these large manufacturing plants into the centers of urban areas. All of the paper mills built in the last 20 years are virgin fiber mills and are not suited either by location or equipment for processing wastepaper. This tremendous capital investment is not portable, and the cost of moving such mills from their present locations would put a disastrous burden on the paper industry. Certainly the construction of new mills designed specifically for recycling should be considered for the future. However, some of the special pollution problems involved in recycling must not be neglected. For example, de-inking of waste papers presents a special disposal problem: the typical de-inking plant loses about 25 percent of its input to waste; the paper itself decreases about 10 percent with each recycling and the lost material escapes as suspended solids in the plant's wastewater. Recycling mills produce more suspended solids in wastewaters than do virgin fiber mills. This is particularly true where magazine stock is recycled. Coated magazine stock such as is used for non-pulp magazines produces as much sludge (ink and clay) as recovered fiber in a recycling process. Obviously the waste problems associated with recycling of paper are themselves quite serious.

The overriding consideration, however, remains the economics of collection. Garden State Paper Company, the nation's largest manufacturer of recycled newsprint, has published forecasts that recycled newsprint will never penetrate more than 10 to 15 percent of the newsprint market. The basis of this prediction is that, first of all, an efficient mill size is a minimum of 350 tons per day of paper production. Since only about 15 to 25 percent of newspapers are expected to be returned in urban areas for recycling, and since the annual per capita newspaper consumption is about 100 pounds, there are only a few locations in the United States that can support an economically feasible mill. Garden State's three current mills are at present located near New York, Chicago, and Los Angeles; this is an obvious demonstration of these economic considerations. However, the situation is not without hope. Several factors make it appear hopeful; one of these is the increased research devoted to the development of an economic method of collection and separation. Another favorable sign is evident in an experiment conducted in Madison, Wisconsin, where through the cooperation of the city government and the populace, approximately 40 percent of all distributed newspapers were collected. Finally, an additional hopeful sign is the consideration by some municipalities of a subsidy to private firms of haulers to collect and remove paper waste before the city collection takes place.

How can paper, especially paper containers, be redesigned to alleviate the solid waste problem? For one thing, if containers were marked conspicuously to show that such contaminants as asphalt tape, wax lining, or pressure-sensitive materials had been used in their manufacture, recycling agencies could more easily separate these contaminated cartons from cartons which are much easier to recycle. Another possibility would be to legislate against the use of these contaminants on corrugated packaging. This, however, is a very complex matter, since the materials which are recycling contaminants do much to enhance the utility of the cartons. Any legislation restricting these contaminants would put affected containers at an economic disadvantage with their competitors: wire bound boxes, steel strapped containers, plastic shippers, etc. Other concepts also must be investigated. One example is the use of paper fiber with water-soluble plastics which after use could be compressed into building blocks. Another possibility would be to convert the wood sugar in paper fibers to a high-protein animal feed; a danger of contamination must also be faced here.

One expedient which is suggested quite often is a simple reduction in the amount of packaging. Does overpackaging exist? This is a complex question to answer. Is an inner stack wrap for crackers, which prevents staling of the last portion, overpackaging? Is a blister card which allows small items to be sold self-service without excessive pilferage also overpackaging? It is questionable whether these materials are indeed overpackaged. It should be noted that packages themselves are not purchased by consumers, but are in fact bought by professional buyers acting for retail outlet chains. These buyers are constantly asking suppliers to reduce the thickness, density, and area of containers.

The most promising use or reuse of paper is the conversion of the material to energy. Paper has a high caloric value, approximately half that of coal, and a third that of fuel oil. In contrast to those materials, however, paper is a replaceable resource. Paper under controlled incineration is a non-polluting fuel, if its by-products, carbon dioxide and water vapor, are both considered natural to the air. Certainly a massive increase in carbon dioxide generation would be considered a polluting mechanism, but carbon dioxide is certainly preferable to some of the more corrosive or toxic gases produced in burning of other materials such as high-sulfur coals. In addition the plastic and wax coatings which render paper difficult for biodegradation certainly enhance its value as a fuel and contribute to its caloric value. Perhaps the idea of the use of paper as a fuel so that energy may be reclaimed in this recycling process is presently one of the most promising reuses of paper.

6-2.2 Recycling Metal

On the subject of metal recycling several general comments may be made. For purposes of discussion, there should be, first of all, some separation of metal into ferrous and nonferrous types. A flow chart showing the recovery route of nonferrous metals is presented in Fig. 6-3. A principal nonferrous metal which is subject to recycling is aluminum. The major source of all recoverable metals is, again, in packaging. Considerable discussion has been devoted to the throwaway beverage can. These cans remain one of the principal sources of recyclable metals, but they are not, by any means, the only source of reusable metal.

The tin can (which in actuality is a tin-coated solder-seam steel can) is familiar as a recoverable item to many people from their recollections of the scrap drives during World War II. During this crisis in our nation's history many persons collected tin cans, and completely cleaned them by removing the labels and the glue by which the label was affixed. To conserve space the ends were often separated from the rest of the can, placed on the inside of the can and the can then flattened for processing. In contrast to the efforts during World War II, modern recycling efforts have tended to decrease. For example, in 1956, 115 million tons of raw steel were produced and of this total 37 million tons came from scrap steel. In 1970, 131 million tons of raw steel

REUSE AND RECYCLING 63

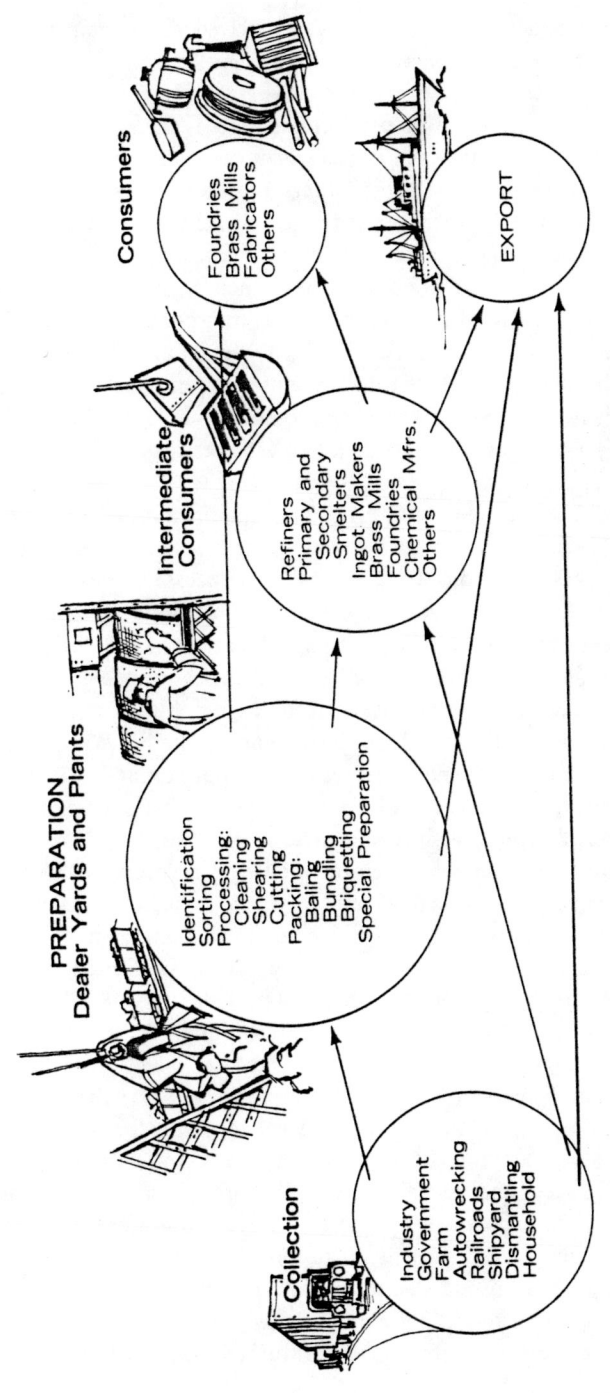

Fig. 6-3 Flow chart of nonferrous scrap metals (*courtesy* NASMI).

were produced and 40 million tons were derived from scrap steel sources. In other words, there has been only a 7½ percent increase in recycling of steel while there has been approximately a 13 percent increase in steel production. One hopeful item for recycling of steel products is the evolution of the modern automobile shredder, which can dispose of a complete automobile body in 30-90 seconds. These automobile shredders represent a capital outlay of from $600,000 to $6 million each depending on capacity and represent one way to recover and recycle steel from the solid waste stream. Several problems exist in the recovery of automobile bodies; these include handling costs for derelict automobiles (who pays for collection and transport of abandoned vehicles), and maintenance of a constant market for the scrap produced. This last item is perhaps the most crucial item in the entire automobile recovery program. A constant steady quality market for automobile scrap has not evolved in this country. The development of such a market is critical to the entire picture of automobile scrap recovery.

The automobile shredder can be regarded historically as an outgrowth of the tin can shredder which was developed earlier. At the present time between 400,000 and 500,000 tons of tin cans are processed each year in shredding operations most of which is used in copper precipitation, principally in the copper mining areas of the Far West. The location of the copper mining areas in the western states points up one of the significant problems in any reuse of tin cans—the cost of transportation from high-population areas on the East Coast to the reuse areas near the copper mines in the Far West. For example, in 1971 many scrap companies were paying approximately $20 per ton for shredded scrap tin cans, delivered. Freight rates in most cases exceeded the price of $20 per ton for the long-haul distances to the Far West. On the other hand, these same tin cans could be used in smelting operations for the steel mills in the Pennsylvania area. As an example, tin cans collected in the Washington, D.C., area and shipped to the Bethlehem Steel Plant in nearby Pennsylvania would have borne the following costs (see Ref. 6-8): freight rate—$4 per ton, baling costs—$8-10 per ton, and collection costs—approximately $7 per ton. Thus the cost of collecting, baling, and shipping these cans approximately equals the price paid by the steel company to the collection agency. In addition to the problem of economics of collection and transport, several metallurgical problems also exist. The tin coating on steel cans and the lead solder along the can seam are detrimental in steel processing. These contaminants produce several effects in the steel and can also affect the processing method itself. The tin present during the processing of steel renders the final product somewhat brittle. Additionally, lead present in the ferrous scrap will settle to the bottom of electric furnaces, invade the refractory material there, and soon render the furnace inoperable. There is some hope, however, for the use of no-tin cans and the removal of the lead solder seam in future cans. If such advances are indeed made and if the consumer can be persuaded to separate and collect steel cans, an economic market will exist for this scrap material. At the present time scrap steel dealers are fully equipped to handle all of the scrap material which would be made available from such steel cans.

A somewhat more hopeful situation exists for the recycling of aluminum cans.

Early recycling efforts were directed primarily at the all-aluminum motor oil can, and as early as 1957 attempts were made to recycle these cans. At the present time motor oil cans are constructed of composite materials that are no longer reclaimed in this way. However, much aluminum is being reclaimed from other containers. Americans are now generating refuse at a daily average of 1 billion pounds. This billion pounds includes approximately 5 million pounds of aluminum, 70 million pounds of steel, 2 million pounds of copper, and a lesser amount of other metal materials. The amount of aluminum currently being wasted amounts to about 20 percent of the total production capacity of the country per day, while the ferrous metals wasted amount to only 10 percent of the daily nationwide production (see Ref. 6-9). Furthermore, aluminum scrap is more easily reprocessed than is other metallic scrap and represents a higher value percentage of the production costs of recycled materials than do other types of metals. An example of the present recycling efforts for aluminum cans in the country today is the Reynolds Metals Company program. Reynolds is currently attempting to reclaim all aluminum beverage cans, through nationwide can reclamation centers, and is also conducting several research efforts directed toward testing the possibility of public participation in the reclamation of aluminum. Since it is felt that a portion of the population would not bring aluminum household scrap to a reclamation center, there is in progress a test program aimed at homeowner separation of aluminum scrap. This program is felt to be worthwhile since the home is the last place in the solid waste stream where aluminum is easily identifiable, separable, and relatively simple to handle. In addition to these programs, a third effort is being undertaken to determine what methods are the most favorable for aluminum recovery from the solid waste materials collected by municipal or private collection agencies.

The can reclamation program began in 1967 in Miami, Florida, and centers were later established in Los Angeles, San Francisco, Phoenix, Houston, Tampa, Jacksonville, Newark, and New York City. At the present time (1973) the aluminum is received, magnetically separated, (see Fig. 6-4), and then shredded for further processing (see Fig. 6-5). It leaves the reclamation center in carload lots, being sent to a smelting operation. The rate of payment (1973) is 10 cents per pound or $200 per ton, which is equivalent to about a half cent for each all-aluminum beverage can. The success for this reclamation program may be judged from the fact that during March 1970 Reynolds' can reclamation centers collected over 1 million pounds of aluminum (equivalent to 20 million cans per month). The system in practice at this time is an economically feasible program.

66 SOLID WASTE MANAGEMENT

Fig. 6-4. The aluminum cans are placed in a hopper and carried along a moving belt through a magnetic separator which removes the steel cans (*courtesy* Aluminum Assoc.).

Of course, these figures must be viewed in the light of another statistic: approximately 5 billion cans are produced each year in the United States.

In addition to can reclamation centers, Reynolds is attempting to recover other large proportions of aluminum consumer scrap through homeowner separation programs. In these programs refuse bags, graphically coded to indicate that they contain aluminum, are given to homeowners, who are asked to separate recoverable aluminum materials, place them in the bags, and set the bags out for collection. These programs are not complete but show some early promise.

Reynolds and other manufacturers have research efforts devoted to the separation of aluminum and other nonferrous metals from refuse. These efforts are being conducted under certain limiting conditions: (1) much of the past work in separation research has been directed toward the reuse of large bulk items; (2) little research has been devoted to the recovery of less bulky but potentially more valuable solid items such as other metals; (3) in order for the recovery of minor bulk items to be practical, this recovery operation must be preceded by salvage, conversion, or destruction of the major bulk items in the refuse; (4) the removal of the major bulk items should not hamper the recovery of minor-

Fig. 6-5 Aluminum cans are usually reduced to dime size chips in a hammer mill for shipping purposes (*courtesy* Aluminum Assoc.).

volume constituents, which have a high salvage value; (5) the separation of metals in municipal refuse is based upon the fact that these metals occur in an uncombined state and therefore the physical characteristics of the individual metals must be used in the separation process.

On the basis of these limitations a procedure for aluminum recovery has been devised by Reynolds. The procedure in essence includes the following steps:

1. Mixed municipal refuse is shredded, dried, and fed into a pyrolysis unit.

2. In the pyrolysis operation the refuse is incinerated under a controlled low-oxygen environment so that aluminum oxidation is minimized (pyrolysis is in essence a destructive distillation operation and is carried out at temperatures of 1,000–1500°F—see Sections 6-4.3 and 11-4 for further information). After the material has been incinerated and leaves the pyrolysis unit it is cooled and mechanically fragmented in a rod mill.

3. After cooling, ferrous and nonferrous materials are size-separated mechanically on screens so that they may be further processed. The magnetic materials are removed for reclamation of the ferrous metal content.

4. The nonferrous materials after being separated into a large fraction and a small fraction on a 2-mesh screen are further processed—the smaller portion con-

68 SOLID WASTE MANAGEMENT

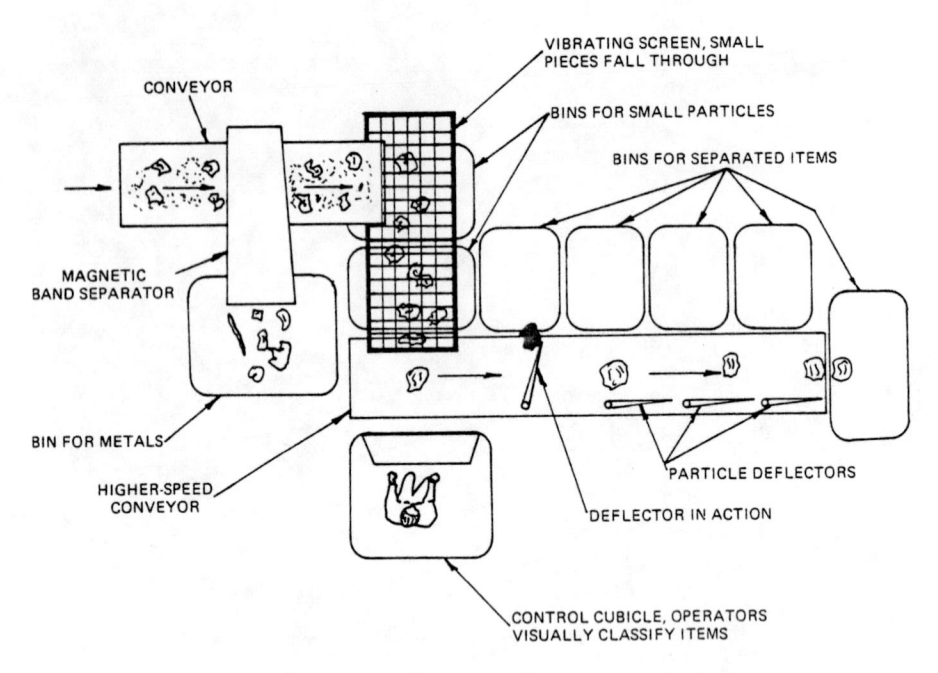

Fig. 6-6 Separation by human coding, schematic diagram.

sisting of jar glass and nonferrous metals is placed in an air classifier which essentially separates the glass and nonferrous metals from the lighter carbonized char. The char is removed at this point for further processing.

5. The material which does not pass the 2-mesh screen, the glass and the nonferrous metals which have been separated by air classification, are worked in a rod mill and subjected to additional processing.

6. After separation on a 35-mesh screen the nonferrous metals are subjected to dense-media separation wherein the specific gravity of the material is used for differentiation so that aluminum is separated from the other nonferrous metals.

7. Glass which has been processed through the 35-mesh screen is subjected to high intensity magnetic separation so that colored glass is separated form clear glass. The end result of this separation procedure is that char, colored glass, clear glass, nonferrous metals other than aluminum, aluminum, and ferrous metals are all separated out of the recycled waste.

It is apparent that aluminum recycling is technically feasible, and there seems to be a market for recycled aluminum. Obviously, high-purity recycled aluminum can be used over again as a substitute for the pure virgin material. One particular benefit from using recycled aluminum such as that recovered from shredded cans is that the energy required to remelt the scrap is low compared to

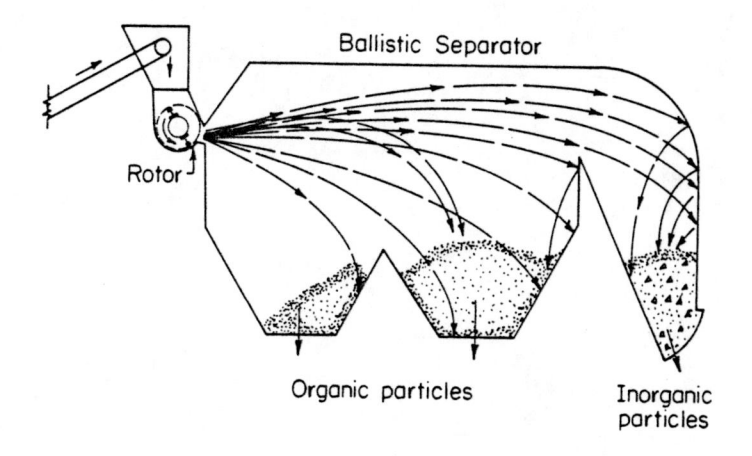

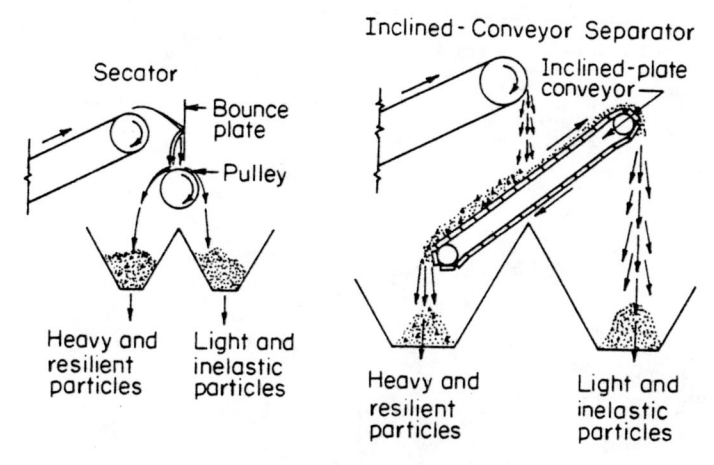

Fig. 6-7 Types of inertial separators (from *Recovery and Utilization of Municipal Solid Waste*).

that required to remelt some of the other common metal packaging materials. Aluminum of less purity, which obviously is cheaper to produce through a recycling operation, is also suitable for some products. For example, the aluminum recycled from municipal refuse is known to contain some other metals, so that the remelted aluminum produced contains certain alloys. These contaminants, such as iron, magnesium, silicon, zinc, and copper, can add strength to the end product of the recycled aluminum. The principal problem as far as aluminum recycling is concerned is participation by the consumer in collecting the cans to be recycled.

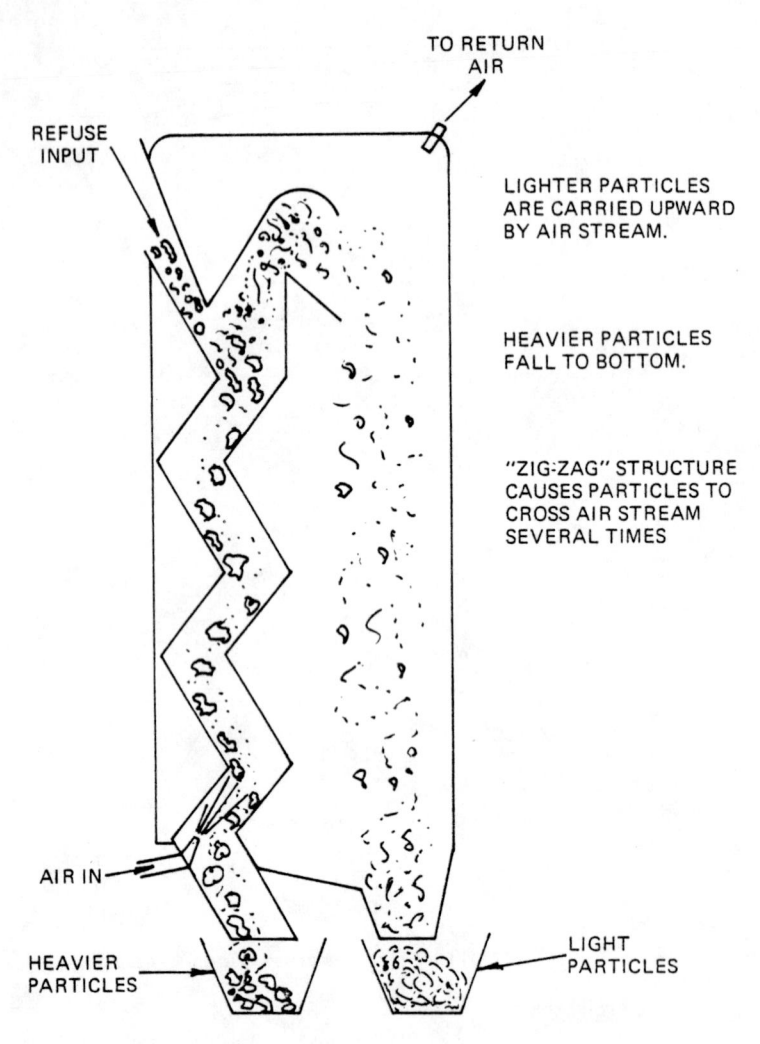

Fig. 6-8 Schematic view of "zigzag" separator.

6-2.3 Recycling Plastics

Another section of the solid waste stream is composed of plastic wastes. Plastics form a rather small percentage of the current waste stream but the use of plastics has increased rapidly in recent years and further accelerated use is predicted. Furthermore, plastics are materials which are basically composed of carbon, hydrogen, and oxygen but are not biodegradable. Therefore, they tend to remain intact in disposal operations such as sanitary landfills. An additional reason for focusing attention at present on plastics is that very little has as yet

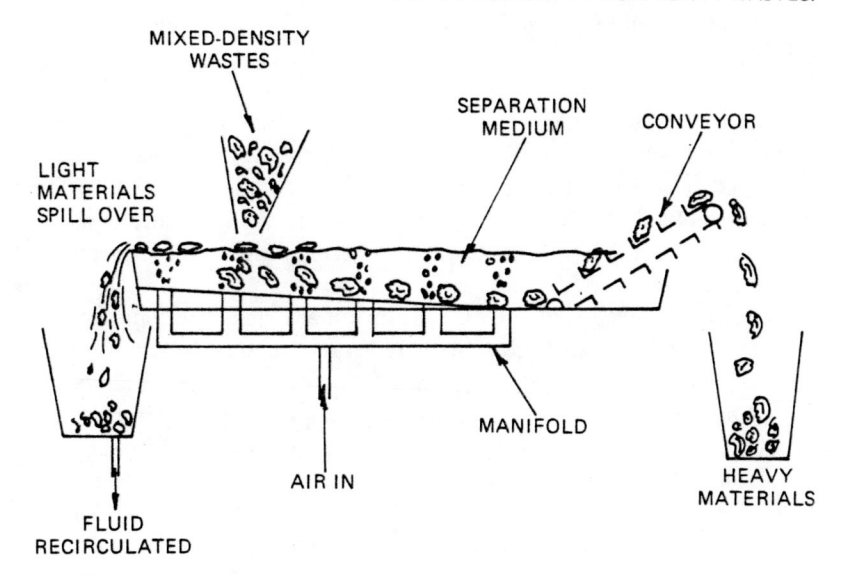

Fig. 6-9 Fluidized bed separator.

been done regarding the recycling of plastic. Plastics recycling may be thought of in three ways:
1. primary recycling, the generation of essentially the same plastic product all over again
2. secondary recycling, the reprocessing and manufacture of an essentially different plastic product of perhaps different composition and in some cases with inferior properties
3. tertiary recycling of plastics, in essence, the complete processing of the plastics to a new form such as in a pyrolysis or incineration operation where the heat value of the material is obtained and the end products are carbon dioxide and water

In examining the plastics in the solid waste stream it is of interest to note that almost 90 percent of all plastics in use today (1973) are thermoplastics which can be reheated, reformed, and possibly used again. In order for any recycling program for plastics to be successful certain conditions are necessary:
1. the plastic product should be made homogeneous—in other words the product should be made of only one type of plastic;
2. the scrap plastic should be intercepted on its way from the consumer to the refuse dump.

Obviously, collectability of scrap plastic will vary according to the form of the plastic product. Plastic films are not as collectible as are rigid plastic containers. Plastic recycling, therefore, could never reach 100 percent. In 1970, over 624 million pounds of plastic were used in bottles and other containers. Eighty-four percent of all this plastic was high-density polyethylene which is easily recyclable. An idea of the recyclability of these containers may be obtained by examination of several operations for recycling and reuse of plastic bottles.

Plastics are a diverse group of products, but all are composed of hydrocarbons and are made essentially from petroleum. Thus, recycling may take the form of reuse as a structural entity or use as an energy source in fuel. Reuse of plastics to form another structural entity is most easily accomplished with fabricator scrap. In the manufacture of plastics generally the primary product comes from a manufacturer in the form of sheets, rolls, or other blank forms and goes to the fabricator who then creates the final container or structural product. During the fabrication operation a certain amount of this plastic is lost. This fabricator scrap is homogeneous and easily separated and therefore may be recycled. On the other hand, after the fabricated article has been used and has entered the waste stream, the scrap is mixed with other materials and is more difficult to recycle. The scrap must be intercepted before it enters the waste stream. In one successful experiment in California, a dairy uses high-density polyethylene bottles for the delivery of milk. These bottles are collected both at home locations and in large containers near supermarkets and returned to the dairy. There they are ground in a Cumberland grinder to produce a flake product. This flake product is homogeneous, contains few contaminants, and has been used as the basis for the manufacture of plastic drainage pipe. One particular problem in the reuse of plastics is that the occurrence of one homogeneous waste plastic is rather rare. Some element or material must be found so that diverse waste plastics may be mixed in a reclamation operation. One such material is chlorinated polyethylene, otherwise known as CPE. The study of compatibilizers such as CPE has been almost all experimental, not theoretical, since the chemical reactions associated with plastics chemistry are not easily quantized. In experiments to date plastics have essentially been used as particles in a mix. Plastics in themselves have low particle adhesion and do not hold together well in mixes. However, the CPE compatibilizer overcomes this difficulty in combining particles of different plastics and produces a usable mix.

In addition to the above-mentioned uses, plastics ground up and shredded can also be used as inert fill material, aggregate for lightweight concrete, and, as mentioned before, fuel.

In addition to the reuse methods mentioned, one other method of recycling of plastic has been proposed: this is the recycling of plastic materials in the form in which they are first manufactured. A prime example of this approach is the returnable plastic milk bottle. Such bottles are in use extensively today.

From only a small beginning in the early 1960s, by the year 1967 over 700 dairies were using polyethylene bottles. However, these bottles made only a single trip from the dairy to the consumer. In several trial experiments, beginning in 1966, the returnable plastic bottle was introduced. The major difficulty in using plastic bottles again as milk containers is the tendency for polyethylene to absorb hydrocarbon contaminants. In other words, milk bottles used in any place in which they come into contact with hydrocarbons are not usable for recycling as milk containers. This problem has been alleviated by the development of an instrument for detecting the presence of harmful hydrocarbons in a milk bottle. Volatile hydrocarbons are detected with this instrument through the following process: first, an air sample is taken from the interior of the bottle; the sample is then mixed with hydrogen and a flame spectrometer test is run to detect the presence of hydrocarbons. Any bottle containing even a trace amount of hydrocarbons is rejected from the recycling operation. This instrument can process 130 bottles per minute. The polyethylene bottles currently in use in the USI Polytrip system generally make more than 100 trips from the dairy to the consumer. The high number of trips results from the fact that the bottles are practically unbreakable, they are safer and quicker in filling and easier to store, and they make for easier distribution. This number of 100 round trips for the high-density polyethylene bottles should be contrasted to an average of 20 return trips for glass containers.

Returnable plastic bottles are in use in only a few localities throughout the country. Plastics now form 10 to 11 percent of all the total milk containers in the country, and of this number only 1 to 1½ percent are returnable bottles. While this method seems to offer a considerable hope for recycling of plastic materials it must overcome several obstacles:

1. supermarkets and other processing agencies are opposed to the returnable bottle with its handling and storage problems
2. there is considerable doubt on the part of the consumer as to the cleanliness and purity of the recycled plastic bottle
3. there has been official reluctance to support this program by the federal Food and Drug Administration
4. finally, there is considerable public apathy toward using a returnable container with a plastic coating

6-2.4 Recycling Glass

The last material toward which attention is focused in this discussion of recycling is glass. Glass is last because from many aspects glass is the least troublesome of the materials in the solid waste stream. For one thing, glass forms only about 6 percent (by weight) of residential solid wastes. Second, glass is made from abundant natural resources such as silica sand, limestone, and soda ash; it is approximately 73 percent silica sand, a material which is in considerable abundance in the world. Third, glass wastes are relatively inert and in many ways are considered a non-polluting waste. Finally, from the economic point of view, glass containers, a major source of glass wastes, are relatively inexpensive: glass containers cost approximately $20 per ton to manufacture, whereas cardboard/paper containers cost approximately $150 per ton, steel $200 per ton, plastic $400 per ton, and aluminum $600 per ton. Glass has some rather unique properties; for example, in some forms it is lighter but stronger than aluminum or steel. Individual strands or fibers of glass have been produced which have a tensile strength of up to 300,000 psi. However, in containers the tensile strength of strands is not utilized. In other words, a glass bottle most of the time fails in tension at a localized zone where the alignment of glass fibers is not parallel with the alignment of tensile stresses. Consequently, in glass containers only about 1 percent of the ultimate strength of glass is utilized. New design practices may put glass fibers in compression in their form as containers so that the latter will thus be much stronger than they have formerly been. The stronger containers will require less protective packaging and will be lighter in themselves. An additional step, chemical tempering, will produce glass which will destruct under grinding to a granular material of the consistency of table salt.

Finally, some attention should be given to current reuse and recycling of glass. At the present time markets exist for almost all of the waste glass which can be collected. Colorless, transparent waste glass can easily be reused as cullet in each new batch of glass produced. Currently about 15 percent of each new batch of glass is obtained from waste sources, usually consisting of in-house scrap. Other uses of waste glass without further processing are in glasphalt or glassbrick.

The principal problem in any reuse of glass is the separation of glass materials from other materials in collected solid waste. Obviously in-house scrap need not be separated from contaminants and is easily recycled. However, even glass scrap from domestic solid waste is being reclaimed at the present time. Reclamation and separation of this glass currently is being done by means of a hydropulper and an optical separator (see Fig. 6-10). The hydropulping operation essentially is a grinding and shredding procedure wherein the solid wastes are mixed with water and ground, shredded, and separated. Various techniques, such as magnetic separation for ferrous metals, ballistic separation for heavy and light materials, and zigzag air separation of materials, are being employed. After the glass has been separated from the other solid waste constituents, a further problem arises in the separation of colored from clear glass. Two methods exist for the separation of the colored from non-colored glass. First, high-intensity magnetic fields will separate the colored glass, on the basis of the metallic coloring agents present, from the clear glass. In another development an optical separator which uses the light transmission properties of the glass itself has been

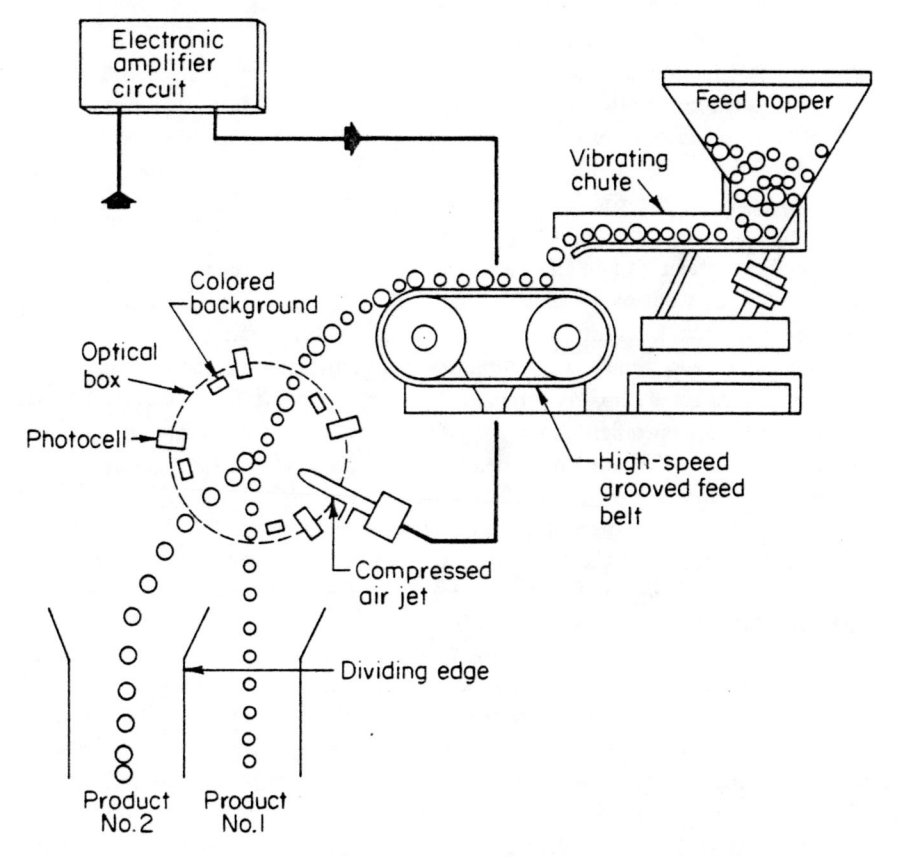

Fig. 6-10 Diagram illustrating the operating principle of the Sortex unit (from *Recovery and Utilization of Municipal Solid Waste*).

developed to separate all colored glass from clear. A further step in the optical separator at present under development will be the separation of green glass from other colors. The separated glass material then can be reused as cullet in a remelting procedure. Separation procedures will be discussed in the last part of this chapter.

Another interesting use of waste glass is in glasphalt. Research in the use of waste as aggregate in asphalt has been conducted at a number of locations over the past two to three years (see Ref. 6-6). For the use of glass as an aggregate in asphalt, only a few criteria must be satisfied. Equidimensional particles are best, but are not absolutely necessary. For another thing, the water deterioration of asphalt made with glass is somewhat higher than water deterioration of conventional asphalt in which crushed stone is used as aggregate. These disadvantages may be overcome rather easily. To balance them there are the advantages of glass in that it is a nonporous, extremely hard material. The disadvantage of water deterioration may be overcome by the addition of hydrated lime $Ca(OH)_2$ in an amount of 1 percent (by weight). With this addition, the wet strength of tested glasphalt is equal to its dry strength. Contrary to expectations by many people, the finished glasphalt product exposes no jagged edges and therefore does not cut tires passing over the surface. In the research conducted to date it has been shown that the skid resistance of glasphalt is a little lower than that of conventional mixes. On the other hand, it has been shown that glasphalt cools considerably slower than does conventional asphalt and so the handling time, the time during which the asphalt may be successfully set in place and compacted, is considerably longer for glasphalt than for conventional asphalt. The major problem in the use of glasphalt is the collection and transport of waste glass in sufficient quantities to supply an asphalt batch operation. For example, in creating a roadway 20 ft wide, 1½ in. thick, and 600 ft long in one test, over 70 tons of ground glass fragments were used. It is doubtful that significant quantities of glasphalt could be economically produced in many areas of the country. However, in areas where waste glass is produced in large quantities, glasphalt may easily be made.

In addition to reuse of glass as now manufactured, other research is being devoted to the development of new types of glass and to the development of new types of glass containers. For example, in research sponsored by the U.S. Public Health Service a type of container has been developed which will consist of a soluble superstructure of a new type of glass covered with an insoluble and impermeable thin film (see Ref. 6-10). This type of development is based upon the philosophy that only the interface between contained materials and container must be durable and insoluble. In the current research development a film approximately 3 mils thick is placed on the soluble superstructure material. The materials considered for the soluble superstructure include sodium silicate glasses, potassium silicate glasses, alkali halides, peptide crystals, and sugar derivative crystals. The peptide and sugar crystals were subsequently discarded as source materials, since some water pollution problems would result in solution of these types of glasses. Alkali halides in essence are too expensive for this use. Of the remaining two materials, potassium silicate glasses are more expensive than sodium silicate types. Therefore most of the research to date has been directed toward the development of a soluble glass made with sodium silicate. The restrictions on any soluble container are the following:

1. the disposable container after solution must not produce a toxic substance
2. no aquatic pollution must result from the solution of the container
3. the containers must be economically competitive with present containers (a 32 oz glass container presently costs 8-10 cents and the additional costs for the new type of container including film practically can increase the cost only $1/3 - 1/2$ cents)

4. finally, the material of the superstructure must dissolve in a relatively short time

Research has shown that the most suitable material for a film is a polymer coating. Presently at Clemson University, research is being conducted in the development of these containers. Sodium silicate soluble superstructures are being coated with polyvinyl hydrogen phthalate (PVHT).

In addition to the research on the films on the surface of the soluble superstructures, additional methods of protection for the superstructures are being investigated. These include ion exchange, so that a less mobile ion would be substituted for the sodium ions on the surface of the soluble container, and surface dealkalinization. There seems to be some potential for the development of soluble glass containers which would alleviate some of the solid waste problem associated with glass.

In addition to development of new types of glass, new types of glass-and-composite containers are also under development. For example, Owens-Illinois Company during 1971 developed two new types of composite containers. The glass composite package (GCP) for beverages is one of the leading items. This container type consists of a glass widemouth globe with a firmly adhered polyethylene base at the lower end. This container will significantly contribute to a saving in the weight of glass used; for example, in a conventional 10 oz quantity container, 7 oz of glass are required for the container. With the new type of container only $2^{3}/_{8}$ oz of glass are used in the globe. In addition, the material is easily recyclable, since returning the container with the adhered polyethylene to the remelting operation results in the burning of the polyethylene and conversion of the plastic to carbon dioxide and water. The heat value inherent in the polyethylene bases may actually aid in remelting of the glass globes. Owens-Illinois's market research has shown that beverage consumers favor these containers, find them easy to use, and in particular like the feature of having a coaster base. Another development by Owens-Illinois is the "plastishield" container. In this container a glass globe is surrounded with a foam polystyrene plastic jacket from the bottom of the container up to approximately two-thirds of its height. The protection afforded the bottle by the polystyrene jacket will make possible a reduction in weight in the glass container itself as opposed to a conventional glass container. These plastishields require 7 oz less glass for a quart size bottle than a conventional container. In addition only a small neck-ring holder similar to the plastic top holder for can beverage containers will be necessary for transport of these plastishield bottles. This development of new types of containers seems to offer some prospect of relief in the solid waste problem, since approximately 75 percent of all beer and 50 percent of all soft drinks are now packaged in convenience packaging, either cans or throwaway bottles. These composite containers can easily be remelted, or can be used as a secondary material. Another idea in the development of these containers is that, although people indicate in opinion polls that they

78   SOLID WASTE MANAGEMENT

prefer returnable containers, in practice few returnable bottles are actually returned. The point of view of the container manufacturer would seem to be that he cannot be a social arbiter and govern consumer practices, but must simply fulfill the desires of wholesale consumers such as breweries and bottlers. At the present time the breweries and bottlers are moving in the direction of composite containers.

6-3. SEPARATION TECHNIQUES

In almost any conceivable recycling or reuse operation, it is necessary to separate mixed refuse into its constituent materials—metals, paper, plastics, wood, etc. Markets simply do not exist for secondary materials that are "contaminated" by other waste materials. Therefore, separation and sorting is an integral portion of any recycling scheme.

The types of sorting or separating mechanisms which could possibly be employed with solid wastes include the following techniques:

1. human sorters separating untreated, unworked refuse, possibly including home separation
2. automated sorters separating untreated refuse
3. automated sorters separating refuse previously processed such as by pulverization, addition of liquid, etc.

All separation techniques must be based on some fundamental property of the constituent materials of which the refuse is composed. Specific gravity, electrical conductance, magnetic response, impact resistance, and infrared reflectance are all examples of such fundamental properties.

Separation may justify significant investments in sophisticated equipment, since income is produced from the sale of salvaged materials. Significant savings are also achieved in disposal operations, since only the residue left after separation must be removed and disposed of.

Home separation has been advocated as the most advantageous and economic sorting means by a large number of persons engaged in solid waste management. Home separation has the following advantages:

1. separation is accomplished before waste materials are mixed, thus forestalling "contamination"
2. labor costs are nonexistent for disposal agencies since the homeowner does the separating
3. lower collection costs may result because of lower volumes of residual (non-resalable) refuse.

However, in most areas certain disadvantages seem to outweigh the advantages listed above:

1. only a few major categories of items will be separated by housewives, while

at least a half-dozen separations should be made if reclamation and reuse is to be economically viable
2. a small number of uncooperative homeowners may ruin the results of all other conscientious separation efforts
3. public apathy exists—many persons will ask, "Why should I separate refuse so that the collector or disposal agency can make a profit?"

At the present time it appears that the only feasible technique is bulk separation at a central location after collection. Very few systems of this kind are currently in operation in the United States (one outstanding exception is Black-Clawson's central Ohio plant); however, considerable amounts of time and money have been spent on research into separation systems, especially automated sorting ones.

In any effort to develop a successful separation system a dichotomy soon appears—preparing refuse for separation and preparing it for collection are conflicting operations. Since collection accounts for a high proportion of all solid waste management costs (80+ percent), any operation which will reduce the collection burden will create significant cost reductions. Such operations include the use of compactor trucks, home compactors, large grinders and compactors in multiple-unit housing systems, and the mixing of refuse constituents (to reduce the number of required collections). All of these operations make for intimate mixing of contaminants and salvageable materials, and thus are directly opposed to separation operation. Because of the economics of collection, however, separation and reclamation processes must almost always be geared for operation on mixed wastes received at a central location.

Separation traditionally has been performed by human labor or by binary sorters. Binary sorters were attuned to the testing of a waste constituent in a yes/no decision situation. For example, electrical conductance testing allows separation of metals (conductors) from nonmetals (nonconductors). In a similar way, electromagnetic testing makes possible the separation of ferrous metals from nonferrous ones. However, such separation is very complicated, since large numbers of sensors are needed for separation into a number of categories. Of much greater utility would be a system employing one or two multiple-output sensors. In such a system, refuse would first be mechanically separated into distinct pieces, and then the individual pieces would be identified and deposited in appropriate receptacles. Identification might be based upon several aspects of the refuse material, such as particle size, bonding strength in particles, and index properties for multiple-output sensing. After size and aggregation separation in a materials-handling system, items would be carried past a row of sensors and could be separated by deflector arms into bins or hoppers for reclamation or ultimate disposal (see Figs. 6-6 to 6-12).

Two multiple-output sensors which have been found to offer promise are both so-called "signature" methods in which an object is excited or driven in a

80 SOLID WASTE MANAGEMENT

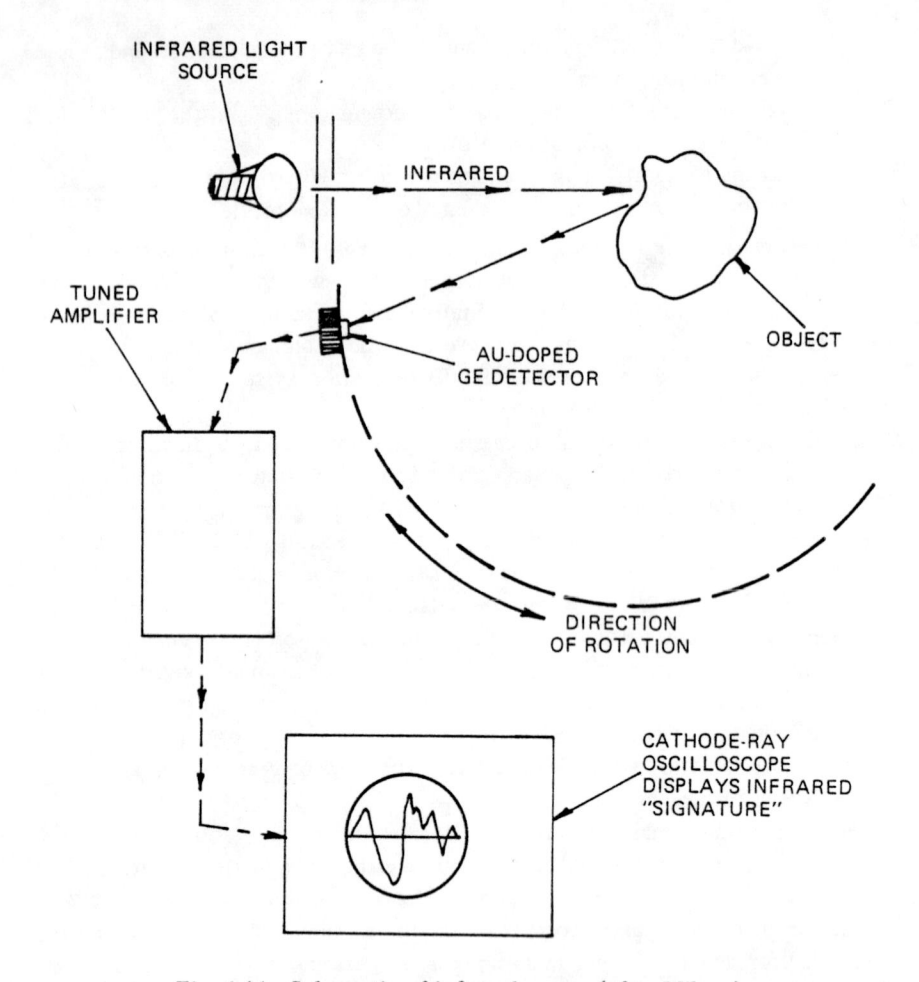

Fig. 6-11 Schematic of infrared sensor (after Wilson).

way so that a distinctive response is achieved. The object is then identified by computerized comparison of the object signature, with the signatures of known materials stored in the computer information memory. Two such methods are developments of research by Dr. David Wilson at the Massachusetts Institute of Technology. One method is based on infrared spectroscopy and the other on impact deceleration.

The infrared spectroscopy technique in brief entails the diffuse reflection of infrared light being monitored for characteristic absorption spectra. Light is reflected from tested specimens in a diffuse form so that even irregularly shaped objects may be examined. The infrared source light is chopped at 1 kHz and a tuned amplifier, synchronized with the source at 1 kHz, is used to strengthen

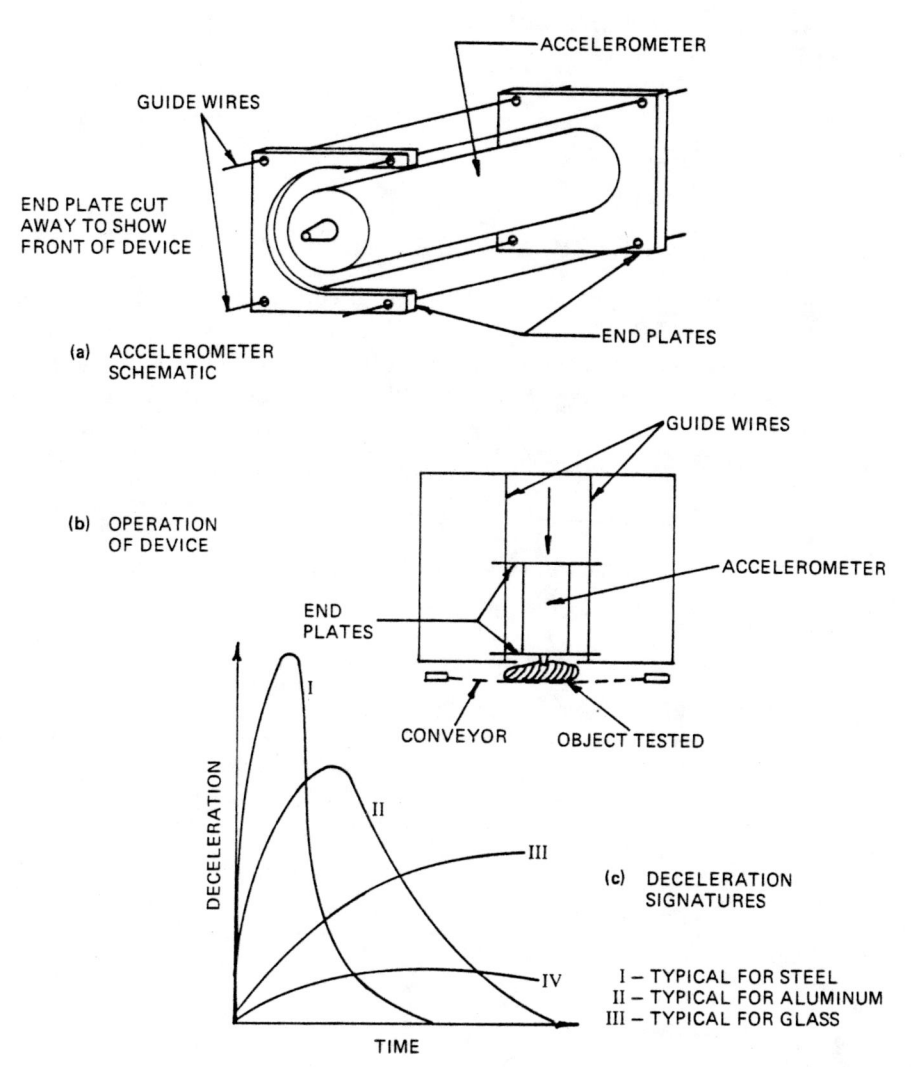

Fig. 6-12 Schematic diagram of deceleration "signature" sensor for separation (after Wilson).

the signal above background light signals. By means of this test, paper, metal, rubber, plastics, and glass may be distinguished and separated.

In the case of impact deceleration, material being tested is struck by a tool on which an accelerometer is mounted. The impact signature is recorded. The signature thus obtained is compared with an information bank of signatures from known samples. Impact signatures for steel plate, aluminum plate, glass,

82 SOLID WASTE MANAGEMENT

and wood are significantly different; unfortunately, however, paper is not amenable to such testing.

Figures 6-6 to 6-12 show the more traditional sorting mechanisms in addition to Sortex classification, zigzag air column separation, and the multiple-output sensors described.

6-4. EMERGING RECYCLING TECHNOLOGY

During the late 1960s and early 1970s there occurred a significant increase in the pilot scale development of new resource recovery systems. Such systems are capable of recovering materials either directly from the solid waste stream (first-order recoverable resources) or indirectly through various available conversion processes (second-order recoverable resources). First- and second-order recoverable resources available in municipal solid waste streams are listed in Table 6-1.

TABLE 6-1 First- and Second-order Recoverable Resources Available in Municipal Solid Waste

First-Order Recoverable Resources	Second-Order Recoverable Resources	
Paper	Heat	Protein
Plastics	Methane	Electricity
Ferrous Metals	Compost	Oil
Nonferrous Metals	Gas	Tar
Glass	Char	Fuel
Miscellaneous materials	Glucose	Yeast

Recently developed resource recovery systems can be broadly classified into: (1) material recovery systems, and (2) energy recovery systems.

6-4.1 Material Recovery Systems

Virtually all material recovery systems have focused on reclamation of the three principal raw materials contained in municipal solid wastes, i.e., paper, metals, and glass. Plastic recovery has not yet proven technically feasible in conjunction with an overall material recovery system. The two most advanced material recovery systems are the Black-Clawson wet-pulping Hydrasposal process and the U.S. Bureau of Mines incinerator residue recovery process.

The Black-Clawson Company has developed a materials recovery system (see Fig. 6-13) designed to accept unsegregated municipal solid waste, remove paper

FLOW SHEET, SOLID WASTE PLANT, FRANKLIN, OHIO

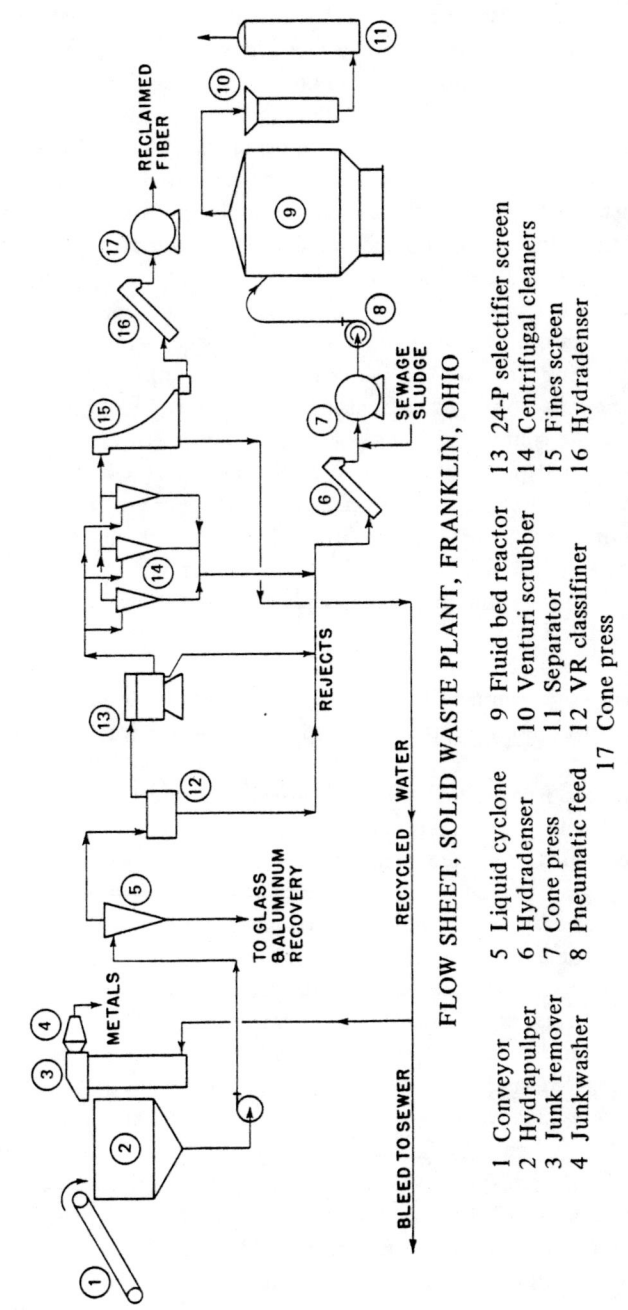

1 Conveyor	5 Liquid cyclone	9 Fluid bed reactor	13 24-P selectifier screen
2 Hydrapulper	6 Hydradenser	10 Venturi scrubber	14 Centrifugal cleaners
3 Junk remover	7 Cone press	11 Separator	15 Fines screen
4 Junkwasher	8 Pneumatic feed	12 VR classifier	16 Hydradenser
		17 Cone press	

Fig. 6-13 Flow sheet of the wet pulping hydrasposal process utilized at the solid waste recycling plant at Franklin, Ohio (*courtesy* Black-Clawson, Inc.).

fiber in the form of reusable pulp, recover metals and glass, and burn the non-recoverable organic fraction. A 150 ton per day demonstration plant of this system has been in operation at Franklin, Ohio, since June 1971. Municipal solid waste arrives at the plant in the "tipping floor," which is covered and enclosed on three sides. The solid waste is transported by an end loader to a conveyor pit which feeds the Hydrapulper. In the Hydrapulper, water is added to the solid waste and a high-speed cutting rotor located in the bottom of the Hydrapulper tub converts the pulpable and friable materials (paper, food waste, plastic, glass, ceramics, and aluminum) into a 3½ percent water slurry. Large, heavy objects such as metal cans, stones, and other non-pulpable materials are ejected through an opening in the side of the tub to a "junk remover." After a preliminary washing the "junker" discharges the materials into a drum washer from which they are conveyed to a magnetic separator. There tin cans and other ferrous metals are removed. The slurry is removed from the Hydrapulper through a perforated plate or sieve at the bottom and sent to a liquid cyclone, where the heavier materials are removed by centrifugal action. The cyclone concentration is then dried, screened, magnetically separated, and air classified into three fractions:

1. Light fraction—aluminum and heavy plastics
2. Middle fraction—glass
3. Heavy fraction—heavy, nonferrous metals

Approximately 80 percent by weight of the material removed in the cyclone is glass, which can be color sorted by an optical separator for reuse in glass container manufacture.

The remaining slurry is passed through a series of screens and cleaners which separate the usable long paper fibers from coarse organics and fine contaminants. The reclaimed fibers are then dewatered to 50-60 percent moisture content in two stages, (i.e., an inclined screw conveyor type thickener and a cone press) and delivered by a screw conveyor to transportation facilities.

The rejected material from the fiber recovery process is essentially non-recoverable organic matter consisting of rubber, textiles, plastics, leather, yard wastes, food waste, paper coatings, paper fines, etc. These non-recoverable organics are dewatered to 60 percent moisture content in a thickener and cone press. The cake from the press is fed to a fluidized bed combustor.

Although energy recovery is not currently practiced at the Franklin plant, such recovery might be feasible based on the fact that 37 percent of the incoming solid waste stream is non-recoverable organic material (see Fig. 6-14) having a heat value of 7,000 to 8,000 Btu/lb. The Franklin plant as currently operated demonstrates a process which accepts unsorted domestic solid waste and produces salable paper pulp, glass cullet, and metals.

Another materials recovery system developed to the pilot plant stage is one created by the U.S. Bureau of Mines for reclaiming and recycling various metal

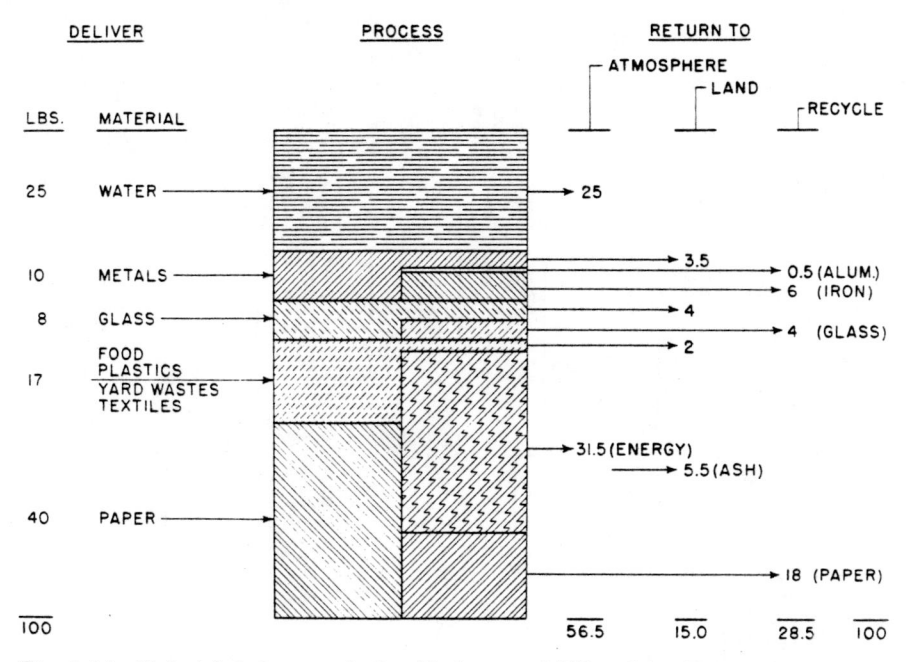

Fig. 6-14 Material balance of the Hydrosposal/Fiberclaim Process (*courtesy* Black-Clawson, Inc.).

and mineral constituents contained in municipal incinerator residues. The Bureau of Mines ½-ton-per-hour pilot plant is located in College Park, Maryland, and has demonstrated the feasibility of recovering metallic iron concentrates, clean nonferrous metal composites, clean fine glass fractions, and fine carbonaceous ash tailings from various incinerator residues. The process as outlined in Fig. 6-15 consists basically of a series of size reduction and separation stages. Initially, wet incineration residue is placed in a trommel which segregates the plus 1¼ in. material from the minus 1¼ in. material. A primary screen separates the minus 1¼ in. material into three size fractions. The larger-size particles from the trommel and primary screen are conveyed to a shredder and secondary screen. The ferrous metals are removed from the larger-size fraction from the secondary screen by means of permanent magnet drum separation. The nonmagnetic material residue is then introduced to a hammermill and tertiary screen, which separates large nonferrous metals for upgrading.

The small particles from the primary, secondary, and tertiary screens are collected and fed to an electromagnetic drum separator to achieve ferrous metal separation. The nonmagnetic residue from the electromagnetic separator is then dewatered in a spiral classifier, conveyed to a rod mill, and screened to accomplish the removal of large nonferrous metals. The fine material is sub-

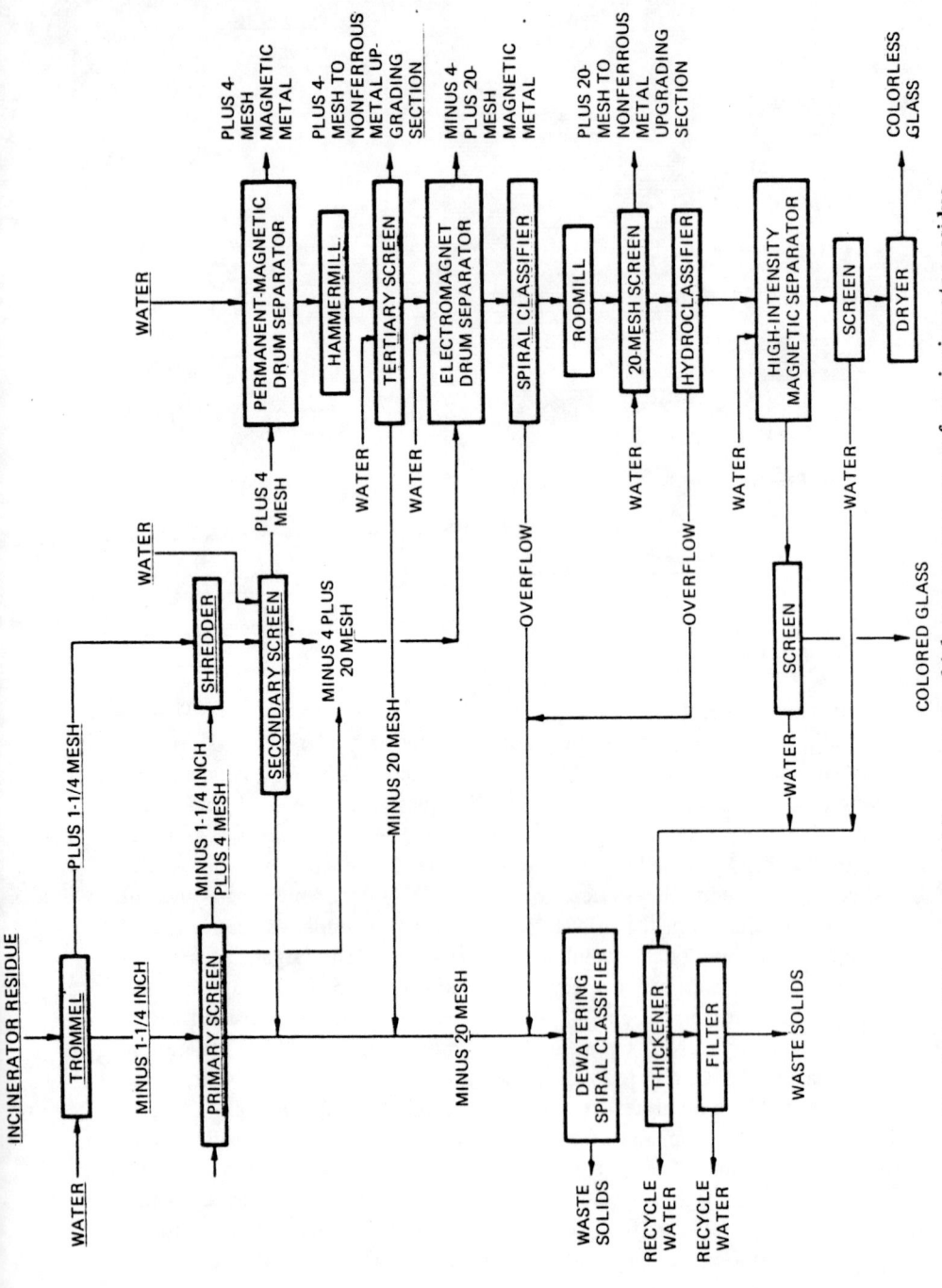

Fig. 6-15 U.S. Bureau of Mines system which recovers resources from incinerator residue.

servently delivered to a hydroclassifier and high-intensity magnetic separator which reclaims the colored and colorless glass. The nonferrous metals are then fed to a sink-float separator which segregates aluminum and copper-zinc scrap. This U.S. Bureau of Mines plant achieves about 80 percent recovery of recyclable materials.

6-4.2 Energy Recovery Systems

Probably the most promising of the emerging energy recovery systems is the Horner-Shifrin-designed fuel supplement system now in use in St. Louis, Missouri. This system is designed to utilize shredded solid waste material as a supplemental fuel for large pulverized coal-fired power plant boilers following ferrous metal removal. It includes a refuse preparation facility as well as the actual power plant. Mixed municipal solid waste is delivered to the refuse preparation facility (see Fig. 6-16) where the raw refuse is conveyed to a

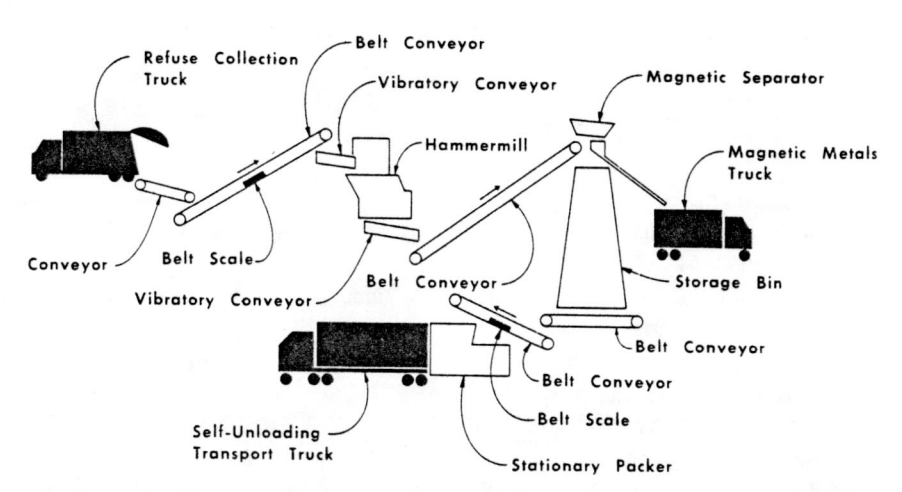

Fig. 6-16 Refuse processing facilities which serve as the "front end" of the fuel supplement system now in use at St. Louis, Missouri (*courtesy* Horner-Shifrin, Inc.).

hammermill. The latter reduces the solid waste to particle sizes under 1½ in., after which it is fed to a magnetic separator to achieve ferrous metal removal. The shredded solid waste is then transported to the power plant and used as supplemental fuel in conventional pulverized coal-fired boilers (see Fig. 6-17). The St. Louis power plant is currently being fed approximately 600 tons per day of shredded solid waste which is more than half the solid waste collected by city-of-St. Louis trucks. The furnace is fired with a 10 percent solid waste-90

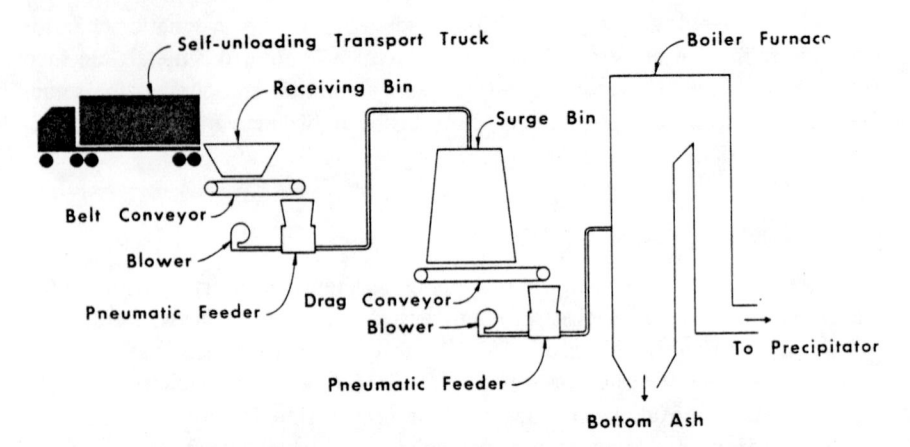

Fig. 6-17 Supplemental fuel receiving and firing facilities utilized at St. Louis, Missouri (*courtesy* Horner-Shifrin, Inc.).

percent coal mixture (based on heat value). It is estimated that fuel recovery during the first year of operation will be equivalent to 150,000 tons of coal.

Major advantages of the St. Louis system include:
1. The process is less expensive then conventional incineration
2. Equipment required for the process is readily available
3. The process has potential applicability in many large American metropolitan areas
4. Air pollution levels will be reduced, since the refuse has a lower sulfur content than does the coal
5. Fossil fuel consumption will be reduced and the ferrous metals recovered
6. Power production costs will be reduced
7. Existing boilers can be retained for longer life as base-load units

Other energy recovery systems in the developmental stage include the CPU-400 (see Fig. 11-2) and the Melt-Zit incinerator (see Fig. 11-1). The CPU-400 system is basically an energy recovery system which utilizes a refuse-fueled gas turbine for electrical power generation. A pilot system in Menlo Park, California, which has a capacity of 40 tons of solid waste per day, is under development by the Combustion Power Company. The Melt-Zit incineration system is operated at a very high temperature (3000°F), which produces complete burning of all combustibles and melting of noncombustibles resulting in 95 to 97 percent volume reduction of the incoming solid waste stream. Flue gases from the incinerator may be passed through a steam boiler. A more detailed discussion of the CPU-400 and Melt-Zit Incinerator processes is presented in Chapter 11.

Several other operational installations throughout the United States utilize a heat recovery type incinerator in addition to separate power generation facilities. One such installation was constructed in 1967 at the Norfolk, Virginia,

Naval Station at which two small heat recovery boilers are fueled with broken shipping crates and other waste wood. Steam from this process is utilized for heating several Naval Station buildings. Similar heat recovery incineration systems are in use in Braintree, Massachusetts; Chicago, Illinois; Munich, West Germany; and Montreal, Canada. The Chicago installation at its construction was the largest in North America, processing 1600 tons of solid wastes per day. While the Munich plant supplies 70 megawatts of electricity to Munich's municipal power system, the Chicago and Montreal plants as of 1972 wasted much of the potentially recoverable heat because of a lack of customers. In most larger cities, the heat that can be generated from the existing solid waste stream can supply enough power to satisfy as much as 10 percent of urban needs. At present, optimum usage of the energy developed in new heat recovery systems is a major problem area. As power demands continue to increase over the next decade, more electric utilities will probably find it worthwhile to make use of this emerging source of energy.

6-4.3 Pyrolysis Systems

The chemical energy and chemical constituent values contained in the typical solid waste stream have suggested recovery of these resources by means of a promising new process—pyrolysis. The pyrolysis process, often referred to as destructive distillation, consists of heating the combustible portion of the solid wastes to a medium high temperature (1000 to 2000°F) in either an oxygen-free or low-oxygen environment. Pyrolysis differs from conventional incineration in that it is an endothermic rather than an exothermic process. Pyrolysis of solid waste matter in a high-temperature and low-oxygen environment results in the chemical breakdown of organic carbon material into three basic components:

1. a gas phase containing primarily hydrogen, carbon dioxide, carbon monoxide, and methane gases (see Table 6-2).

TABLE 6-2 Gases Evolved by Pyrolysis

Constituent	Percent by Volume at Indicated Temperatures (°F)			
	900	1,200	1,500	1,700
H_2	5.56	16.58	28.55	32.48
CH_4	12.43	15.91	13.73	10.45
CO	33.50	30.49	34.12	35.25
CO_2	44.77	31.78	20.59	18.31
C_2H_4	0.45	2.18	2.24	2.43
C_2H_6	3.03	3.06	0.77	1.07
accountability	99.74	100.00	100.00	99.99

(See Ref. 6-20).

90 SOLID WASTE MANAGEMENT

2. a tar or oil phase containing simple organic acids such as acetic acid, methanol, and acetone; this tar or oil produced in the pyrolysis process exists as a liquid at room temperature
3. a char phase made up of virtually pure carbon and inert materials (metals, glass, rock, etc.) not consumed in the process. Table 6-3 outlines the relative quantities of various materials found in pyrolysis char.

TABLE 6-3 Proximate Analysis of Pyrolysis Char

Percent	\multicolumn{4}{c}{Pyrolyzing Temperatures (°F)}	Pennsylvania Anthracite
	900	1,200	1,500	1,700	
Volatile matter, %	21.81	15.05	8.13	8.30	7.66
Fixed carbon, %	70.48	70.67	79.05	77.23	82.02
Ash, %	7.71	14.28	12.82	14.47	10.32
Btu per lb	12,120	12,280	11,540	11,400	13,880

(See Ref. 6-20).

Experimental results to date indicate that the relative proportions of pyrolysis end products is strongly dependent on the temperature of the process (see Table 6-4). An economic analysis of the pyrolysis process, as applied to solid waste conversion to commercial grade acetic acid, indicated a net operational cost of about $5.70 per ton of raw refuse (see Ref. 6-21).

Of the various pilot scale pyrolysis units in operation in the United States today, the most completely developed device is the Lantz Converter. The Lantz Converter accepts mixed municipal solid waste which is fed continuously through a hammermill operation to a revolving stainless steel drum. Natural gas is used to initiate the pyrolysis process by elevating equilibrium temperatures to

TABLE 6-4 Pyrolysis Product Yield

Temp. (°F)	Refuse (lb[a])	Gases (lb)	Pyroligneous Acids and Tars (lb[b])	Char (lb)	Mass Accounted For (lb)
900	100	12.33	61.08	24.71	98.12
1200	100	18.64	18.64	59.18	99.62
1500	100	23.69	59.67	17.24	100.59
1700	100	24.36	58.70	17.67	100.73

[a]On an as-received basis, except that metals and glass have been removed.
[b]This column includes all condensables and the figures cited include 70 to 80 percent water.
(See Ref. 6-20).

approximately 2000°F in an air-devoid environment. If 70 percent of the gases produced in the process are fed back into the gas burners, the system becomes self-sustaining. The excess gases are currently wasted; however, eventual usage of these is contemplated. At present only charcoal recovery has been studied extensively.

In summary, it would appear that the pyrolysis process possesses great potential for solid waste management, since a variety of valuable end products can be produced. Several large pilot pyrolysis units have been operated with relative success, and the process has the possibility of being self-sustaining. Specific applications of the pyrolysis process to particular solid waste disposal problems are discussed in Chapter 11.

6-4.4 Integrated Recovery Systems

Eventual incorporation of the individual resource recovery units described in this chapter into waste disposal systems must occur if planners and designers are to logically solve their overall solid waste management problems. Several integrated resource recovery systems are on the drawing boards and plans to construct and operate them are currently being formulated. One of these total recycling systems has been sponsored by the Aluminum Association. This recycling system includes various proven recycling unit processes and consequently could be installed in different combinations to meet the specific needs of individual communities. The plant as currently envisioned (see Figs. 6-18 and 6-19) would convert 60 percent of the incoming solid waste material to heat, and recover the remaining 40 percent for recycling. Basically, incoming solid waste material would be shredded and magnetically separated. Noncombustibles would be screened, air classified, and forwarded to the Bureau of Mines unit mentioned previously for recovery. Combustible materials would be forwarded to a refuse storage pit and incinerated or pyrolyzed with energy recovery, and then sent to the Bureau of Mines unit. Materials recovered would include ferrous metals, nonferrous ones, clear glass, colored glass, and sand. Additionally, dry fiber reclamation could be incorporated into the process with little change in overall operation. A 500-ton-per-day plant (the quantity of solid waste generated by approximately 200,000 people) located in the Washington, D.C., area and utilizing the above flow scheme would be expected to produce an annual income of $1,450,334 (see Fig. 6-20) with a net annual profit to the operator of $22,000 to $133,000. Such a plant would require approximately 10 acres of land and a capital investment of $15 million.

Although such integrated recovery systems could still be regarded as in the developmental stage, several plants will probably be constructed during the 1970s.

Fig. 6-18 An artist's rendering of the Aluminum Association's total recycling plant (*courtesy* Aluminum Association).

6-5. RECYCLING PROBLEM AREAS

All basic problems associated with increasing the percentage of recycled material usage focus upon the development of markets for secondary raw materials. Collection and separation of constituents from the solid waste stream merely represent the movement of solid waste material from one point to another, unless such market outlets exist. The processed solid waste cannot be considered recycled until it is utilized by agricultural, industrial, or private consumers. Utilization outlets for recycled materials, therefore, must be expanded with the cooperation of government, industry, and the private sector.

One of the greatest problems related to increased utilization of secondary raw materials is tax equalization. Currently, virgin materials industries benefit from depletion allowances and capital gains allowances which indirectly serve to place secondary raw materials in a weaker competitive position. Legislation has re-

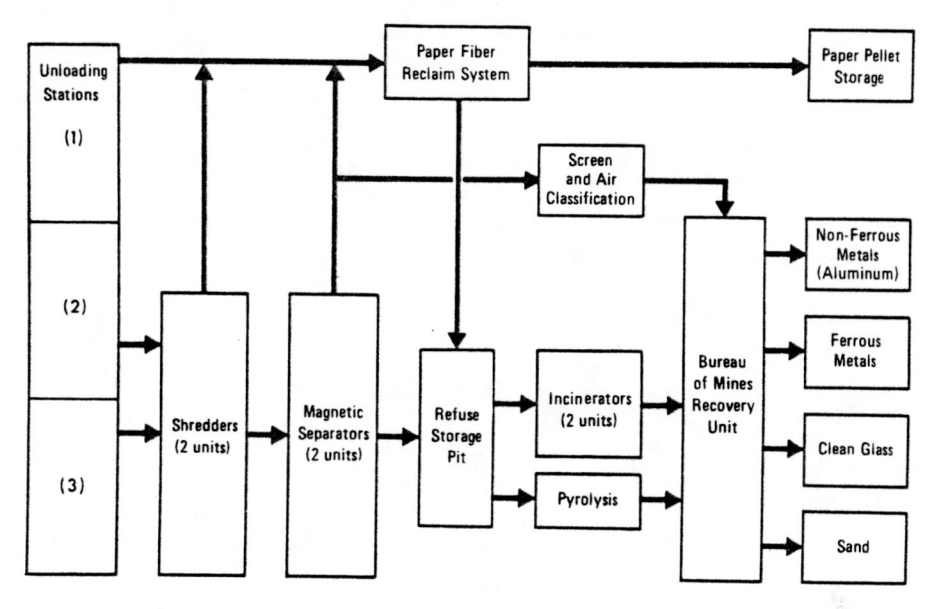

Fig. 6-19 Flow chart of the Aluminum Association's proposed recycling plant (*courtesy* Aluminum Association).

cently been proposed that would provide a tax incentive for the utilization of recycled materials, thereby offsetting existing income tax advantages for primary industries.

Equalization of transportation rates is another area of high priority. As of 1972, many recyclable materials carried higher domestic and international transportation rates than similar primary commodities. Congressional action is required to equalize freight rates for secondary and primary raw materials if solid waste recycling is expected to expand in the future.

Other major problem areas include government procurement practices and zoning and licensing regulations. Until very recently many national, state, and local governmental procurement agencies restricted their purchases to primary material products. However, New York City and the federal General Services Administration have initiated changes in procurement policies to guarantee increased usage of recycled materials. It would seem apparent that additional changes in governmental and industrial procurement policies are required to broaden market opportunities for recycled materials. Regional and municipal zoning and licensing regulations have also served to impede the recycling effort at these levels. In many areas, recycling systems are not allowed to locate in close proximity to manufacturing and retail operations which could utilize secondary materials. Transportation costs for recycled goods are therefore ele-

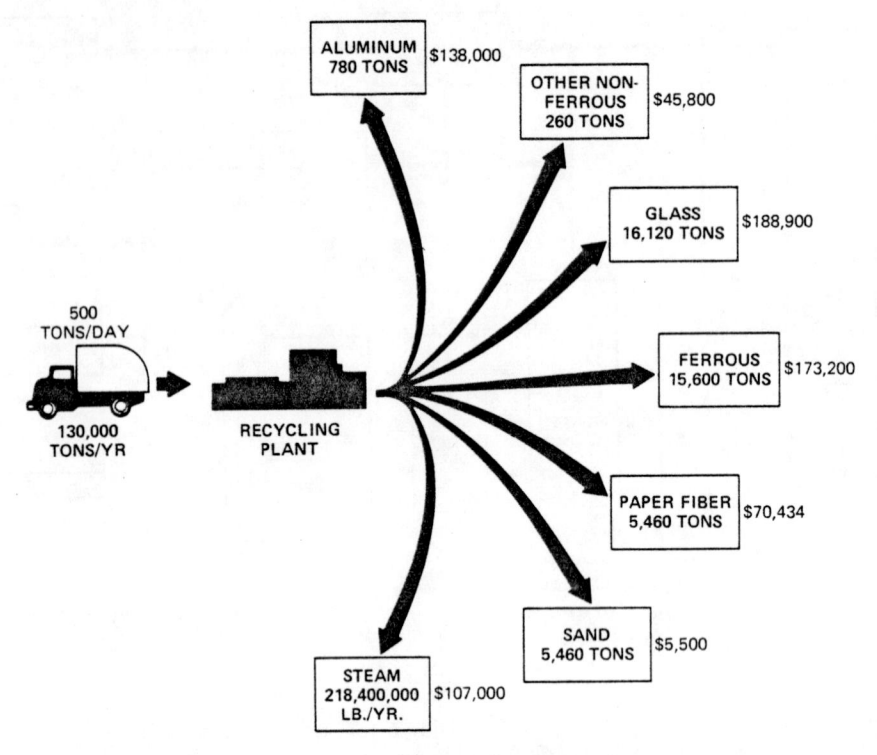

Fig. 6-20 Expected annual income from the Aluminum Association's proposed recycling plant (*courtesy* Aluminum Association).

vated following forced location of recycling activities at distant points. Various licensing requirements and record-keeping rules also serve to discourage resource recovery operations. Major effort needs to be expanded toward the development of model codes for solid waste management activities.

Other primary priority actions required to promote greater usage of recycled solid wastes as outlined in a 1972 Battelle Memorial Institute study (Ref. 6-13) include:

1. the development of methodology required to utilize lower-grade solid wastes including unsegregated materials and mixed refuse
2. refinement of solid waste recycling techniques to permit more economical operations
3. increased consumer education with regard to the quality of products manufactured with secondary materials and the indirect effect on environmental quality control

Other secondary priority actions also exist with regard to increasing the marketability of secondary materials; however, the emphasis of future corrective action should be focused on the above-mentioned problem areas during the next several years.

6-6. RESOURCE RECOVERY OF PACKAGING WASTES

It has been estimated that 90 percent of all packaging materials enters the solid waste stream. Furthermore, it is predicted that by 1976 over 75 million tons of waste packaging material will be produced in the United States each year. For example, in 1956 approximately 404 lb/cap/yr of packaging materials were produced; by 1976 this figure is expected to approach 661 lb/cap/yr of packaging material. The composition of the material used for packaging at that time is expected to be as follows (see Ref. 6-1):

Paper and board	57%
Glass	18%
Metals	13%
Wood	7%
Plastics	5%

6-6.1 A Specific Packaging Waste Problem—Litter

In a 1969 highway litter survey prepared by the National Academy of Sciences and in which twenty-nine states participated in an attempt to categorize and quantize the litter which is found along the nation's highways, it was found that each mile of primary highway in the United States receives about 1300 pieces of litter each month, not counting such small items as cigarette butts, etc. However, this same study showed that only one-sixth of the litter consisted of no-deposit beverage cans or bottles. Surprisingly, the study showed a ratio of 26 returnable bottles to 37 no-deposit bottles found along a typical mile of highway. Furthermore, the ratio of beer cans to soft drink types was 4 to 1 in highway litter, in contrast to the sales ratio of beer to soft drink cans which is about 3 to 2.

Of course, highways are not the only places where litter is deposited. It appears that beverage containers also constitute a large portion of the litter found in other areas, while statistics also show that beverage containers have increased proportionately in the litter produced since the emergence of no-deposit containers. This is due to two factors. First, soft drink sales have tripled in the last 10 years (see Ref. 6-1). Second, the percentage of beverages consumed outside taverns or soda fountains has increased from 25 to 65 percent. Still, research has shown that about 95 percent of all soft drinks and beer is being consumed indoors, in taverns or at home. Only 5 percent is consumed in parks, at drive-ins, or at other public places. Thus, any sort of restriction placed upon sale or use of no-deposit cans or nonreturned bottles to prevent litter would be punitive to a large portion of the population who very legitimately and innocently dispose of these containers in a proper way. Of course, returnable bottles are preferable to throwaway containers for purposes of recycling and reuse. Resource conservation and recovery must also be considered.

However, an additional factor is consumer apathy. People just do not want to return bottles. For example, before World War II beverage bottles sold for home use averaged 30 trips back to the bottling location before they were lost or broken. At the present time such containers average about 4 trips.

6-6.2 Control of Packaging Wastes

In a research study completed in 1969 (Ref. 6-7), the Midwest Research Institute listed five possible mechanisms for controlling the solid waste problems created by packaging. These methods were: (1) regulation; (2) taxes; (3) incentives and subsidies; (4) education; and (5) research and development. Which is the best of these alternatives? The Research Institute concluded that regulation had the best chance to improve the overall situation. However, there are many pitfalls in this method of control. Many individuals have suggested restrictive measures for various products such as plastics, cans, wrappings, etc. There is no guarantee that if this were done on a national level it would alleviate the problem. Some situations may be cited; for example, there is the question of the returnable bottle versus the no-deposit bottle. Many persons believe that the no-deposit bottle is a large factor in the solid waste problem and, in particular, in the litter problem which today has reached large proportions.

A second measure to improve the situation as far as packaging is concerned would be the use of special taxes, such as a packaging-use tax or a depletion tax. However, the former tax would reduce the depletion of natural resources only slightly and would in no way make disposal of waste containers any easier. It appears that regulation has a better chance of success than a tax.

Another method for control and disposal of packaging that has been suggested is the use of incentives, for example, preferred use by large consumers such as the federal government, or a reduction in the depletion allowance of certain natural resources such as oil or timber. Much more study is required of incentive measures so that a detailed picture of their full impact may be obtained. However, at the present time one incentive measure is being put in practice in the form of federal laws which require that the government through its purchasing agencies obtain paper stock materials (for use in governmental publications) which contain at least 15 percent recycled paper pulp material. Thus, a large market for recycled paper has been created.

A fourth measure, also suggested by the MRI report, is the use of education programs in industry, in the private sectors, and in intragovernmental groups. Most of these programs are based on the assumption that today's citizen is ignorant of the solid waste problem and will behave in such a way as to alleviate that problem if better informed. There are some indications that this may be a rash assumption. For example, a few years ago the Pepsi Cola Company introduced over 14 million returnable bottles into the New York City area, publicized the effort considerably, and raised the deposit value per bottle from 2 to 5 cents as an additional incentive for the consumer to return the bottle. The experiment was a dismal failure because the bottles were not returned.

A fifth measure for reducing the solid waste problem due to packaging is the use of research and development to create new materials, new methods and techniques for manufacture, and for transport and disposal of wasted items. As far as the materials themselves are concerned, there is some question as to the near-term success which is foreseeable through the use of different materials. Under development are several types of biodegradable or soluble plastics and glasses. Among this last group are polymers which when exposed to solar energy will depolymerize and become brittle, thereby being more easily ground and shredded in a disposal process. In the area of disposal technology, some hope is held for advanced techniques of salvage and reuse. An example is the use of glass as cullet for remelting and use in making new glass, or as aggregate in glasphalt as mentioned previously. Another salvage mechanism is the wet pulverization described in Section 6-4.1; from 1,000 tons of solid waste, approximately 200 tons of paper pulp, 80 tons of ferrous metals, and 20 tons of glass bits may be recovered by the Black-Clawson Hydrapulping operation. In contrast to these favorable developments there are also negative ones on the horizon. For example, automated supermarkets and the increased use of portion packaging will create a much larger disposal problem. It is doubtful whether research can gain on these adverse developments. After all things are considered, it appears that regulation still may hold the best hope for a solution.

REFERENCES

6-1. Alexander, J. H., "Banning the Can Won't Clean Up the Mess," statement before a state of Michigan Special House Committee studying disposable beverage containers, American Can Co., 1971.

6-2. Burgess, K. L., "Reuse of Plastic Milk Container Materials," *Proc.*, Conference on Design of Consumer Containers for Reuse or Disposal, Columbus, Ohio, 1971.

6-3. Emich, K. H., "Returnable Plastic Milk Bottles," *ibid*.

6-4. Abrahams, J. H., "Utilization of Waste Container Glass," *Waste Age*, Vol. I, No. 4 (1970).

6-5. Cheney, R., "Design Trends in Glass Containers," *Proc.*, Conference on Design of Consumer Containers for Reuse or Disposal, Columbus, Ohio, 1971.

6-6. Malisch, W. R., Day, D. E., and B. G. Wixson, "Use of Salvaged Waste Glass in Bituminous Paving," paper presented at the Centennial Symposium, Technology for the Future to Control Industrial and Urban Wastes, University of Missouri (Rolla), 1971.

6-7. Midwest Research Institute, "The Role of Packaging in Solid Waste Management 1966 to 1976," USPHS Publ. No. 1855, 1969.

6-8. Story, W. S., "Ferrous Scrap Recycling and Steel Technology," *Proc.*, Conference on Design of Consumer Containers for Reuse or Disposal, Columbus, Ohio, 1971.

6-9. Testin, R. F., Bourcier, G. F., and K. H. Dale, "Recovery and Utilization of Aluminum from Solid Wastes," Reynolds Metals Co., Richmond, Va., 1971.

6-10. Hulbert, S. F., Fain, C. C., Cooper, M. M., Ballenger, D. T., and C. W. Jennings, "Improving Package Disposability," paper presented at the First National Conference on Packaging Wastes, San Francisco, 1969.

6-11. National Association of Secondary Material Industries, "Recycling: A Guide to Effective Solid Waste Utilization," 1972.

6-12. Ness, H., "Recycling as an Industry," *Environmental Science and Technology*, Vol. VI, No. 8 (Aug. 1972), p. 700.

6-13. Battelle Memorial Institute, Columbus Laboratories, "A Study to Identify Opportunities for Increased Solid Waste Utilization," June 1972.

6-14. Resource Planning Associates, Inc., "Resource Recovery Systems Catalog," Report No. A-72-5 (P), April 1972.

6-15. Bendersky, D., Park, W. R., Shannon, L. J., and Franklin, W. E., "Resource Recovery from Municipal Wastes—A Review and Analysis of Existing and Emerging Technology," presented at an International Meeting on Pollution: Engineering and Scientific Solutions, Tel Aviv, Israel, June 1972.

6-16. Anon., "A Solid Waste Recovery System for All Municipalities," *Environmental Science and Technology*, Vol. V, No. 2 (February 1971), p. 109.

6-17. Engdahl, R. B., "Solid Waste Processing—A State of the Art Report on Unit Operations and Processes," USPHS Publ. No. 1856 (1969).

6-18. Belknap, M., "Paper Recycling: A Business Perspective," New York Chamber of Commerce, Sept. 1972.

6-19. Wisely, F. E., Sutterfield, G. W., and Klumb, D. L., "St. Louis Power Plant to Burn City Refuse," *Civil Engineering–ASCE*, Jan. 1971.

6-20. Drobny, N. L., Hull, H. E., and Testin, R. F., "Recovery and Utilization of Municipal Solid Waste," USPHS Publ. No. 1908 (1971).

6-21. Porteous, A., "Towards a Profitable Means of Waste Disposal," presented at Winter Annual Meeting and Energy Systems Symposium, American Society of Mechanical Engineers, November 1967.

6-22. Hoffman, D. A., and Fitz, R. A., "Batch Retort Research on Pyrolysis of Solid Municipal Refuse," *Environmental Science and Technology*, Vol. II, No. 11, (November 1968), p. 1023.

6-23. Eggen, A., and Powell, O. A., Jr., "Feasibility Study of a New Solid Waste System," University of Hartford Report DUST/TR November 1967, p. 701.

6-24. "Segregation Systems for Processing Heterogeneous Solid-Waste Materials for Resource Recovery," Victor Brown *Proc.* Natl. Indus. SW Management Conference, Houston, Texas, March 1970.

6-25. M. L. McKenna, "Sorting Technology and Its Application to Materials Management," ASME paper 69-MH-14, 1969.

6-26. Richard A. Boettcher, "Air Classification for Reclamation Processing of Solid Wastes," ASME paper 69-WA/PID-9, 1969.

6-27. David G. Wilson, "Present and Future Possibilities of Reclamation from Solid Wastes," *Proc.*, New Directions in Solid Wastes Processing Institute, Framingham, Mass., 1970.

6-28. David G. Wilson, *et al.*, "New Sensors for the Automatic Sorting of Municipal Solid Wastes," Amer. Inst. Chem. Engrs. Conference, Cincinnati, 1971.

7

Solid waste disposal microbiology

7-1. INTRODUCTION

The oldest biological waste treatment system utilized by man involves the decomposition of solid waste in soil by microorganisms. For centuries the wastes of man and other animals, including their bodies, and the tissues of plants have been disposed on or buried in the soil. Eventually all of these wastes have disappeared, being transferred into some of the substances that make up various soils. It is the soil microorganisms that produce these changes by breaking down complex organic compounds into simple compounds, that in turn make up the nutrient material of the plant world. Currently there are at least two general categories of solid waste disposal accomplished by microbes: sanitary landfilling and composting. This chapter will briefly explain the specific microbial interactions which are responsible for solid waste disposal in soil.

7-2. THE SOIL ENVIRONMENT

The fertile soil environment provides an ideal habitat for a virtual microscopic menagerie, consisting of a wide range of microorganisms varying in number from a few per acre to billions per gram of soil (see Table 7-1). Soil microorganisms

TABLE 7-1 Soil Population (Number of Organisms per Gram in a Fertile Agricultural Soil)

Bacteria:	
Direct count	2,500,000,000
Dilution plate	15,000,000
Actinomycetes	700,000
Fungi	400,000
Algae	50,000
Protozoa	30,000

Source: A. Burges, *Micro-organisms in the Soil*, Hutchinson & Co., Ltd, London, 1958.

may be thought of as those forms of life which generally cannot be seen without the aid of a microscope, including bacteria, protozoa, rickettsia, viruses, algae, and fungi. Simply stated, bacteria are typically single-celled vegetative organisms; protozoa are single-celled animals; rickettsia are similar to both bacteria and protozoa but are classified under a single genus; viruses are ultramicroscopic, considered by some to be proteins capable of multiplication; algae are microscopic plants containing chlorophyll; and fungi are microscopic plants devoid of chlorophyll.

The wide variety of these microorganism species makes enumeration of total soil populations difficult, since most cultural methods will reveal only those nutritional and physiological types which are compatible with the specific cultural environment. Theoretically, direct microscopic counts would permit enumeration of all microorganisms except viruses; however, the use of this technique is understandably limited. Microbial analyses of soil systems is most often focused on the identification of specific types of organisms. It should be apparent that no single analytical method available is capable of revealing the absolute total microbial population when one considers the variety of microorganisms harbored in the soil.

The various microbial species, in reality, change the surface layer of the soil environment from an aggregate of mineral particles to a mass teeming with microbes and honeycombed by visible channels. Between the soil particles are invisible channels and spaces of various size, intricate networks of microscopic grooves whose surfaces are coated with colloidal slime, and a variable water solution, transporting mineral and organic materials. Microorganism populations are not evenly distributed through the pores of soil systems. For example, bacteria commonly occur in colonies or clumps of a few to many thousands of individuals scattered along the walls of pores or channels in the soil or over the surfaces of the soil particles.

The microbial flora existing in a specific soil will depend almost exclusively on the composition and structure of the earth in question. The major constituents which make up all soil systems may be broadly categorized in terms of solids, liquids, and gases. The solid fraction of soil may be subdivided into biological systems, organic residue, and minerals. Fertile soil is usually inhabited by large numbers of microorganisms, many varying animal forms (rodents, worms, insects, etc.), and the root systems of higher plants. The leaves, plants, remains of animal bodies, and solid waste deposited on or in the soil contribute organic substances, referred to as humus, in the last stages of decomposition. The mineral constituents of soil range in size from particles as small as bacteria to large pebbles and gravel. The chemical nature of these mineral particles is comparable to that of the parent rock.

The liquid phase in soil consists almost exclusively of water which in turn serves as the vehicle of nutrient transmission. Various inorganic and organic constituents are dissolved in water and are thus made available for plants and microorganisms.

The gaseous character of soil is comprised largely of carbon dioxide, oxygen, and nitrogen which may be utilized by various plants and animals in performing their metabolic functions.

7-3. SOIL BACTERIA

The bacterial population of soil exceeds the population of all other microorganism groups in both number and variety, with reports of direct microscopic counts as high as several billion per gram in soil containing the proper food supply, moisture, temperature, and physical conditions. However, plate counts from the same soil samples yield only a fraction of this number (millions), since no single cultural medium will provide an environment nutritionally adequate and physically satisfactory for the growth of all physiological types present.

Ordinarily bacterial generations in soil are very short-lived, although particular individual bacteria may survive for long periods under optimal conditions. The numbers of bacteria present in any situation are continually changed by the death of myriads of organisms and their replacement by others. Every gram of soil may contain billions of dead microorganisms and consequently the protein determination in the typical soil analysis is largely composed of microbial remains.

The most numerous bacteria found in arable soil are those capable of utilizing a wide variety of food materials. They degrade complex organic substances (solid waste) to simple compounds such as carbon dioxide and ammonia. Certain bacteria initiate the decomposition process and others complete it, except where the nature of the material is such as to resist attack (nonbiodegradable).

Bacterial types likely to be found in soil include aerobes (require oxygen for growth), anaerobes (grow in an oxygen-devoid environment), facultative (exist either aerobically or anaerobically), psychrophiles (cold-loving bacteria—exist at temperatures as low as $0°C$), mesophiles (survive in a temperature range from $25-40°C$), thermophiles (heat-loving bacteria which exist at temperatures as high as $50°C$), autotrophs (those microorganisms that utilize carbon dioxide as a carbon source and inorganic nutrients for energy), heterotrophs (those microorganisms that utilize organic compounds, such as solid waste materials, for both a carbon source and energy), cellulose digesters, sulfur oxidizers, protein digesters, and nitrogen fixers.

The great bulk of the decomposition of solid waste in soil is carried on by facultative organisms. Initially, facultative bacteria degrade complex organic wastes to simple organic acids under anaerobic conditions. The resulting water-soluble organic acids are then converted either aerobically to carbon dioxide and water, or anaerobically to methane. The compositional makeup of solid waste material will largely determine the speed with which decomposition will proceed, since starches, sugars, and proteins will be easily decomposed, whereas cellulose, lignin, and other related carbohydrates are very resistant to biological attack.

Millions of actinomycetes may be present in each gram of dry, warm soil imparting the characteristic musty or earthy odor associated with a freshly plowed field. These bacteria require little nitrogen and are therefore quite amenable to typical solid waste material, which usually has a low nitrogen to carbon ratio. Consequently, actinomycetes are capable of degrading many complex chemical substances thereby playing an important role in solid waste disposal in soil systems.

7-4. SOIL FUNGI

Many different species of mold or fungi inhabit the soil, existing predominantly near the surface, where aerobic conditions are likely to prevail. Since they may exist in either the mycelial vegetative filamentous stage or in the spore stage, it is difficult to estimate their population accurately; however, enumerations ranging from thousands to hundreds of thousands per gram of soil have been reported. The accumulation of mold mycelium imparts a physical soil structure or "crumb structure" which is of considerable agricultural importance. This binding together of fine soil particles is accomplished by the penetration of the mycelium through the soil, forming a network which entangles small particles making water-stable aggregates. Among the fungi are many curious forms especially adapted to restricted conditions, but more important by far are cosmopolitan genera, present in soils everywhere, which participate in the decomposition of organic matter. Fungi are active in decomposing the major constituents of plant tissue such as cellulose and lignin and therefore play an important role in the stabilization of solid wastes in both landfills and composting processes.

Yeasts are not abundant in soils except where environmental conditions favor their growth, and in general are not involved in solid waste decomposition to a large extent.

7-5. SOIL ALGAE

The population of algae in soil is generally smaller than that of either bacteria or fungi, the major types present consisting of the green algae, the blue-green algae, and the diatoms. Their photosynthetic nature accounts for their predominance on the surface or just below the surface layer of soil. In a rich, fertile soil the biochemical activities of algae are dwarfed by those of bacteria and fungi; however, in some instances algae perform prominent and beneficial changes. For example, on barren and eroded lands they may initiate the accumulation of organic matter. This activity has been observed in some desert soils. Algal growth also contributes to soil structure and is beneficial for erosion control. Their role in nitrogen fixation, particularly in paddy soils of the kind used for cultivation of rice in the Far East, is extremely significant. The basic physiological characteristics of algae, however, inherently endow them with a minimal role with regard to solid waste disposal in either sanitary landfills or in the composting process.

7-6. SOIL PROTOZOA

In a soil population of millions of organisms to the gram, the one-celled animals (protozoa) constantly appear, primarily as destroyers that prey on bacteria. Certain of the smaller species have been found to be represented in soils of widely differing physical and chemical composition and under most varied climatic conditions. Most soil protozoa are amoeba or flagellates, the number per gram of soil ranging from a few hundred to several thousand. Of academic interest is the fact that they exhibit a preference for certain microbial species. Because of their primary activity of feeding upon other organisms, the protozoa may be regarded as an important agent in maintaining the equilibrium of microbial flora in solid waste disposal systems.

7-7. SOIL VIRUSES

Bacterial viruses (bacteriophages), as well as some plant and animal viruses, also exist in soil systems. Bacteriophages can conceivably exert some influence on the microbial ecology of the soil; however, this influence has not been fully established.

7-8. INFLUENCE OF SOIL CONDITIONS ON MICROBIAL FLORA

The region at which soil and roots interface is designated the rhizosphere. Its microbial flora are considerably greater than that of root-free soil. Bacteria growth may be enhanced by nutritional substances (amino acids, vitamins, etc.) released from plants, while plant growth may be influenced by various bacterial end products. Although the bacteria flora of the rhizosphere are known to be physiologically active because of the tremendously complex biological system, solid waste decomposition in landfills or composting processes is usually accomplished by means of non-rhizospheric organisms.

In addition to the effect of root systems on microbial content in soils, other growth influencing soil parameters include soil composition and physical characteristics, temperature, pH, available moisture, nutrients, oxygen content, variations in climatic conditions, and interactions between various microbial species.

These environmental soil parameters will indirectly affect solid waste decomposition in landfill operations because of their direct effect upon the soil flora. Those parameters which can be artifically controlled should be maintained at proper levels to insure optimal environmental conditions for biological degradation of the solid wastes.

7-9. ACTIVITY OF MICROORGANISMS IN SOLID WASTE DISPOSAL SYSTEMS

Most solid waste exists in the form of complex organic materials, and may be deposited in various types of solid waste disposal systems to be stabilized by

microorganisms. The end product of this biological activity is humus, a dark-colored, amorphous substance composed of residual organic material virtually nonbiodegradable in nature. Humus may contribute several important aspects to soil systems which include:
1. desirable alterations in soil texture
2. contributions to the buffering capacity of the soil
3. nutrient source origin for plants and microorganisms
4. improvement of the water-holding capacity of the soil
5. increased availability of minerals

The complexity and diversity of the soil microflora have their parallel in the biochemical changes which they accomplish in solid waste disposal systems. While complex organic solid wastes are dissimilated, inorganic compounds such as sulfates and phosphates are being produced in a process referred to as mineralization. Acids resulting from degradation reactions may dissolve minerals, making them available to plants and other organisms as nutrients. The metabolic potential of microorganisms in some solid waste disposal systems is relied upon to dissipate not only the "normal" solid wastes but also herbicides and insecticides which are being discarded in ever-increasing amounts. The complexity of the biochemical activities of microorganisms in the solid waste disposal environment cannot be overemphaiszed. Only a limited discussion is presented here, relating the important phases of soil microbiology to solid waste degradation directly involving the cyclic transformations of carbon, nitrogen, sulfur, and compounds thereof.

7-10. TRANSFORMATIONS INVOLVING CARBON COMPOUNDS

Compounds of carbon are involved in a series of chemical changes of a cyclic nature. These biochemical events extend from reactions involving carbon dioxide to the complex polysaccharide constituents of solid wastes. Changes involve both degradation and synthesis.

Solid wastes deposited in the soil contain carbon as a constituent of organic compounds, some of which is not available for plant growth. This carbon must be released, or returned as carbon dioxide, before it can again be utilized by plants. Microorganisms bring about this transformation.

The major organic carbon-containing substances present in solid waste include cellulose, starch, sugar, lignin, and pectin. Glycogen from dead animals, as well as the lipids and proteins of all animal cells, also contain carbon in complex organic organization. Through a variety of integrated microbial enzymatic reactions, the organically bound carbon in these substrates is ultimately released as carbon dioxide. Microbes capable of breaking down long-chain cellulose molecules (cellulolytic microorganisms), including many species of bacteria and fungi, are common in landfill or compost systems. These organisms excrete an enzyme (cellulase) which hydrolyzes the long cellulose chain into smaller glucose molecules. This glucose is readily metabolized by a multitude of microbes via various biochemical pathways. Aerobic bacteria are likely to produce carbon dioxide and cell substance, whereas anaerobes produce methane gas, a variety of fatty acids, alcohols, and other neutral products.

Examples of carbon dioxide transformation or incorporation into organic compounds by bacteria include:

1) Utilization of carbon dioxide by autotrophic bacteria. Carbon dioxide represents the sole source of carbon for these organisms and is transformed by a reduction reaction to carbohydrates as shown below:

$$CO_2 + 2H_2 \longrightarrow (CH_2O)_x + H_2O$$

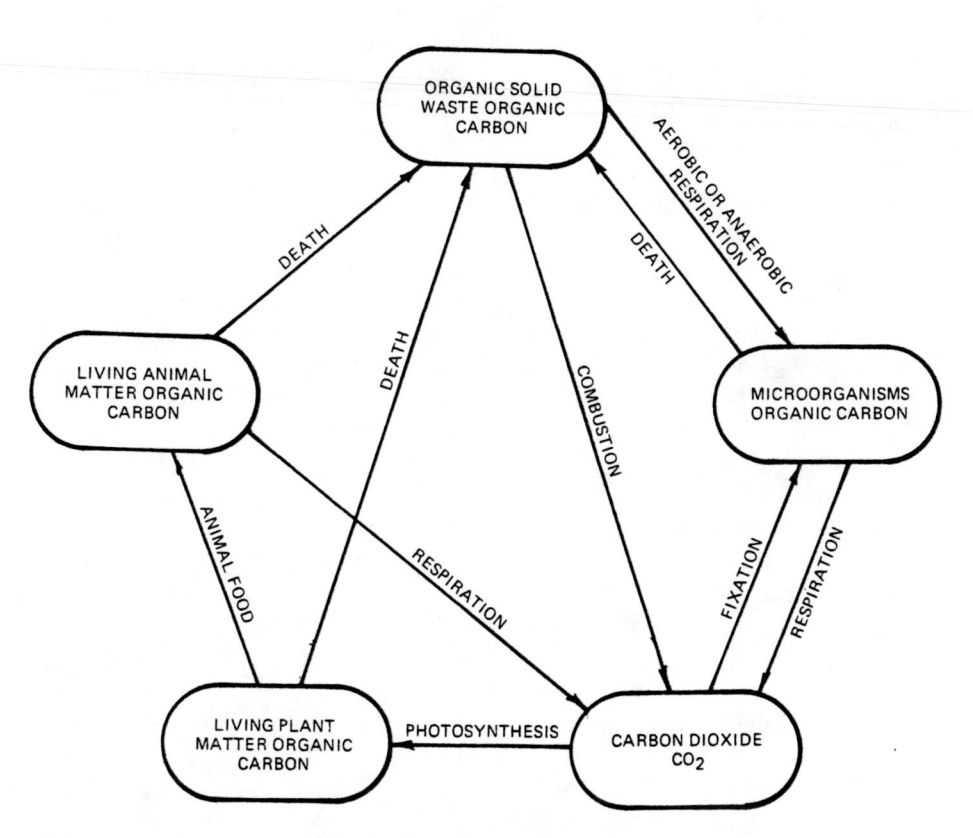

Fig. 7-1 The carbon cycle.

2) Many heterotrophic bacteria are capable of "fixing" carbon dioxide into a preexisting organic compound. A specific example of this type of reaction is

$$CH_3COCOOH + CO_2 \longrightarrow HOOCCH_2COCOOH$$

(pyruvic acid)   (oxalacetic acid)

Microorganisms are capable of synthesizing a variety of carbohydrates from simple carbon compounds, some of which are integral parts of the cell structure, capsular material, or products secreted into the soil environment as slime. The slimes released into the soil are beneficial in the formation of water-stable aggregates of soil particles, a desirable type of soil structure. A general summary of the carbon cycle is shown in Fig. 7-1.

7-11. THE NITROGEN CYCLE

Nitrogenous substances are found in almost every type of organic solid waste. Proteins, a constituent of all living cells, are complex nitrogenous compounds having an average nitrogen content of approximately 16 percent. Other complex organic nitrogenous substances occurring in solid wastes include nucleic acids, purine and pyrimidine bases, and amino sugars. The simplest form of nitrogen involved in biological transformations is gaseous elementary nitrogen. The overall transformations in which microorganisms are involved range from nitrogen gas to protein and other complex organic nitrogenous compounds. Many intricate enzymatic reactions are involved in bringing about these changes, and not all the steps have been clearly elucidated. Some of the changes include:

1. Proteolysis. The nitrogen present in proteins (as well as nucleic acids) may be regarded as "unavailable"; insofar as synthesis of nitrogenous compounds is concerned it is "locked" and not available as a nutrient to plants. In order to set this organically bound nitrogen free for recirculation, the first process that must occur is the enzymatic breakdown of proteins, or proteolysis. This is accomplished by microorganisms that release extracellular enzymes (proteinases) which convert the long-chain protein molecule to smaller units of amino acids (peptides). These peptides are then attacked by different enzymes (peptidases), resulting ultimately in the release of individual amino acids. Relatively few bacterial species possess large amounts of proteolytic enzymes; however, many fungi and soil actinomycetes are extremely proteolytic and consequently of great value in solid waste disposal systems.

2. Amino acid breakdown. The ultimate products of proteolysis are amino acids, which may be utilized as nutrients by microorganisms or degraded by microbial attack through various metabolic pathways. The liberation of nitrogen from these compounds is accomplished by the removal of the amino group (deamination). Although several variations of deamination reactions are ex-

hibited by microorganisms, one of the end products is always ammonia (NH_3). The fate of the ammonia produced varies, depending upon the soil conditions. Possibilities include accumulation and utilization of the ammonia and plants by microorganisms and its oxidation to nitrate (NO_3).

3. Nitrification. The oxidation of ammonia to nitrate is referred to as nitrification. It is one of the most important activities of autotrophic bacteria from the standpoint of soil fertility, since the reaction product (nitrate) provides a form of nitrogen most available to plants. Nitrification is carried out in two stages by specific bacteria, as shown below:

1. Oxidation of ammonia to nitrite by *Nitrosomonas*:

$$2NH_3 + 3O_2 \longrightarrow 2HNO_2 + 2H_2O$$

2. Oxidation of nitrite to nitrate by *Nitrobacter*:

$$HNO_2 + \tfrac{1}{2}O_2 \longrightarrow HNO_3$$

Few genera of bacteria have been found capable of performing these reactions, and only *Nitrosomonas* and *Nitrobacter* have been studied in detail. Organisms of both these genera are obligate autotrophs and strict aerobes and will not grow upon the usual media employed for cultivation of heterotrophs. They grow only in a mineral salts medium containing ammonia or nitrite. Nitrification was discovered to be a biological process in 1877, whereas isolation of the bacteria responsible for the process was not accomplished until 1890.

4. Reduction of nitrate to ammonia. Several heterotrophic bacteria are capable of converting nitrates into nitrites or ammonia under anaerobic conditions. The oxygen of the nitrate serves as an acceptor for hydrogen. The process involves several reactions, and the overall result is as follows:

$$HNO_3 + 4H_2 \longrightarrow NH_3 + 3H_2O$$

5. Denitrification. Certain microorganisms are capable of transforming nitrates to nitrogen gas or nitrous oxide. This process is called denitrification, and the change leads to a net loss of nitrogen from the soil. Organisms incriminated in this reaction include both autotrophs and heterotrophs. Denitrification does not occur to any significant degree in well-aerated soils with moderate amounts of organic matter and nitrates. It is most likely to occur in anaerobic soils saturated with water and containing an abundance of organic substances.

6. Nitrogen fixation. A number of microorganisms are able to utilize molecular nitrogen in the atmosphere as their source of nitrogen. The conversion of molecular nitrogen into nitrogenous compounds is known as nitrogen fixation. Two groups of microorganisms are involved in this process: (1) nonsymbiotic microorganisms, those living freely and independently in the soil, and (2) symbiotic microorganisms, those living in roots of plants.

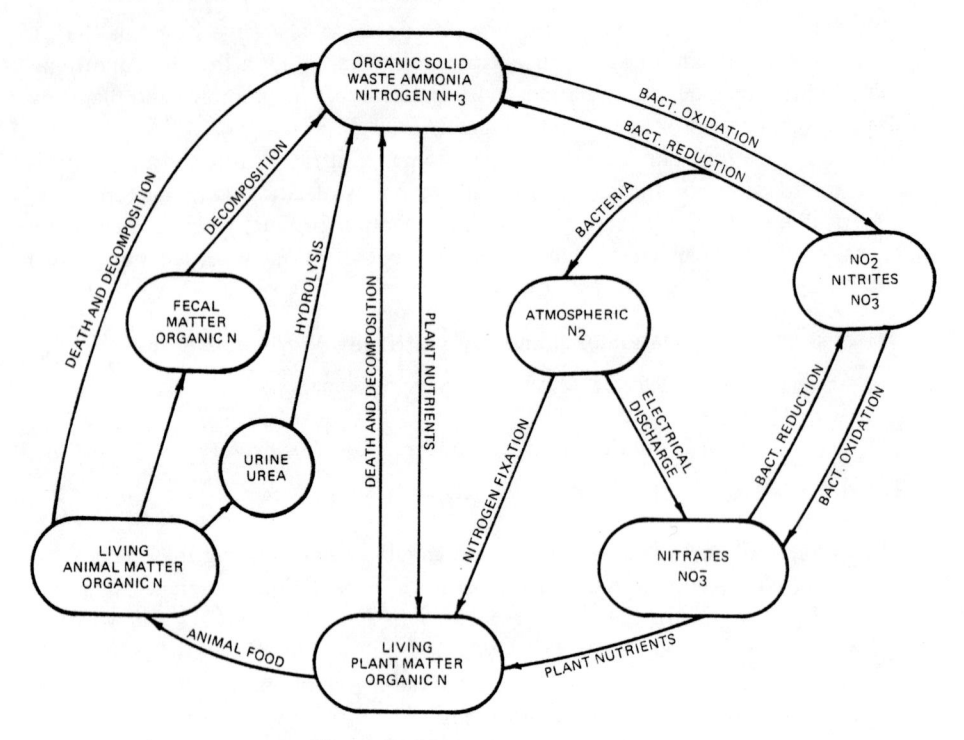

Fig. 7-2 The nitrogen cycle.

A summary of the various nitrogen transformations known as the nitrogen cycle is shown in Fig. 7-2.

7-12. SULFUR TRANSFORMATIONS

Sulfur, like nitrogen and carbon, passes through a cycle of transformations mediated by microorganisms which oxidize and reduce various sulfur compounds (see Fig. 7-3). Some of the biochemical changes performed by microorganisms involved in this cycle may be summarized as follows:

1. Sulfur in its elemental form cannot be utilized by plants or animals. Certain bacteria, however, are capable of oxidizing sulfur to sulfates as shown below:

$$2S + 2H_2O + 3O_2 \rightarrow 2H_2SO_4$$

2. Sulfate is assimilated by plants and is incorporated into proteins. Degradation of proteins (proteolysis) liberates amino acids, some of which contain sulfur. This sulfur is released from the amino acids by the enzymatic activity of various heterotrophic bacteria.

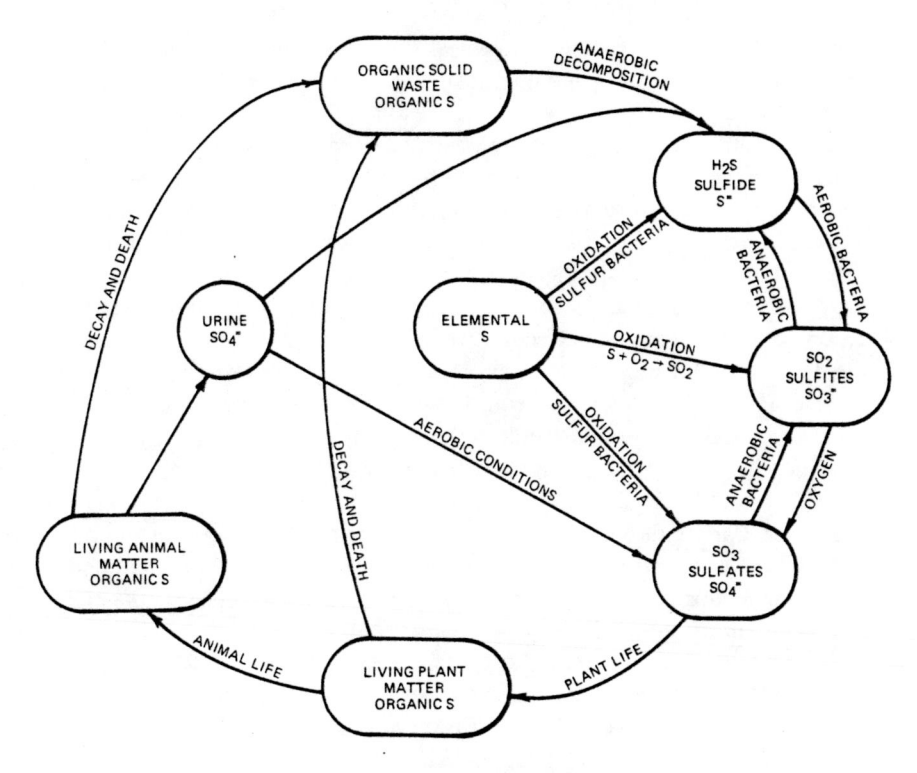

Fig. 7-3 The sulfur cycle.

3. Sulfates may also be reduced to hydrogen sulfide by soil microorganisms as follows:

$$4H_2 + CaSO_4 \longrightarrow H_2S + Ca(OH)_2 + 2H_2O$$

4. Hydrogen sulfide resulting from sulfate reduction and amino acid decomposition is oxidized to elemental sulfur. This reaction is characteristic of certain pigmented (photosynthetic) sulfur bacteria.

The basic biochemical reactions discussed in the last several sections, dealing with carbon, nitrogen, and sulfur transformations are a small sampling of the microbial activity occurring in solid waste disposal systems.

7-13. MICROBIOLOGY OF SANITARY LANDFILLS

A sanitary landfill can be described as engineered burial of solid wastes which are subsequently degraded by soil microorganisms. Following burial, the microorganisms slowly degrade the organic portion of solid wastes to stable compounds. Compared to other biological treatment systems (activated sludge, anaerobic digestion, etc.), the microbial degradation of solid wastes proceeds at a very slow rate.

Complex organic solid wastes are degraded primarily by aerobic or facultative bacteria and fungi. Facultative organisms are of importance in a landfill ecosystem since they can survive in both oxygen or non-oxygen environments. This is significant, since air does not penetrate a well-compacted landfill to any extent, and oxygen inside the fill is utilized rapidly by aerobic microorganisms as they decompose organic solid wastes. When this oxygen supply is depleted, decomposition by anaerobic facultative microorganisms begins and accounts for the degradation of most of the organic solid wastes in the landfill. Water soluble organic acids are produced, through the hydrolysis of complex organic solid wastes, by these anaerobic facultative bacteria. These organic acids enter the water media and are then able to diffuse through the landfill soils where fungi and other bacteria aerobically metabolize this organic matter to carbon dioxide and water. Carbon dioxide produced inside a landfill can dissolve in groundwater making the water weakly acidic. Limestone and other rock media coming in contact with this groundwater can be dissolved, thereby increasing the dissolved solids content of the water.

Anaerobic methane bacteria can occasionally accumulate in large quantities in landfill systems thereby discharging sizable amounts of methane gas through the soil. Aerobic bacteria can utilize a portion of this methane as it diffuses through the landfill; however, most of the methane is lost to the atmosphere. Buildings can trap escaping methane, creating an explosion hazard; and this, plus potential settlement of the fill, are the two factors of concern in erecting buildings on or directly adjacent to sanitary landfills. Obviously, any excessive production of methane in landfill systems is a potential fire hazard.

Water is essential for biological degradation; therefore landfilled solid wastes containing less than 60 percent moisture will be difficult for bacteria to decompose. Groundwater often serves to moisten solid wastes, allowing microbial activity, and also serves as a medium for end product diffusion; however, an excessive amount of water may fill the air voids, which then makes aerobic metabolism impossible.

Proper landfill design and operation is required to avoid difficulties with regard to gases and groundwater pollution. Current research is directed at determining the movement of intermediate products of microbial decomposition into groundwater and bacterial pollution of groundwater. Other studies have demonstrated the feasibility of pumping air into landfills to achieve aerobic conditions. Aerobic landfills are desirable in that they exhibit more rapid settling characteristics (i.e., more waste disposal capacity) than anaerobic ones, since aerobic metabolism proceeds at a faster rate. Also any explosive potential will be greatly reduced in aerobic landfills because of the elimination of excessive amounts of methane gas. Much of the future research regarding landfill systems will be largely microbiological in nature since such research will be directed at upgrading the knowledge of microbial decomposition products in landfills. Such knowledge will facilitate a more thorough understanding of the landfill stabilization process, thereby enabling optimal process control.

7-14. MICROBIOLOGY OF COMPOSTING

Composting operations consist of controlled microbial reactions yielding a stable end product much sooner than can be obtained using sanitary landfill methods. Composting produces a stable material than can be utilized to recover or rejuvenate wasteland. In countries having an abundance of mineral fertilizers (such as the United States), composting has never been used to any appreciable extent.

Fungi and actinomycetes are the predominant group of microorganisms in composting, since these organisms are favored in the semimoist conditions that prevail in the process. Bacteria will begin to predominate under high-moisture conditions. Little is known about the microbiology of composting, other than that natural mixtures of bacteria, actinonmycetes, and fungi will produce a compost just as efficiently as so-called special pure cultures.

The predominance of fungi in the compost will depend largely on the type of organic solid wastes being decomposed. Various fungi species have been reported in composting operations. These include *Penicillium*, *Aspergillus*, *Mucor*, and *Rhizopus*. Bacterial species found in most waste water treatment systems can also be found in composting operations.

The major microbiological control parameters for optimum composting include:

1. Temperature. Optimum thermophilic composting occurs at approximately 140°F. Microbial decomposition proceeds at a faster rate at elevated temperatures, thereby speeding up the composting process.

2. Moisture. The desired range of moisture content is from 40 to 70 percent with an optimum content of 55 percent. Moist conditions will favor bacteria over fungi or actinomycetes, thereby enhancing the rate of the composting process.

3. Hydrogen ion concentration (pH). Satisfactory composting occurs in a pH range from 4.5 to 9.5 with optimum conditions at pH 6.5. This pH range is required for optimal microbial metabolism and consequently proper composting.

4. Nutrients. The desired carbon to nitrogen ratio is 40:1 while the optimal carbon to phosphorus ratio is 100:1. Nitrogen and phosphorus must be present in the compost at these ratios to ensure an adequate nutrient content for microbial growth.

5. Air. It has been observed that 10 to 30 cu ft of air/day/lb of volatile compost solids is required for proper aerobic biological metabolism.

6. Particle size. Solid waste particle sizes ranging from ¼ to 1 in. provide an optimal surface area for the microbes to begin the degradation process.

Little is known about the chemical constituents of compost which might be considered essential plant nutrients, excepting the three primary nutrients (phosphorus, nitrogen, and sulfur). There currently exists no common agreement on the point at which degrading organic solid wastes becomes "compost." A fully satisfactory method for determining the extent of solid waste degradation has not been developed and consequently research on the chemical and microbiological changes that occur in the composting process has been hindered. Plant design is also hampered, since it should logically depend on the optimal or required degree of waste treatment.

Composting potentials are much greater than those of basic landfill operations; however, composting will not become a major method of solid wastes disposal in this country until composting economics become more favorable.

REFERENCES

7-1. Burges, A., *Microorganisms in the Soil*, Hutchinson & Co., Ltd, London, 1958.

7-2. Ehlers, V. M., and Steel, E. W., *Municipal and Rural Sanitation*, McGraw-Hill Book Co., New York, 1965.

7-3. McKinney, R., *Microbiology for Sanitary Engineers*, McGraw-Hill Book Co, New York.

7-4. Alexander, M., *Introduction to Soil Microbiology*, John Wiley & Sons, New York, 1961.

7-5. Pelezar, M. J., Jr., and Reid, R. D., *Microbiology*, McGraw-Hill Book Co., New York, 1965.

7-6. American Chemical Society, *Cleaning Our Environment: The Chemical Basis for Action*, 1970.

7-7. Waksman, S. A., *Soil Microbiology*, John Wiley & Sons, New York, 1952.

8
Disposal methods—composting

8-1. INTRODUCTION

In Chapter 6 the potential for recycling and reuse of solid waste was discussed. In a summary analysis it appears that a certain amount of wastes will always remain for disposal and, moreover, until one or more of the proposed recycling methods are implemented, means must be at hand for disposal of the wastes currently being generated. For these reasons, it is imperative that the various disposal methods which are now in use be examined and alternative methods of disposal which have been proposed be investigated. The discussion in the following chapters deals with the principal methods of solid waste disposal currently in use in the United States. Additionally, some discussion is devoted to methods under development and to research efforts in the area of solid waste disposal, in Chapter 11.

The methods of solid waste disposal discussed in this and the next two chapters are refuse composting, incineration, and sanitary landfilling. An effort will be made to present the technological basis of each method, to discuss the economic advantages inherent in each, and to evaluate the environmental ramifications of each such form of waste disposal.

8-2. COMPOSTING AS A DISPOSAL METHOD

8-2.1 General

Refuse composting is defined as the aerobic, thermophilic degradation of putrescible material in refuse by microorganisms. There is no clear distinction made in practice between the refuse being processed and the final end product called "compost." There is no general agreement, moreover, on the composition of the material which is known as "compost." However, certain definitions or criteria must be met when speaking about compost. For example, the entire process must be aerobic; in other words, the refuse stabilization must take place in an oxygen-rich atmosphere and no further degradation should take place under anaerobic conditions during subsequent storage. Composting should not be confused with other biodegradation processes for wastes.

8-2.2 Background

For many years decomposition of waste material has been utilized in the Orient for the production of soil additives; however, this decomposition is primarily anaerobic. During the twentieth century in western Europe aerobic composting was introduced for the treatment of refuse and solid waste. There has been a considerable amount of experience accumulated since the 1920s and 1930s in the practice of refuse composting, but European practice and European experience are not directly applicable to the refuse disposal situation in the United States because there is a significant difference in refuse composition; for example, there is more paper in American refuse.

In the refuse in the United States a major portion of the putrescible material can be stabilized in from 5 to 7 days if forced aeration techniques are used. Research studies by Wiley and Schultze (Ref. 8-1) have demonstrated this fact. On the basis of commercial endeavors, a period of 5 to 6 days has generally been selected as the average decomposition time for ground-up refuse in American mechanical composting procedures. On the other hand, where the composting action occurs through a spreading of the material in windrows which are periodically turned, a much longer composting period is required. From two weeks to three months may be required for adequate stabilization under such conditions.

Since composting is referred as a "thermophilic" degradation process it is obvious that relatively high temperatures are implied; during the average mechanical composting operation the temperature should exceed 140°F for a minimum of 96 hours in order for adequate stabilization to occur. To accelerate the degradation of putrescible materials by the active microorganisms, the refuse should first be ground to a particle size of less than an inch (average particle size). Furthermore, the moisture content of the ground-up refuse should be established at about 55 percent (on a total weight basis), so that the biological activity may proceed at an optimum rate. The carbon-to-nitrogen ratio should be established at approximately 40 to insure the most rapid stabilization. In practice the requirement that a carbon-to-nitrogen ratio of approximately 40 be maintained will cause some process difficulties. Average mixed refuse in the United States has a very high paper content which ordinarily yields a carbon-to-nitrogen ratio in excess of about 70. For the composting procedure to progress at the most rapid rate, either sewage solids (high in nitrogen content) or nitrogen solutions must be added to the composting refuse, in order that the carbon-to-nitrogen ratio may be adjusted to a value of about 40.

No general statement can be made concerning the composition of the refuse handled in the composting operations now existent throughout the United States, because this refuse, as mentioned previously, has a very heterogeneous character and is also time-variable in composition.

8-3. COMPOSTING SYSTEMS IN THE UNITED STATES

8-3.1 Operations

In general any composting operation consists of three basic operations: refuse preparation; stabilization of biodegradable materials; and upgrading of the final product. Preparation of the refuse includes receiving, sorting, separation and salvage operations, grinding, and the above-mentioned addition of both moisture and nitrogen if required. The subsequent aerobic digestion process can be accomplished in several ways, for example, open-field windrow techniques or mechanical plant techniques. The upgrading of the final compost product may consist of final grinding of the material, enrichment with selected additives, granulation in some cases, shipment, and marketing. A detailed discussion of each of these steps follows.

8-3.2 Refuse Preparation

In all of the composting operations currently operated in the United States a certain amount of hand and/or mechanical sorting of the incoming refuse is practiced. This separation and sorting procedure is required because certain materials in domestic refuse are actually non-compostable; while certain other bulky items such as major appliances may actually retard or interfere with subsequent composting operation and must be removed. Of course, separation may be done for the purpose of salvage, and at the incoming end of a composting operation a certain amount of salvage should be practiced. Most composting operations include hand sorting of some materials (for example, corrugated paper) from a slowly moving conveyor belt, and subsequent magnetic separation of the ferrous metal content within the refuse. Some systems now operating (and certainly some systems planned for future use) include inertial or ballistic separation in salvage plans for other materials in addition to the ferrous metals.

The second step in refuse preparation consists of grinding, which is necessary in order that the composting operation be efficient. Grinding may be accomplished by a number of different apparatus: hammermills, chain mills, rasp grinders, or wet pulpers, followed by some sort of dewatering apparatus. Wet pulping followed by dewatering is generally not suitable for most composting operations in the United States today. The grinders used in this refuse-reduction step require a significant amount of power; from 5 to 30 hp/ton/hr may be required. Plants are now being constructed with grinders that are sufficiently large so that the entire input per day to the plant may be ground in one 8-hr shift. Thus a plant, as initially designed, could be tripled in capacity by the addition of more digesting capacity and the operation of the grinding units on a 24-hr, 3-shift basis.

Grinding is usually performed in a 2-stage operation: course grinding to produce a particle size from 2 to 3 in., and second-stage grinding to produce a particle size of approximately 1/4 to 1 in. After the material is ground to the proper size, moisture and nitrogen in the form of a dilute ammonium nitrate solution is added. The second phase of the composting operation, the digestion phase, follows. The discussion of the second phase of the procedure must be broken down into separate discussions, since several types of composting digestion operations are presently being utilized.

8-3.3 Digestion Methods

In windrow composting, the refuse is brought into a plant, hand sorted, ground and moistened, and is then conveyed to an outdoor decomposition area where it is placed in windrows. The windrows are turned once or twice a week during a composting period of approximately 5 weeks. During these 5 weeks thermophilic microorganisms degrade the organic material present in the refuse. After the composting digestion phase, the material is commonly cured for from 2 to 4 weeks of additional time to insure stabilization. Windrow composting of this type has been practiced in many locations throughout the United States. The process requires a moderately large area for the composting plant, since the windrows are outside and the material must be retained at the plant site for a time period of from 30 to 90 days. For a city of 200,000 population a typical windrow composting plant would require approximately 55-60 acres of land.

Obviously this type of operation would be suitable for small cities with adequate land available close by and having a suitable nearby market area for the final product.

8-3.4 Mechanical Composting Systems

In addition to the windrow composting procedure, certain mechanical composting systems have been developed within recent years in the United States. Three main systems have proved successful: the Fairfield system; the International Disposal Corporation (IDC) system; and the MetroWaste system. The principal advantage of mechanical composting systems is that the land required for such plants is much less than that required for windrow plants of comparable capacity. A mechanical composting plant occupying approximately 8-10 acres could serve a city of 200,000 population (in contrast to 60 acres required for a windrow plant).

The first system to be discussed is the Fairfield system. A plant using the Fairfield system has been operating for many years in the city of Altoona, Pennsylvania. This plant processes approximately 25 tons of separated refuse per day. A schematic diagram of the Altoona plant is shown in Fig. 8-1. The initial operation at this plant is primary grinding with no prior hand sorting. Garbage and rubbish are collected separately in Altoona and thus no hand sorting is required; the delivered material goes immediately into a Williams hammermill. After initial grinding, secondary grinding is performed in a hydropulper. In this particular unit, sewage solids are added to the refuse to increase the moisture content and to enrich the material for the composting operation. After the material leaves the hydropulper it passes through a bar screen which removes film plastics, tin cans, and other non-compostable items. It is then fed into a circular digester. Before entering the digester, the wet pulp goes through a screw press, which reduces the moisture content (by compressing the material) to approximately 55 percent. After the material is placed in the digester, air is blown through the perforated bottom of the circular tank to keep the mixture aerobic. Differing amounts of air may be fed to various sections of the digestor. Agitation of the refuse is accomplished by a number of augers which operate on a revolving arm; these rotating augers continually mix the material and immediately blend the fresh wet pulp into the already composting mixture present. After a nominal 5-day digestion period, the material is removed and cured in windrows for about 3 weeks; the processed and cured material is moistened with a starch suspension, granulated, and dried. The operating costs for this digester are relatively high, since the agitation by the rotating auger must be continuous. One additional disadvantage is that expansion of the plant requires complete new construction of an additional digesting tank.

Another system is that of the International Disposal Corporation system. The

120 SOLID WASTE MANAGEMENT

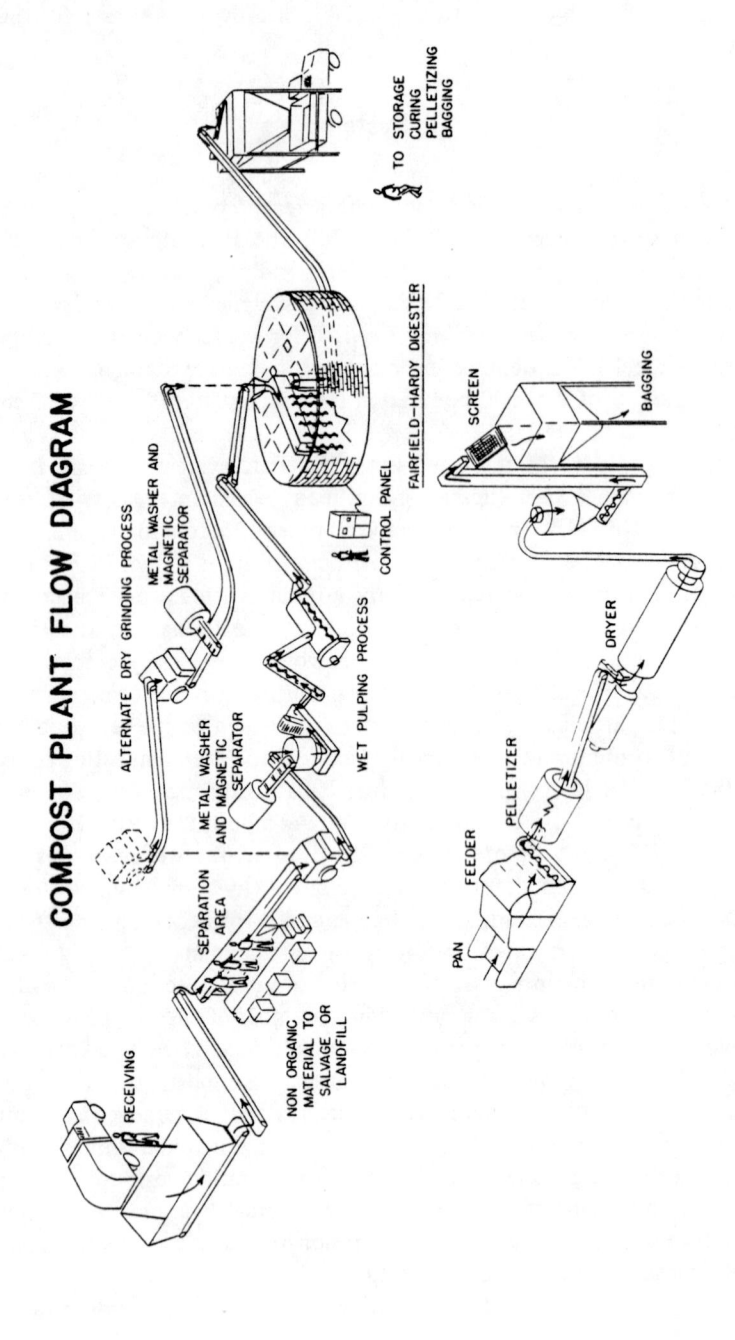

Fig. 8-1 Typical design for Fairfield hardy digestor installation and related equipment (*courtesy* Fairfield Engineering Co.).

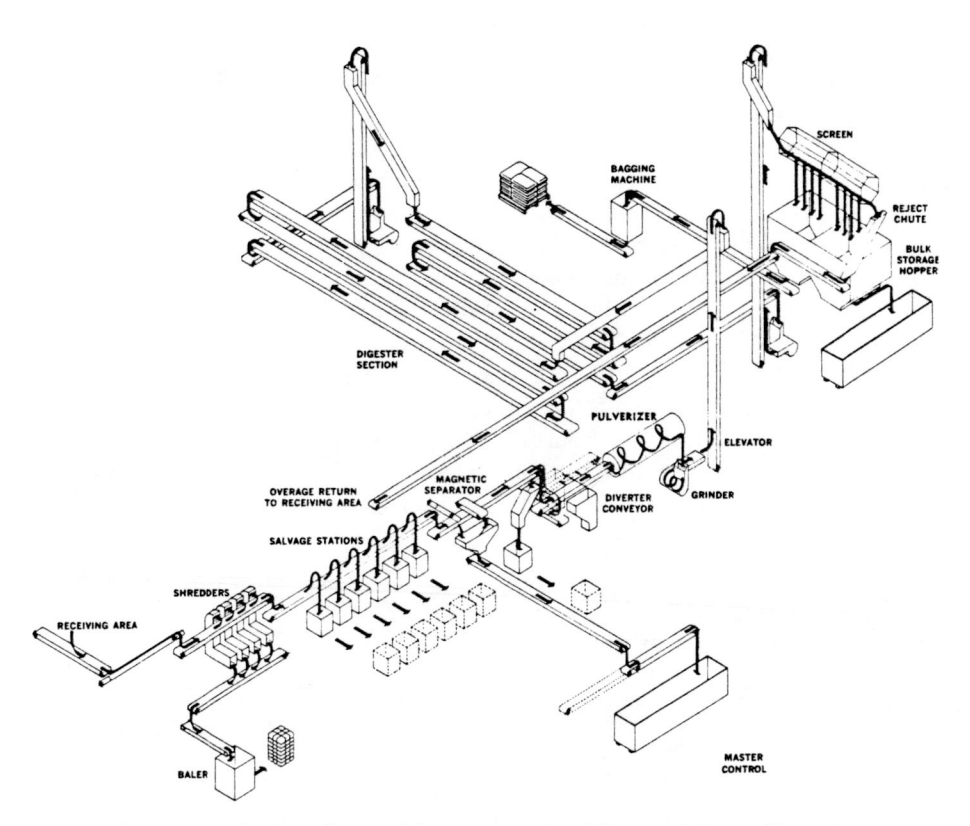

Fig. 8-2 Typical plant designed by International Disposal Corp. (from *Recovery and Utilization of Municipal Solid Waste*).

mechanical composting systems developed by International Disposal Corporation are typified by the plant shown schematically in Fig. 8-2. The initial operation at this kind of plant is sorting to remove non-compostable items from the incoming refuse. The sorted refuse is then processed through a magnetic separator to remove ferrous metals. The unit operating after the magnetic separator is a rotary mixer called a pulverator; the refuse is fed into this unit and a moistening agent, an ammonium nitrate solution, is added. After leaving the rotary mixer the moistened refuse enters a flail mill grinder which shreds the refuse but does not remove or shred plastic items or rags. These latter enter the composting process virtually intact. The next unit in the schematic diagram is a plugged-flow digester housed in a vertical building; the unit consists of horizontal, moving belts on which the ground refuse decomposes. Air is blown into the refuse pile just above the belt in order to maintain the decomposition process as an aerobic operation. Temperatures within the refuse are approximately 140°F, or in the range which is suitable for thermophilic microorganisms. After 2 days of proc-

122   SOLID WASTE MANAGEMENT

essing the material is reground and reinserted into the compost conveyor system. At the end of a detention time (total) of 5 days, the composted material is removed and is passed through a screen with ¾ in. openings. The screen separates non-compostable materials such as rags and plastic from the compost. The separated compost is then ground again and conveyed to outdoor curing piles. An additional 10 days is allowed for adequate curing of the composted material, after which it is sold in bulk, or is enriched and bagged for retail sales. This process is similar to the Fairfield process in that any expansion of the plant's capacity would require construction of a completely new digestion unit. The only other way to increase the capacity of this unit would be to reduce the detention time of the refuse in the digestion units; however, such practice may result in dangerous survival of pathogens in the improperly stabilized refuse.

The third mechanical composting system to be discussed is the MetroWaste system. A typical MetroWaste plant was the 150 ton per day MetroWaste plant in Gainesville, Florida. A schematic drawing of a MetroWaste system is shown in Fig. 8-3. The initial operation in this process is hand sorting of the incoming

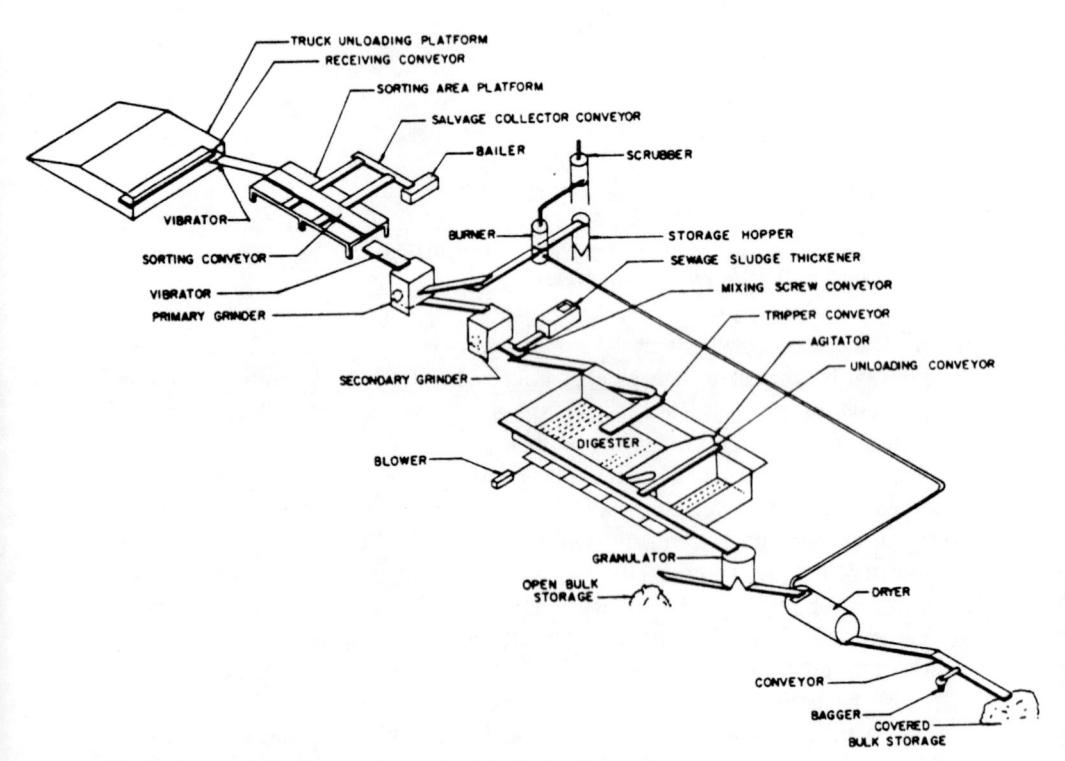

Fig. 8-3 Typical Metro-Waste installation (from *Recovery and Utilization of Municipal Solid Waste*).

refuse; after hand sorting the refuse is ground in either a hammermill or a centriblast unit, which also provides inertial separation. After grinding and separation, the material is then passed through a magnetic separator, a secondary grinder, and is moistened prior to composting; moistening may be done with either sewage solids or with a nitrogen solution. The digesters used in the Metro-Waste system are batch digesters consisting of horizontal tanks with perforated bottoms. The ground refuse is placed in the tanks and is held there for 4-6 days (the length of detention time depending upon the plant operating conditions). Aerobic conditions are maintained by blowing air through the bottom of the tanks either on a periodic cycle or continuously. A special agitator-unloader is used to mix and unload the composting material at the completion of the detention time. For convenience, the digester tanks are usually built in pairs, with a center belt serving both tanks for feed and takeoff. Such an arrangement allows one agitator to be used for both tanks with a periodic shifting of the agitator from one tank to the other.

Several experiments have been carried out with the MetroWaste composting system in an attempt to reduce the detention time. In these experiments oxygen

Fig. 8-4 Conveyor belt feeding the preparation stage of a conversion process. Manual separation is accomplished at the top of the conveyor with the salvaged paper fed directly into balers. Baled news and corrugated papers are shown at the right (*courtesy* Metropolitan Waste Conversion Corp.).

Fig. 8-5 Overall view of the shredded materials ready to be fed into the digestor. In the background the compost is being turned to accomplish a mixing operation (*courtesy* Metropolitan Waste Conversion Corp.).

enrichment during the first half-day to day of composting has significantly reduced the time required for the biodegradation to reach thermophilic temperatures. An increase in the oxygen content of the inlet air to a value of about 30 percent (by volume) has reduced the detention time by 1 to 2 days.

One advantage of this system is that the capacity of the plant can be increased by simply adding to the digester length while retaining and using the same agitator. This capacity increase in the MetroWaste system provides the cheapest additional capacity of any of the three mechanical systems discussed.

After the material has completed the composting cycle it passes through a series of secondary grinders, is screened and either cured or granulated for sale. An additional feature of the MetroWaste system (which is unique to it) is the use of air suction on the discharge side of the primary grinders; this air suction removes film plastic, and also some of the drier paper and glass fragments. The removed materials are burned in a suspension drier and provide heat for the drying of the material which has been composted.

The relative costs of these three mechanical composting systems in 1970 are compared in Table 8-1. The labor costs are quite variable and the labor requirements for operation of composting plants may vary from 0.06 man per ton of refuse processed per day to as much as 0.15 man per ton. An additional cost which has not received adequate attention is the cost of hammer wear and maintenance from the grinding operations. This cost is estimated (1970) to vary from 65 cents to $1.25 per ton of refuse processed (Ref. 8-2).

TABLE 8-1 Estimated Capital Costs, Energy and Manpower Needs, Three Mechanical Compost Plants (1970 Figures)

Capacity T/day	Fairfield			MetroWaste			IDC		
	$ × 10^6	hp	Labor	$ × 10^6	hp	Labor	$ × 10^6	hp	Labor
100	1.4	900	8	0.9	1250	12	1.4	600	20
200	2.1	1400	11	1.2	1700	17	2.1	800	28
300	2.5	1700	14	1.5	1900	25	2.7	950	36
400	3.2	2500	20	1.6	2000	30	3.2	1100	45

Source: Ref. 8-2.

8-3.5 Product Upgrading

The third basic step in composting operations is the upgrading of the final product. Upgrading operations which follow digestion may consist of any or all of the following: curing, grinding, screening, pelletizing or granulating, drying, magnetic separation, and bagging.

8-3.5.1 Curing. Storage of refuse after it has been stabilized through mechanical composting will result in a slow curing or maturing process. The net result of this slower curing is the production of a darker-color material with a shorter fiber length. Such material is esthetically more desirable. Most compost plants operating in the United States cure material for 10-60 days. The final compost product may be very salable even if this curing is omitted if the carbon-to-nitrogen ratio of the material is adjusted to insure a minimum of 1½ to 2 percent nitrogen content in the material. During any curing operation the temperature in the curing pile will remain near 140°F; properly stabilized compost will not produce odors, since the material can be piled and left without agitation for a considerable period of time without any anaerobic action taking place. After curing in such a pile, the material will be very dark brown in color and will function well as a soil conditioner or fertilizer filler.

8-3.5.2 Granulation. The best example of an operating granulation system is the Altoona, Pennsylvania, plant previously described. At this installation an attractive granular product has been produced. The moisture content of the material is reduced in this process from the 40-50 percent which is common in the regular compost product to about 10 percent for ease of handling.

All product upgrading operations are intended to make the final product from the composting plant more readily marketable.

8-4. COMPOST MARKETING

The traditional markets for compost have been truck farms relatively close to urban areas, luxury and specialty agriculture, home gardening and landscaping, highway right-of-way and city park use, and some use as filler for fertilizers. The potential for development of these traditional markets is somewhat limited. For example, New York City produces approximately 23,000 tons of refuse per day, or approximately 7 million tons annually. Since the reduction ratio in compost plants is approximately 2 to 1, this means that composting of all of New York City's refuse would produce about 3½ million tons of compost per year. There is little likelihood that this compost could be sold commercially in the New York City area. The entire fertilizer market within a 200-mile radius of New York City now utilizes approximately 4½ million tons a year. During the first few years of an expanded compost marketing program, only a small portion of this market could be captured—probably as little as 5-10 percent of it. Thus between ¼ and ½ million tons of compost could be sold while approximately 3½ million tons of compost could be manufactured (Ref. 8-7). There is thus a great discrepancy between predicted sales and potential production.

The present development of composting as a waste disposal system has been primarily directed toward semirural areas where a rather small-population city is surrounded by an area with high potential sales for the final compost end-product. Little likelihood exists for greater development of composting as a means of handling refuse in very large urban areas.

The present costs of finished compost product are $20 to $24 per ton purchased at the plant site. These costs are for bulk sales of the compost. Bag sales of compost have not as yet proved successful in the United States. The best potential market for compost is in bulk use as a building material in the fertilizer industry; this market is likely to grow, since there is an increasing trend to the use of organic fillers in fertilizers.

8-5. POTENTIAL FOR COMPOSTING OPERATIONS

8-5.1 General

Certainly for moderate-to-large size communities where urban space is a problem and surrounding agricultural activities are extensive, composting should be considered as a waste disposal operation. In these circumstances composting can probably compete effectively with incineration as a disposal operation, particularly if the operators of the compost system can develop a market for the compost and the items salvaged from the incoming refuse. An additional and advantageous situation for refuse composting is the possible combination of composting with sewage treatment.

8-5.2 Use of Sewage Sludge in Composting

It has been estimated that a city can save approximately 30 percent of the cost of its sewage treatment by pumping the raw sludge to a compost plant for use as a moistening agent and as a source of nitrogen in the compost (Ref. 8-4). The addition of raw sewage sludge to the composting operation certainly enhances the process. The greater use of paper and disposable packaging materials has led to an increase in the amount of paper and paper products in refuse and a corresponding percentage decrease in the amounts of putrescible materials such as garbage. Paper and paper products now constitute 50-60 percent of the weight of combined refuse and the content of garbage has decreased to about 10-15 percent. Consequently the refuse is much drier and bulkier than formerly and contains smaller proportions of nutrients. The addition of sewage sludge to the refuse speeds up the decomposition process and improves the quality of the finished compost by increasing its nutrient content. Sludge can replace water in adjusting the moisture of the composting mixture and will also improve the color and consistency of the compost. The moisture and nutrient contents of different types of sewage sludge are shown in Table 8-2.

TABLE 8-2 Moisture and Nutrient Contents of Several Types of Sewage Sludge

Type of Sludge	% Moisture	N	P_2O_5	K_2O
Primary				
raw	95-98	3.0-4.0	1.0-3.0	
digested	87-95	1.3-3.0	1.5-4.5	0.3-0.50.5
Primary + trickling filter				
raw	95-98	3.5-5.0		
digested	90-95	1.5-3.5	2.8-4.5	
Activated				
raw	98-99.5	4.3-6.4	4.0-7.0	0.3-0.7
digested	93-97	2.5-4.8	2.5-4.8	0.3-0.6

Nutrients, % Day Weight

Source: Ref. 8-4.

The most critical nutrient in any aerobic decomposition process is nitrogen. Refuse after separation and grinding as it commonly occurs today may contain only about $\frac{1}{2}$ to $\frac{9}{10}$ percent of nitrogen. A low nitrogen content requires that the microorganisms acting during the decomposition process recycle this nitrogen through many generations, gradually building up its percentage, as carbonaceous material is decomposed in the aerobic decomposition process to carbon dioxide and water. This process is extremely slow. On the other hand, raw sludges are high in nitrogen content and therefore can greatly speed up the composting process. Unfortunately these raw sludges are considerably more dilute than digested sludges. Furthermore, any type of sludge may be expected to contain pathogens in greater concentrations than are found in refuse. This means that much care is required in order that plant workers and public users of the compost are protected from disease. In addition to providing a benefit for the composting process itself, the addition of raw sewage sludge to compost will result in a reduction in the problem of sludge processing and disposal. The processing of sludge and its disposal presents problems to any sewage treatment operation. The amount of suspended solids being removed from sewage is constantly increasing with the cost of sewage treatment and increase in disposal of the sludge. Increased use of garbage grinders will cause additional difficulties in that the amount of sludge produced per capita is likely to increase. Final sludge disposal methods include use as fertilizer or soil conditioner; landfill and lagooning; combustion to ash and dilution in large bodies of water. These processes involve significant additional costs. Disposal of the final sludge product in a composting operation thus may save considerable amounts of money and alleviate significant problems in the disposal of the sludge itself. The joint processing of the refuse and sewage sludge by a composting operation should also produce a significant savings in the construction and operation costs of the combined plant over the cost of separate treatment and disposal facilities. The greatest savings obviously could be realized in a community where new sewage treatment plants are being planned and where the operations could be planned for composting and sewage treatment plants on adjoining sites. The conventional sludge digestion and drying operations could be eliminated, with the raw sludge simply thickened and pumped to the composting plants. The thickening of the sludge to approximately 88 percent moisture content can be obtained rather inexpensively by gravity filtration through cloth. For a slight increase in the net cost of composting the refuse, a considerable savings in the sewage treatment cost can be realized through the processing of the sludge with the refuse. The composting material will be increased in volume by only 6-10 percent, while at the same time the addition of the sludge will speed up the composting operation and produce a better final product from the point of view of nutrient contents. Thus, in addition to the savings in the sewage treatment plant operation, a greater market value for the end product will result.

8-6. FINANCING COMPOSTING OPERATIONS

8-6.1 General

The final operation in projecting any composting development in a community is the planning of financial support for the proposed construction. Since the composting operation is a municipal refuse disposal function it should be underwritten by dumping fees, so that these latter cover the disposal costs of the operation. Thus, amortization costs on all capital expenditures, an equipment replacement fund, operating costs including the cost of transporting compost to an ultimate disposal site during market development, and an additional cost for plant downtime operation should be covered by the dumping fees. During compost plant downtime an alternate method of disposal must be available. All of the above-mentioned items should be covered by a guaranteed minimum dumping fee for the composting plant. Furthermore, a realistic escalation clause for the effects of prices and costs inflation should be included in any contract for waste disposal, so that increased labor and operating costs may be met. It is suggested that in the financing of the compost plant no credit balance should be estimated for sale of salvage material or for sale of the final end composted product. Furthermore, an incineration cost should be included in the estimate of disposal charges so that plastics and other non-compostable but combustible items may be eliminated.

8-6.2 By-product Financing

The financing of the plant should also include a second phase, a by-product phase. In this financing operation the cost of final grinding, upgrading, marketing, etc. should be met by planned revenue received from the sale of the end product. Any revenue over and above the cost of product upgrading could represent profit for a private concern undertaking this phase of the compost operation. By thus separating the financing of the composting operation into distinct phases—a disposal phase supported by dumping fees, and a final product phase supported by sale of compost—a realistic approach to the problem can be achieved. Under this scheme of financing, when the additional savings in sewage treatment costs achieved by combined sludge/compost operations are included, composting economics become more attractive.

8-7. SUMMARY ON COMPOSTING

In summarizing the composting operation, it may be said that composting consists essentially of aerobic degradation of biodegradable materials. This oversimplified statement must also include a mention of the additional operations required: materials preparation, digestion itself, and end product upgrading.

Additional mention should also be made of some other problems associated with composting. A significant problem is the survival of pathogenic organisms in the compost. For example, viruses, bacteria, protozoa, fungi and other organisms capable of causing human, plant, or animal disease may persist in the compost. The possible hazards involved during production or use of the compost have also not yet been fully assessed. Further research in this area must be undertaken. Some control of these pathogens appears to be possible but no definite statement can be made at this time.

In addition to the survival of pathogenic organisms in compost a problem may also exist with disease vectors. Flies are attracted by and may develop in raw compost, particularly when the composting operation is of the windrow type. A further problem may be the existence of rodents and roaches in compost heaps if the composting operation does not include adequate housekeeping. However, some amelioration of the rodent problem has been attained through the use of grinders which intermingle the edible portions of the compost intimately with inedible portions and reduce the size of the material to the point where rats and other large animal vectors are unable to secure harborage within the material.

In addition to these problems, there are those of odors, noise, and visual nuisances. However, with proper operation most of these nuisances can be controlled. It must be anticipated that a certain amount of downtime will occur, that certain breakdowns will occur during the life of the plant, and that emergency alternative measures must be available. The public eye being on environmental problems today, little sympathy will be extended by the public to a composting operation which offends the senses with objectionable odors, noise, or unsightly appearance. Finally, the visible inclusions in the end product compost may put it at somewhat of a disadvantage for sales. For example, plastic films, tinfoil, and other inclusions are very expensive to remove from the composted material. However, all of these materials serve to remind a user of the original source of the compost and serve to prejudice the user against the composted material. All of these considerations should be kept in mind when considering composting as a means of community disposal of refuse. At the same time it must be realized that the composting operation may save considerable volume in a landfill disposal operation and may provide cover material for an existing and operating landfill. The composted end product may also find considerable usage in areas which are currently being explored. These areas of possible extensive usage of compost include reclamation of open-mined territories such as the strip-mined areas in the Appalachian region and in the denuded forest areas in the western states. The use of end product compost as a soil conditioner and partial fertilizer could alleviate some of the current problems of land reclamation. Finally, it must be realized that the entire composting operation uses land as an acceptor of materials which have originally been drawn from the land; therefore the composting process constitutes in truth a recycling operation. This recycling may not draw the most benefit from the recycled materials, but likewise it is not harmful and in some cases may have definite benefit.

REFERENCES

8-1. Wiley, John S., *Composting of Organic Wastes—An Annotated Bibliography*, PHS, Feb. 1958; Supplement 1, June 1959. Supplement 2, April 1960.

8-2. Harding, Charles I., "Recycling and Utilization," Proc., Surgeon General's Conference on Solid Waste Management for Metropolitan Washington, D.C., Govt. Printing Office, 1967.

8-3. Hart, Samuel A., "Solid Waste Management: Composting/European Activity and American Potential," USPHS Publ. No. 1826, Govt. Printing Office, 1968.

8-4. Wiley, John S., "A Discussion of Composting of Refuse with Sewage Sludge," paper presented at the 1966 APWA Public Works Congress, Sept. 13, 1966, Chicago.

8-5. Shell, G. L., and Boyd, J. L., "Composting Dewatered Sewage Sludge," Report SW-12c, HEW, USPHS, 1969 (Bur. SW Management).

8-6. "Gainsville Compost Plant—An Interim Report," HEW, USPHS 1969 (Bur. SW Management).

8-7. Varro, Stephen, "Composting As a Feasible Waste Disposal System," *Proc.*, New Directions in Solid Wastes Processing Institute, Framingham, Mass., 1970.

9

Incineration

9-1. GENERAL

The following discussion of incineration is by no means intended to be all-inclusive. Complete volumes, copious research articles, and numbers of general interest articles have been written on this subject. Therefore the intent of the following discussion is simply to present the basic principles of the incineration process and to discuss the mechanics of planning, design, construction, and operation of refuse incinerators.

9-1.1 Introduction

Incineration is a waste disposal process by means of which solid, liquid, and gaseous combustible wastes are converted through controlled combustion to a residue which contains virtually no combustible matter and to gases which are released to the atmosphere. As a disposal process, incineration is not a complete operation; the end products of incineration, the residue and materials in the gas exit stream, must be taken into account. In addition to the gases which are produced, a significant amount of particulate matter is entrained in the gas stream, and a significant amount of effluent process water is produced. Despite the incompleteness of the process, incineration has many advantages as a waste disposal operation. If land suitable for solid waste landfilling operations is not available within economic haul distances from the sources of the refuse, then incineration is quite advantageous. An incinerator should be positioned centrally with respect to the source areas of the wastes to be processed and in general will cause no disturbance or disruption of normal activities, if it can be located in areas which are zoned for industrial development. The basic advantages of incineration are that the solid waste is reduced in weight and volume; and, if the operation is carried out correctly, the residue after incineration is free of degradable material and offers no source of nuisances. For an adequately designed incinerator, a variety of different wastes may be processed to produce a nuisance-free material as an end product.

The major disadvantage is that an incinerator requires an investment of a significant amount of money. In addition to a large capital expenditure, operating costs are higher in an incinerator operation than the corresponding costs for operation of a sanitary landfill; one reason for these higher costs is that skilled labor is necessary for operation, maintenance, and repair of the installation. In addition to these disadvantages, consideration must be given to the attitude toward the incinerator in the surrounding community; adverse community reaction may be a principal deterrent to the construction and operation of an incinerator in any given area and in particular in a residential area.

Volume reduction by incineration of municipal solid waste is approximately 80-90 percent. Weight reduction in the combustible material portion is usually from 98-99 percent, since the combustible materials are converted to carbon dioxide and water vapor. On the basis of the total weight of incoming refuse, the weight reduction is generally 75-80 percent. If compaction of the resultant residue from an incineration operation is practiced, an even greater volume reduction may be achieved.

On the other side of the question, certain materials are not suitable for incineration and therefore the volume reduction of these wastes is practically zero. Generally the materials which are not suitable for municipal incinerators are those which are either too large to insert into the incinerator, those which burn too slowly, or those which contain some structural component which could interrupt the operation of the charging systems or the residue removal system. Bulky, burnable wastes include logs, mattresses, large appliances, tree stumps, tires, etc. To overcome the problem of bulky wastes, in some localities shredding operations have been added as a preliminary treatment for prior processing before the bulky material reaches the incinerator. In other localities, bulky wastes have been reduced in special incinerators which are designed and constructed to handle just such material.

Incineration is most advantageous when wastes are separated prior to or during collection. For materials which contain no bulky wastes, approximately ten times as much volume reduction is attained through incinerator processing as compared to compaction within a landfill; in other words, the ratio of volume reduction by incineration to that in landfilling is approximately 10 to 1. If, however, the wastes are the more typical domestic-municipal-commercial-industrial kinds which are collected in today's cities, a certain portion of the refuse will be bulky items. If the latter make up approximately one-fifth of the total volume of wastes to be processed, the volume reduction ratio may be as low as 2 to 1 when comparing incineration to landfill.

9-1.2 Fundamental Design Information

In order for an incinerator to be designed properly certain basic information must be furnished to the designer; this includes such items as data on the population currently served by the projected facility and the population to be served in the future. It also includes an estimate of the characteristics, composition, and amounts of waste which will be brought to the incinerator for disposal. Finally, some consideration must be given to control measures which will be necessary to insure that no pollution of air, land, or water may occur. While this information is needed for the design of an incinerator, few localities in the United States have gathered and obtained such detailed data. Therefore in most cases it is necessary for the designer to also undertake the collection of these basic facts.

Standard demographic techniques are available for estimating current and future population, but certain considerations should be kept in mind by the designer in gathering population data. These data will serve several significant purposes; for the most economical operation of the incinerator it should be located preferably at a central point with respect to the population to be served; further, the estimate of required incinerator capacity will necessarily be based upon an estimate of the current populace *and* of the projected growth of that populace.

It is exceedingly important that some estimate of the characteristics of the refuse to be incinerated be made, since the design of furnaces, grates, feed mechanisms and auxiliary equipment will depend in part upon the characteristics of the charged refuse. The characteristics of refuse generated in the United States have changed considerably in recent years, both in chemical and physical properties. As the amount of household garbage in the solid waste stream has decreased, the resultant overall moisture content of refuse has decreased and the combustible content and heat value have increased. This heat value increase is caused principally by an ever-increasing use of paper and plastic products, which because of their own high heat value have increased the average heat value of the solid waste. The overall effect of these changes in characteristics of the refuse has been that greater furnace volumes and more air for combustion are required in order that the rated burning capacities of existing incinerators can be maintained.

The composition of municipal solid wastes will vary considerably as a function of time, locality, and income level of the surrounding community. In general, municipal solid waste weighs between 200 and 300 lb/cu yd. After compression and compaction in a collector truck the solid waste increases in density to approximately 500 to 700 lb/cu yd; however, after the refuse is dumped in an incinerator storage pit the density of the material will decrease to approximately 400 to 500 lb/cu yd. The constituents of the refuse are highly variable, as shown in Table 9-1. In addition to the materials shown in Table 9-1, as mentioned previously, bulky items may make up as much as 20 percent by volume

TABLE 9-1 Range in Composition of Incinerator Charge

	Minimum	% (by Weight) Maximum	Average
Food waste	7.5	34.6	22.3
Garden waste	1.6	12.1	4.4
Paper products	29.8	61.8	47.6
Metals	6.6	10.9	8.3
Glass and ceramics	4.6	11.0	6.8
Plastics, rubber, leather	2.5	5.8	2.8
Textiles	1.4	4.8	2.2
Wood	0.4	2.2	1.4
Rocks, dirt, ashes, etc.	0.2	12.5	4.2

Source: Office Solid Waste Management Programs

of the total waste stream. The problem of bulky wastes has been overcome to a certain extent in some localities by the use of separate collection techniques or by size reduction at the incinerator itself. In addition to the information contained in Table 9-1 on composition of domestic solid wastes, the approximate analysis, ultimate analysis, and heat values of the different components of solid waste are shown in Table 9-2. Of particular interest is the ultimate analysis contained in Table 9-2, since this analysis provides the basis for calculating a materials balance of incoming and outgoing matter in the combustion operation and thus forms a basis for furnace design. In the ultimate analysis, moisture content is of particular importance in the design of the furnace.

Heating value is another characteristic of solid waste which is of considerable

TABLE 9-2 Physical and Chemical Characteristics of Charge

Proximate Analysis		Ultimate Analysis	
Constituents	% by Weight	Constituents	% by Weight
Moisture	15-35	Moisture	15-35
Volatile matter	50-65	Carbon	15-30
Fixed carbon	3-9	Oxygen	12-24
Noncombustibles	15-25	Hydrogen	2-5
		Nitrogen	0.2-1.0
		Sulfur	0.02-0.1
		Noncombustibles	15-25

"Higher" heating value
3,000-6,000 Btu/charged lb

Source: Office Solid Waste Management Programs

importance in the design of an incinerator. The heat values shown in Table 9-2 are "higher heat" values; i.e., the higher heat value is the total amount of heat which is released during the combustion process per unit weight of combusted material. As mentioned, this increase in heat value is attributed to the greater use of paper and plastic packaging materials and other combustible products. The present trends in the solid waste stream indicate that the values of heat from the waste given in Table 9-2 will probably increase by 500 to 1000 Btus per pound within the next decade.

In addition to the characteristics and composition of refuse, other important data for the design of an incinerator include the determination of the quantities of the solid waste to be anticipated from the serviced community. Quantities will vary with climate, season, community characteristics, relative prevalence of commercial, industrial, and residential areas, and with the amount of usage of food waste grinders, home compactors, and on-site incinerators. Table 9-3 shows the wide ranges of per-capita quantities of solid waste produced in the United States. It is apparent from Table 9-3 that local surveys of quantities produced will be necessary. Much of the variation in Table 9-3 is attributed to seasonal fluctuations; in general the maximum quantities of solid waste will be generated during the summer months. However, the amount or magnitude of

TABLE 9-3 Solid Waste Collected for Incineration, lbs/cap/day

Item	Quantity
Residential (domestic)	1.5-5.0
Commercial (stores, restaurants, businesses, etc.)	1.0-3.0
Incinerable bulky solid wastes (furniture, fixtures, brush, demolition, and construction wastes)	0.3-2.5

fluctuation in the per-capita quantities will vary significantly from community to community. It has been demonstrated that the four-week averages in waste generation within any given area commonly will range from -10 to +10 percent of the average weekly quantities; also, the weekly variation in any year can exceed one-fourth of the average weekly quantity for the entire year. These large fluctuations in solid waste quantities will have considerable bearing on the size of the designed incinerator. For economy the incinerator should be sized on the basis of a certain weekly quantity to be incinerated, with the storage pit designed to hold daily peaks in quantity. Several methods or techniques of kiln sizing are available. In one method, the highest estimated four-week-period quantity is determined for the design year; the average weekly delivery is then estimated from this highest four-week period and the incinerator sized to handle that average weekly input. Other methods include the use of frequency diagrams, so that the incinerator size is determined in such a way that its capacity will be exceeded only during a certain given percentage of time during the entire year. In any case, the incinerator should be of sufficient size so that provisions for preventive or routine maintenance and repairs may be met. The downtime necessary for these operations varies considerably from incinerator to incinerator. Newer model incinerators permit continuous operation for periods of up to 90 days. For such an incinerator the estimated downtime for repairs and maintenance will be approximately 15 percent; good designs of incinerators will also include a number of furnaces, so that the 15 percent downtime may be accounted for by the provision of an extra furnace. If the number of furnaces is not increased, so that any one unit may be shut down for repairs without exceeding the capacity or amount of incoming refuse, some alternative disposal methods must be provided.

Finally, an increasingly important aspect of design information which must be considered in the planning of any incinerator is the existence of regulations to prevent pollution of the environment and to safeguard the health and safety of the operating personnel within the incinerator plant itself. In addition to these considerations, standard building code requirements must be met, as must the zoning regulations of the community in which the incinerator is constructed. It is also reasonable to assume, in the present state of affairs in the United States today, that more stringent regulations concerning environmental pollution will be brought into existence in the immediate future.

9-1.3 Incinerator Costs

The expenditures for incinerators, as for any other solid waste disposal apparatus, include capital costs for the construction of the facility, and operating costs which include both direct and indirect operational and maintenance expenditures. Included in capital expenditures are such items as furnaces, cranes, scales, air pollution control devices, fans, process residue treatment and recycling equipment, instrumentation, waste heat recovery equipment, steam delivery equipment, etc. Furnaces and their appurtenances account for 60-65 percent of all capital costs; other components and their relative costs are physical plant, 20-30 percent; air pollution control equipment, 8-10 percent. Capital costs, as established by the 1968 National Survey of Community Solid Waste Practices of the Public Health Service, were approximately $6200 per ton. Almost two-thirds of the incinerators studied had average costs which were less than the $6200 per ton figure. However, 15 of the 170 plants surveyed had capital costs above $11,000 per ton and one plant had a capital cost of approximately $30,000 per ton.

Owning and operating costs were investigated at the same time that these capital costs were established. The average cost of operating municipal incinerators was found to be $5 per ton of incoming refuse. However, almost three-fourths of the incinerators surveyed had operating costs below $5 per ton; on the other hand, 4 of the 78 incinerators had reported operating costs above $10 per ton. This wide variation in costs has been attributed to several sources: variations in the amount and type of pollution control equipment installed in the particular incinerator; variable regional labor rates; differences in utility costs; differences in alternate disposal costs (for residue); and the amount of automation within the plant. In addition to operating costs and capital outlay, an additional cost factor is the amount which must be paid each year in financing the incinerator; these costs include depreciation and interest payments. Such financing costs have been estimated at between $1 and $2 per ton.

9-1.4 Physical Plant Location and Design

Several important factors should be considered whenever an incinerator facility is being planned. Among these factors are the characteristics of the site, the planning of the physical plant layout, and the actual design of the structure itself.

One of the most important criteria for selecting a site for an incinerator is acceptance by the public of the location. Public acceptance of an incinerator at a particular site will be much more readily attained if certain general principles are followed: (1) in general, the incinerator should be located in an area which has been zoned for commercial or industrial use and, usually, industrial land is more suitable than commercial; (2) the incinerator should be located so that there is no interference with the operation or appearance of other facilities (for example, an excellent location for an incinerator is near an existing or planned sewage treatment plant, since service facilities have already been established at this particular location and some joint operation of both plants may be attained); (3) a comprehensive program should be initiated to establish good public relations in the area surrounding the proposed incinerator—included in such a program should be public meetings, during which the proposed neighbors for the facility are acquainted with the economics of locating an incinerator in their neighborhood and also the possibilities for screening the facility from view. Additionally it should be made clear to the people in the affected neighborhood that the adverse effects of an incinerator have been exaggerated in the past and that the exteriors of many operating incinerators have been architecturally designed and have proved to be assets to the particular area in which they have been located.

Physical factors about the site itself which are important to the overall success of the facility include: the building code restrictions for that area and site; the foundation conditions, topography, and drainage provisions for the site; the ready access to utilities at the site; and the overall meteorologic conditions for the area. Topography, in addition to affecting the design of the structure itself, can play an important role in the dispersal of gases and particulate effluents from the incinerator into the atmosphere. Topography and meteorological conditions must be considered together when the air pollution hazards of the particular site are being evaluated. Foundation conditions and drainage will play an important part in the structure design since an incinerator with its furnaces and other appurtenances constitutes heavy concentrated loads for a supporting substratum. Since electricity, gas, and water must be supplied to the facility, and since sewage and process water must be carried away, ready access to utilities plays an important part in the selection of any given site. In addition to these particular properties of an individual site, its overall location with respect to the collection area is considerably important. The overall traffic flow pattern of the collection vehicles must be determined in advance and the incinerator must be located as close as possible to the center of that resultant flow pattern. In the actual layout of the physical plant and the design of the structure itself certain facilities must be planned, designed, and constructed: adequate traffic flow facilities must be provided, since an incinerator may be visited by hundreds of public and private collection vehicles per day; the physical plant should be designed so that it will harmonize well with the character of the surrounding neighborhood and be so situated that any unsightly portions such as the receiving areas are not visible to passing persons; facilities for the incinerator personnel must also be provided, and these facilities should include dining facilities, locker rooms, showers, toilet areas and, possibly, separate washroom facilities for collection personnel; other facilities within the structure itself must include a control room, administrative offices, a weighmaster's office, a laboratory, and maintenance and repair facilities. The facilities for incinerator personnel must receive adequate consideration, since they promote efficiency and cleanliness and lead to efficient operation; whereas insufficient facilities will create an unsuitable attitude on the part of the working personnel. An indifferent and inefficient working crew may, willingly or not, lead to the overall failure of an incinerator facility.

The external features of the plant itself should also receive consideration.

Such features will include: adequate lighting for the entire area; landscaping, including perimeter planting; fencing; traffic control equipment such as signs and lights; and also roadways, sidewalks, and parking areas. As a final observation, it is important that the roadway system be constructed so that one-way traffic flow is established. One-way traffic flow will insure that only one entrance and one exit to the facility is used and therefore careful monitoring of incoming material can be practiced.

9-2. COMPONENTS FOR MUNICIPAL INCINERATORS

The following discussion presents a brief description of the necessary components for operation of a municipal incinerator.

9-2.1 Utilities

The utilities required for a municipal incinerator include: supplies of potable water; electricity for power operations and lighting; sewerage systems for handling both process water and sewage from personnel; telephone service; storm sewers or drains; plant heating, and hot water supplies. The amounts of water required for incinerator operation vary from approximately 300 to 2,000 gallons per ton of waste incinerated. The cost for this water, assuming a rate of 20 cents per thousand gallons as a water rate would amount to 7 to 40 cents per ton of waste processed. Incoming water which is supplied to the plant will be changed in temperature, chemical characteristics, and solids content and will require subsequent treatment. Much attention is presently being directed toward water recycling and reuse in incineration operations.

The electric power requirements of incinerators will vary according to the equipment used within the plant; this equipment may include fans, pumps, hoists, cranes, air pollution control devices, grate drives, stoker mechanisms, etc. Electric power costs may amount to as much as 75 cents per ton of wastes generated. At the present time all the electric power used in an incinerator facility is usually supplied by outside sources, since generation of electrical power through the use of waste heat recovery within the incinerator is not being practiced as a general method within the United States.

Some additional comments may be made about other equipment necessary for good operation of an incinerator: (1) both storm and sanitary sewers should be provided; (2) communication within the incinerator plant should include public address systems, intercoms, bells, and other apparatus suitable for communications in areas with high levels of background noise; (3) in addition, in new incinerator installations closed circuit television monitoring of the entire operation should be considered; (4) finally, adequate fuel must be provided for water heating, building heating, and plant processes.

A further need for fuel may exist if auxiliary fuel is required.

9-2.2 Weighing Equipment

In the operation of an incinerator it is imperative that accurate weight records be obtained so that the actual operation of the incinerator may be managed correctly, future facilities may be planned, and reasonable fees and charges for incoming refuse may be established. The most efficient overall operation for waste disposal will be achieved when the weight records for an incinerator are analyzed so that the collection and disposal systems may be properly integrated. The scheduling and routing of collection vehicles then may be done in such a way that the incinerator facility is furnished with a uniform flow of incoming materials and no serious overload will result. Furthermore, measurement and recording of the weights, characteristics, and sources of the incoming solid wastes will be of vast assistance to planners in developing plans for any new incinerators or other waste disposal facilities.

The type of scales available for use in the weighing operation vary considerably from incinerator to incinerator. Smaller, older incinerators with capacities of less than about 100 tons/day generally may use platform scales which are mechanically operated, and records may be kept manually by the scale operators. However, for the operation of a large modern incinerator it is virtually mandatory that automatic weighing systems be installed. Such systems will include the use of load cells within the scales themselves, electronic information channeling from load cells to recording apparatus, and automatic printing of the scale weight results. In general, the scales should be designed so that they can accommodate the largest and heaviest vehicles which are anticipated for use in the community. Usually, a 50-ft-long platform scale can be used for weighing most semitrailers and large trucks; the capacity of such a scale should be at least 30 tons. In general, scale accuracy of about ± 1 percent is justified.

Some additional considerations for scales and other weighing apparatus are design and construction of the platform; the scale pit, which must be designed to facilitate cleaning and maintenance; operational procedures for the scales; establishment of tare weights; and maintenance, including periodic inspections. The operation of the scale should be designed to handle as many trucks as can reasonably be expected to enter the facility during normal operations. Tare procedures are also important for scale operation, since the fees assessed the collection vehicles are based upon weights of refuse delivered, which represent the difference between total weights and tare weights of the incoming vehicle. For operations where only a small number of vehicles come into the plant, the easiest method of establishing tare weights is to weigh the vehicle before and after the refuse is dumped. For other operations, especially where municipal collection agencies are involved, the easiest procedure may be to weigh all of the regularly incoming trucks one time and then simply keep a record of these tare weights.

Generally, maintenance problems are not severe for modern design platform scales. However, periodic inspection should not be neglected, and if competent employees are not available to act as maintenance personnel for the scales, a separate contract should be let for repair and maintenance. Some problems may occur in what should be a very simple operation, the weighing of the incoming refuse. For example, loaded trucks may sometimes bypass the scale during peak loading periods when confusion in traffic may occur. In addition, misplacement of the truck on the platform can also cause errors in weighing. Finally, dirt, water, ice, and other debris may accumulate on or under the scale deck and can cause errors in the weights, in addition to providing hazardous driving conditions. Most of these problems can be eliminated by adequate design of the gate system for the scales, curbings and markings for the entrance and exit areas to the scales, and adequate housekeeping procedures.

9-2.3 Materials Handling Systems

One of the aspects of incinerator operation which is of considerable importance to the overall efficiency and success of the operation but which is frequently neglected is the design, construction, and operation of the in-plant system for handling of solid waste. The materials handling system which transfers the waste from the dumping or incoming area through the furnaces and out into the residue removal area is of primary importance. Such a system can be thought of as consisting of several different parts.

Refuse is brought into the plant and encounters the materials handling system first in the dumping or tipping area. The dimensions of the tipping area and the adjacent refuse storage pit should receive careful consideration in the planning and design of the incinerator. For example, the width of the dumping area should exceed the turning radius for each truck which will use the dumping area. Single-chassis trucks generally have a turning radius of less than 35 ft; for semitrailer trucks the turning radius generally is 40-50 ft. Therefore, the width of the dumping area should be 50-70 ft, so that easy turning is available for all trucks. Other considerations for the design of the dumping area are: the vertical clearance for all portions of the dumping area must be sufficient to provide the necessary clearance for all the trucks which will use the dumping area—at least 24 ft of overhead clearance is recommended; in other portions of the entrance area the vertical clearances should be at least 18 ft; warning devices such as hanging chains should also be employed as precautionary measures to indicate vertical clearances; individual dumping spaces along the dumping area should have a width of from 10 to 12 ft; the floor slabs in the dumping area should slope away from the area to provide easy drainage and cleaning, and should be rough surfaced to provide traction; enclosing the dumping area should be considered for a number of reasons—adverse climate may dictate that the area be enclosed or the reduction in noise, odor, and dust may make it necessary;

finally, the drainage system for the dumping area should be designed so that large volumes of wash water can be accommodated and so that large objects in the wash water can be prevented from going into the drainage.

The next item to be considered in the waste handling system is the storage pit; proper design of the pit is essential for efficient operation of an incinerator. The storage pit should be designed so that easy access is obtained for transfer cranes; pits are usually rectangular in plan so that the overhead cranes may easily reach each portion of the pit. Pits should be designed to have a capacity equal to a minimum of 250 percent of the daily 3-shift incinerator capacity to insure operations over the weekend period; extra capacity in the storage pit is particularly essential if the incinerator operation is designed to furnish waste heat to other facilities. Any lag in refuse delivery would produce an interruption in heat delivered if adequate capacity is not available in the storage pit. Pits should be designed for a current average unit weight of waste equal to about 350 lb/cu yd, but ample consideration should be given to future changes in unit weight as a result of increasing use of packaging materials, etc. Several other considerations are of importance in the design of a storage pit: (1) obviously, structural design should be adequate to withstand any external pressures such as water and soil pressure, as well as the internal forces of the solid wastes and any water in the pit; (2) since it is relatively easy for sparks, live coals, or spontaneous combustion to ignite the stored refuse, the pits should be designed so that fires may be dealt with easily; (3) the pit area should also be equipped (in addition to the fire equipment available) with dewatering apparatus so that water used in fighting fires may be easily removed; (4) for ease of dewatering, the entire pit area should be made watertight and the floor of the pit should slope to drains; (5) finally, facilities should be provided so that the pit may be easily cleaned out in the event that the furnace or any auxiliary equipment breaks down for a prolonged period. Cleanout facilities for the storage pit are essential, since the wastes cannot be left indefinitely in a storage pit until any broken-down facilities are repaired.

The wastes may be removed from the storage pit to the grate and furnace area by a variety of mechanisms; the type of waste transfer device generally varies from incinerator to incinerator. In small incinerators, front end loaders or other smaller rubber-tired equipment may be used to transfer wastes. In larger types, the general practice is to transfer wastes by means of overhead cranes. The latter should be designed on the basis of the planned capacity of the processing furnaces. An adequate number of cranes of sufficient capacity should be provided so that the furnace units of the incinerator are never forced to remain idle. For some installations this may mean the inclusion of an extra crane, so that in the event of downtime for one crane an extra one is available to provide continuous feed of the furnaces. Frequently, the furnace feed operation is done by means of charging hoppers or charging chutes. The waste is dumped by the crane operators or the loader operators into the upper, wider end of the truncated pyramid hopper. The gates at the lower end of the hopper provide an air seal for the furnace and grate system. Frequently, the waste is transferred from the hopper into a chute so that the waste can be charged into a particular area of the grate system. The design and construction of these handling systems has become fairly standard throughout the country.

In summary, it should be noted that the entire operation of transfer of waste through an incinerator should be designed for maximum convenience and ease of transfer. From the dumping area the wastes contained in the storage pit should then be transferred to the combustion area by a system designed specifically for the materials which will be transferred; i.e., the materials handling system should be designed specifically for handling solid waste in an incinerator and should not be simply a waste handling system adapted from another industry.

The next section of the incinerator system is the combustion area—the furnaces and associated appurtenances, including grates for charging the furnaces.

9-2.4 Furnaces and Accessory Apparatus

Since the primary operation in incineration is the controlled combustion of solid, liquid, and gaseous wastes, very careful consideration must be given to the design, construction and operation of the furnaces, grates, and other equipment involved in the combustion process itself. The combustion operation should be thought of as occurring in several stages: first, moisture is evaporated; second, the combustible portions of the wastes are vaporized; and, third, the vaporized constituents are oxidized. The end products of the combustion operation obviously must include the noncombustible portions, and the by-products of the combustion itself: carbon dioxide, water vapor, and exhaust gases.

At the present time no single furnace design for municipal incinerators can be considered to be superior in all cases to any other design. The types commonly used in refuse incineration are the rectangular furnace, the multicell rectangular furnace, the vertical circular furnace, and the rotary kiln furnace.

The rectangular furnace is the type most installed in newer incinerators (see Fig. 9-1). This type is amenable to several different types of grate systems; charging of the furnace by the grate system usually includes agitation of the waste as it drops from one level to the next level of grates; then waste is carried through the primary combustion chamber on a horizontal grate. Secondary combustion in this type of furnace is generally carried out near the exit port for the exhaust gases. This type of furnace is very commonly refractory-lined.

The multicell rectangular furnace contains two or more cells set side by side, and within each cell rectangular grates move the solid waste (see Fig. 9-2). The

INCINERATION 145

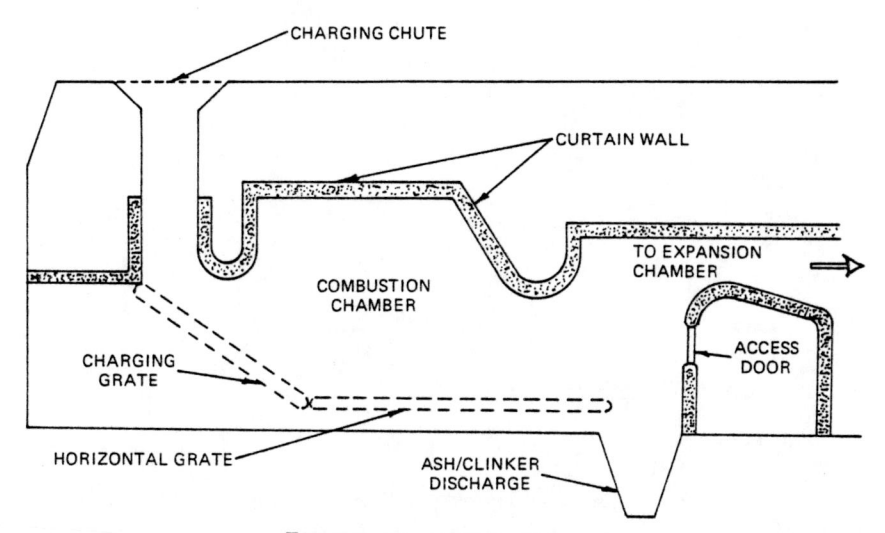

Fig. 9-1 Rectangular furnace.

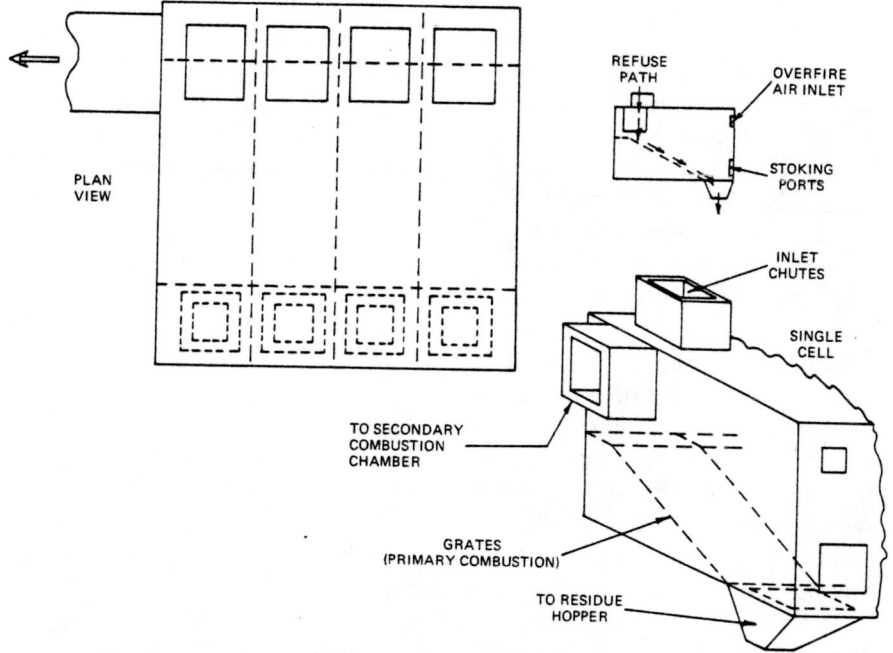

Fig. 9-2 Multicell rectangular furnace.

146   **SOLID WASTE MANAGEMENT**

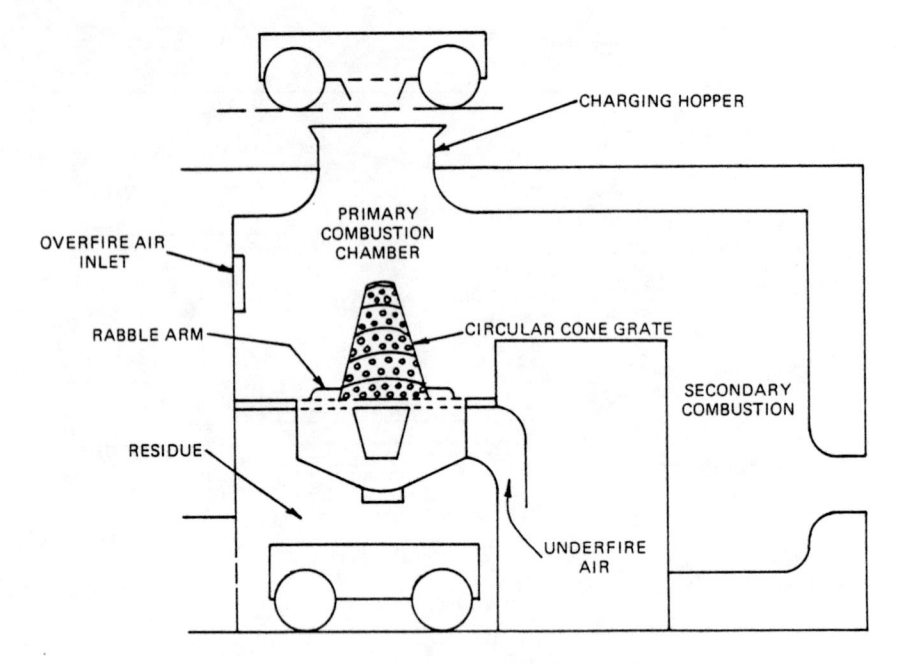

Fig. 9-3   Vertical circular furnace.

multicell type furnace is also known under the name "mutual assistance" furnace. Wastes are charged into each cell individually through the top of the cell, but the cells generally have a common secondary combustion chamber and also have a common residue disposal device. Multicell rectangular furnaces may be either refractory-lined or water-cooled.

In the vertical circular furnace solid waste is inserted through a door or entry port in the upper portion of the furnace and drops onto a cone-shaped central grate and a surrounding circular grate (see Fig. 9-3). Air is forced up through the grate system. As the central cone and attached arms rotate slowly over the surrounding circular grate, the solid waste is agitated and the residue after combustion is shifted to the sides of the circular grate. When the residue has reached the outer portions of the circular grate, it is discharged through a dumping grate. Generally secondary combustion is accomplished in an adjacent chamber; overfire air may be added to assist in this combustion. Vertical circular furnaces are usually refractory-lined.

The rotary kiln furnace generally is constructed to consist of a rectangular primary furnace and a slowly revolving kiln inclined to the horizontal as a second unit in the system (see Fig. 9-4). Generally the purpose of the rectangular furnace is to dry the solid waste and to only partially burn it. The partially burned solid wastes are then fed into the rotating kiln, where the tumbling

INCINERATION 147

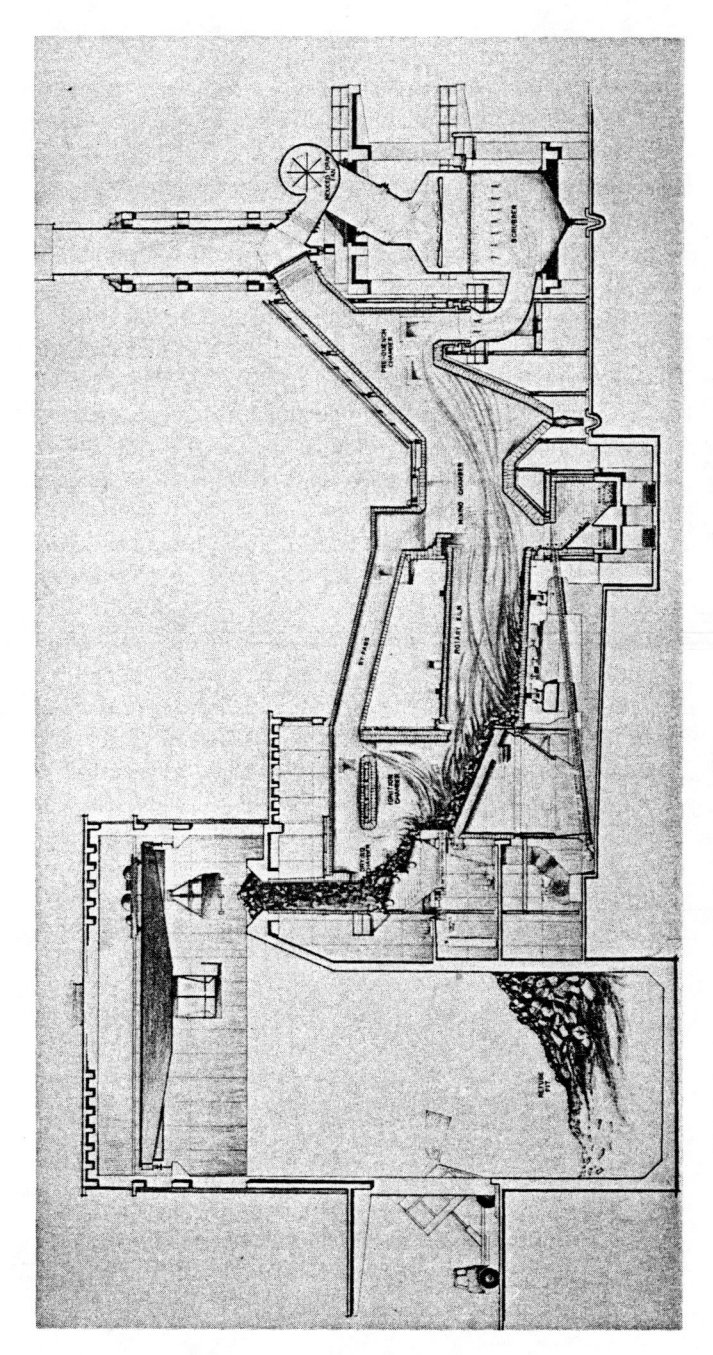

Fig. 9-4 A rotary kiln furnace incinerator (*courtesy* International Incinerators, Inc.).

action within exposes any unburned wastes. Complete combustion is assured through the final combustion of gases and suspended particulate matter in the mixing chamber which follows the kiln. Refuse is removed from this type of furnace by the force of gravity; i.e., the residue simply falls from the end of the kiln to a lower water-filled quenching trough.

Of significant importance in the operation of any system are the grates which are used to charge the furnace and which are used to transfer the wastes through it. The design of the grate system must satisfy several seemingly conflicting requirements. For example, combustion will be promoted by agitation of the wastes and passage of ample air up through the grate system. However, excessive tumbling of the wastes or excessive flow of underfire air will lead to entrainment of particulate matter in the exhaust gases. At times inert materials such as glass particles and bits of metal may aid combustion, since their presence tends to increase the porosity or amount of voids in the fuel bed; on the other hand, excessive amounts of inert material may clog the air inlet openings and inhibit combustion. Finally, grate systems must be designed to operate in high-temperature environments and must be adequate to withstand abrasion, clogging, impact loads, and thermal shocks. The necessary area of the grate system may be obtained by considering that between 50 and 70 lb of waste can be transferred per square foot of grate per hour.

In the design of the system attention must be given to the functions performed by the grates: the incoming refuse is first dried, ignited, and then completely burned. To fulfill these functions a variety of mechanical grates have been developed. In the United States grates in incinerators will fall into one of the following categories: traveling; reciprocating; rocking; rotary kiln; and circular. Some of these are shown in Figs. 9-5, 9-6, 9-7, and 9-8. Traveling grates in themselves do not promote combustion, since a significant amount of agitation is not achieved on a simple type. Therefore, several traveling grates should be used together to promote agitation by dropping refuse from one elevation down to a traveling grate at a lower elevation.

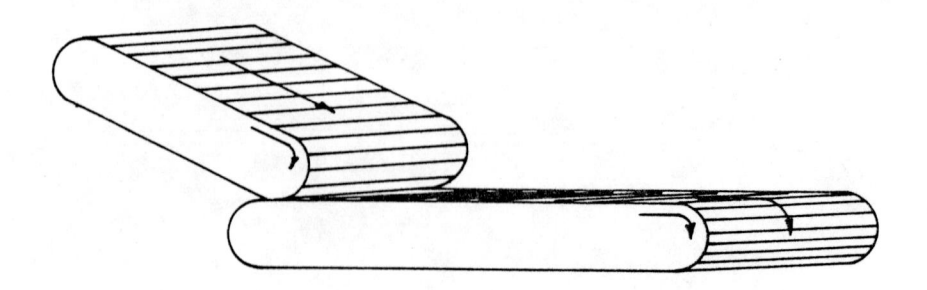

Fig. 9-5 Traveling grates.

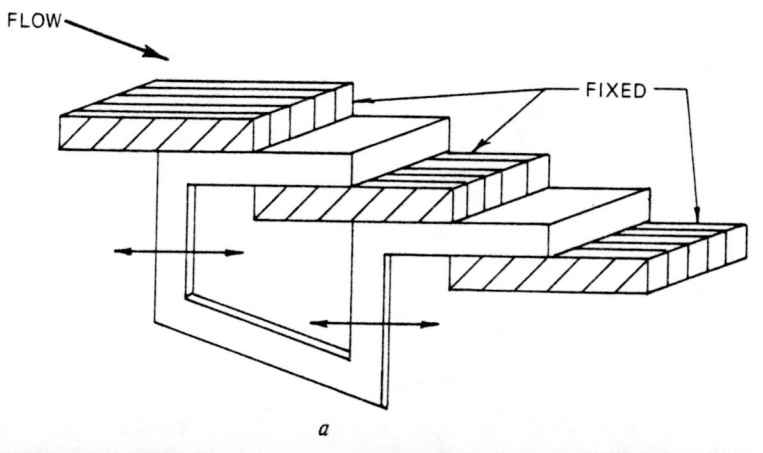

Fig. 9-6 (*a*) Reciprocating grates; (*b*) typical installation of reciprocating grates in Volund Incinerator (*courtesy* International Incinerators, Inc.).

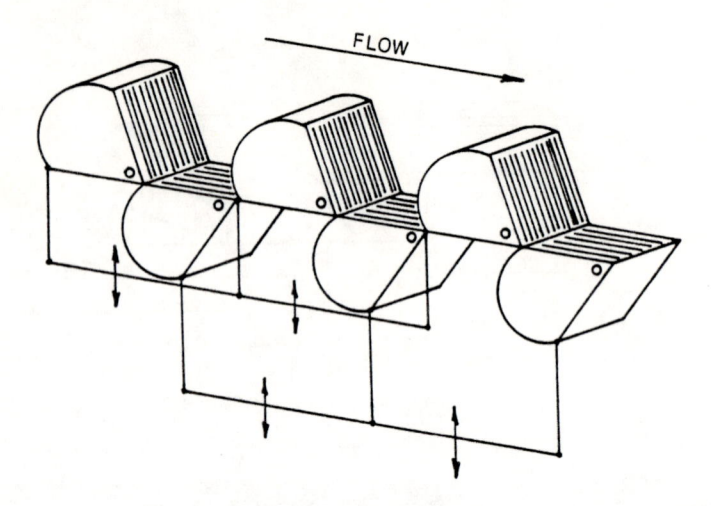

Fig. 9-7 Rocking grates.

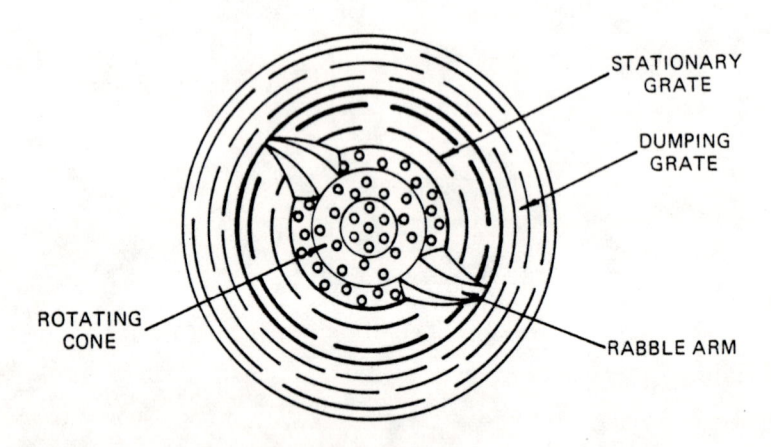

Fig. 9-8 Circular grates.

Wastes are charged to the grate system either in batches or continuously. In a continuous feeding operation the wastes are fed directly by means of a chute that is kept filled at all times so that an air seal is maintained. In the batch feeding process, the wastes are fed into the furnace intermittently through a hatch. The hatch is kept closed at all other times so that the air seal for the furnace is maintained. Most modern incinerators have been designed for a continuous feeding operation, since this minimizes irregularities in the combustion itself. Batch feeding generally causes fluctuations in combustion and can lead to thermal shocks because of the introduction of large quantities of cool air into the furnace. Temperature irregularities of this sort can be quite detrimental to the material of the furnace itself.

Another aspect of the grate system which should receive consideration is the removal of materials which filter down through the grates themselves; these materials are generally called "siftings" and may contain metal fragments, glass fragments, partially burned combustible materials, and pieces of ash. Siftings should be removed and returned to the furnace or removed and transferred to a residue area. The removal of siftings is important, since any combustible materials in them may ignite and cause damage to the grate systems through which they have fallen.

After the solid wastes have been processed in the furnace, any noncombusted residue must be processed in some way. The residue from the combustion process generally includes ash, clinkers, rock, glass, unburned organic materials, and tin cans. Various systems for refuse residue removal have been installed and are in use in incinerators throughout the United States. Refuse removal systems may be either continuously operating or may operate in batches. Whatever the means of residue removal, certain aspects of the residue should be considered in design of the removal system. For example, the residue itself is generally quite abrasive and rapidly produces wear in the system. In addition to its abrasive characteristics, it should be noted that the water used in quenching the residue quite commonly is highly corrosive and may lead to rapid deterioration of metal parts of the system. In many large incinerator installations, the residue is discharged by a continuous feed operation into a common trough connected to all furnaces; a drag conveyor submerged in the bottom of the water-filled trough then removes the residue. The submerged drag conveyor system has the advantage that an air seal is maintained between the removal ports and the interior of the furnace.

In any discussion of the design and operation of furnace systems within an incinerator, some consideration should also be given to the combustion process itself. In general, the governing factors of this process are time, temperature, and turbulence. As mentioned previously, the combustion process may be thought of as occurring in three stages. During the first stage, surface and internal moisture are evaporated from the combustible material. During the evaporation phase, the heat input into the wastes is used to remove moisture, and the temperature of the material does not exceed the vaporation temperature of water, 212°F. After the moisture has been removed, the temperature of the waste rises and, with increasing heat input, the ignition temperature of the materials is attained. During this process, for the larger particles of materials, the outer surfaces may be dried and ignite before complete drying of the inner portions of the material is attained. Since with large particles outer burning and inner moisture removal may be occurring at the same time and could lead to incomplete combustion, many modern incinerators employ some grinding and shredding apparatus to reduce the particle size of the incoming wastes.

Within the furnace chambers themselves, the combustion operation occurs in two phases: primary and secondary combustion. During the former the waste is dried, the combustible components are volatilized, and the volatile materials ignited. During secondary combustion the gases and particulate matter entrained in the gas stream after primary combustion are oxidized. Secondary combustion is essential to the operation of the incinerator since in it all unburned furnace gases are oxidized, odors from incompletely burned materials are eliminated, and the carbon suspended in the exhaust gases from the primary combustion chamber is also burned. To insure adequate primary and secondary combustion, sufficient air must be supplied so that both oxygen and turbulence requirements are satisfied. The turbulence imparted by incoming airstreams generally serves the purpose of mixing gases with sufficient amounts of air for complete combustion of the volatile matter present and any particulate matter suspended in the gases. Turbulence must be sufficiently high so that complete combustion occurs within the secondary combustion chamber, where temperatures are high.

Air is supplied to a furnace via openings in the bottom of the furnace area beneath the grates and also through upper openings; air admitted through the lower ports is called underfire and air admitted to the area above the grates is called overfire air. Underfire air generally supplies the necessary oxygen for the combustion process and will aid in suspending the fuel materials in a fluidized-bed operation. Overfire air is added to provide the necessary turbulence. The amounts and proportions of underfire and overfire air will depend upon the design of the incinerator. Quite often the correct proportions of underfire and overfire air will be determined by a trial-and-error process. In general, underfire air amounts to from 40-60 percent of the total amount of air entering into the furnace. This proportion for the underfire air generally provides enough air for complete combustion in the fuel bed and also provides adequate cooling of the grates themselves. Since mixing of the combustible materials with oxygen is not always perfect, and since a certain amount of turbulence is desirable, air should be supplied in the combustion process in excess of that amount necessary to provide sufficient oxygen for combustion of the wastes. In other words, only a certain amount of air would be required to provide sufficient oxygen for perfect theoretical combustion of the waste materials present; however, an amount in excess of this theoretical figure must be supplied for the purposes of complete combustion turbulence. Generally, refractory furnaces require 150-200 percent excess air and water wall furnaces require 50-100 percent excess air.

During the combustion process, gases obviously are produced. Adequate provision must be made for exhausting these. If reference is made to the ultimate analysis of the wastes charged to the incinerator, a calculation may be made of the amounts of gaseous by-products from the combustion of that waste. The gas exit passages, incinerator stacks, and any air pollution control devices may be designed to accommodate this quantity of exit gases. Moreover, the gas exit facilities may be designed so that gas velocity is maintained at the proper level in order that entrained material is precipitated from the effluent gas flow stream at a desired location. The design of the stacks and air flow systems for the incinerator must take into account various considerations: the topography of the surrounding countryside; climatic conditions, including average wind direction and wind velocity; Federal Aviation Agency regulations; local regulations pertinent to air pollution; distribution of population surrounding the incinerator facilities both present and future; the architectural considerations of the incinerator facility itself; as well as any other design constraints. One observation may be made: to control the combustion process it is necessary to control the draft through the furnace. Since natural draft stacks generally will provide an irregular or erratic draft to the furnace, induced draft fans should be used to provide a regular, easily controlled air supply.

As mentioned previously, for proper incinerator operation, time, temperature, and turbulence are all important. Sufficiently high temperatures must be maintained so that the oxidation reactions are completed within the furnace. However, if the combustion temperatures are too high, severe difficulties may result. Equipment and other structural damage may be caused by such excessive temperatures; for example, refractories may spall off within the furnace area if they are overheated and consequently expand excessively. An additional consideration in temperature control is that excessively high heat leads to the production of nitrogen oxides, which are extremely troublesome air pollutants. Careful consideration must be given to this aspect of incinerator operation. However, control of the temperatures within the furnace and auxiliary areas is very difficult, since the charged material varies considerably as to composition and characteristics. The incoming air will vary from ambient outside temperatures to approximately 200-300°F if preheating is practiced. After the air enters the furnace, in the primary combustion chamber, temperatures of the burning gases reach between 2100 and 2500°F. For certain materials temperatures of up to 2800°F may be reached in some portions of the furnace area. After the gases leave the combustion chamber and mix with exit air, the general temperature of the mixture is decreased to from 1400 to 1800°F. In the secondary combustion chamber and through the stack, the temperatures of the exit air must be carefully controlled, because of the presence of auxiliary mechanisms such as air pollution control devices. Generally, the gases enter the stack at a temperature of approximately 1000°F. If induced draft fans or other mechanical devices such as electrostatic precipitators are used, the temperature of the exit gases must be reduced. It is necessary to cool the gases to a temperature between 500 and 700°F to avoid detrimental effects within air pollution control devices or induced draft fans.

Temperature control is attained in the combustion chamber and in the gas effluent areas through one or more of several methods: use of excess air, evaporation of water, and heat exchange. The use of excess air is practiced because of requirements for oxygen and turbulence within the furnace itself, and, obviously, the same excess air may be used to lower the temperatures of the gases. In any refractory furnace, use of excess air is the only means to control the temperature of the gases. It appears that another method of cooling hot gases also holds promise for use in incinerators. This method uses water tubes in the walls of the combustion chambers. In addition to providing cooling capabilities for the exiting gases, the heated waters within the tube walls may be used for a variety of purposes, including the generation of electric power. A further advantage of such heat exchange is that no additional gases or vapors are added to the gas flow during the temperature-reduction process. Thus a significantly smaller gas volume is produced in a water-tube wall furnace than in a refractory-lined one, where excess air is used to cool exit gases or water is sprayed in the exit gas stream to lower the temperatures. Water injected into the hot exit gases will cool these flue gases through the evaporation of the water and the absorption of heat during superheating of the water vapor. However, the cost of the water added to the system must be considered in the overall economics of the process. Even though water vapor is added to the stream of exit gases, the total volume of water vapor plus cooled gases will be less than the volume of high-temperature gases before cooling.

In summary, perhaps the best method of cooling exhaust gases is through the use of heat exchange and water-tube walls, since with a smaller volume of exit gases the sizes of collection devices, fans, gas passages, and other pollution control equipment may be reduced.

However, since most of the furnaces currently in operation and many of those planned for the immediate future will not include the use of water-tube walls, and will be built using refractory materials, some discussion should be devoted to the latter themselves. These materials are used to resist the effects of heat, and consist of such ingredients as alumina, magnesia, and silica with certain other additives. Often these ingredients are intermixed in a grinding operation with the clay mineral, kaolin, to produce a combined refractory product. Such materials are then combined and cast as bricks which may be laid in a conventional bricklaying operation with mortar. On the other hand, refractory materials may also be used in the form of a dry powder which is then mixed with water and cast in place in forms. A continued high degree of emphasis in incinerator design must be placed upon the properties and characteristics of the refractories used, since trends in modern incinerators have been toward higher and higher combustion chambers and thinner walls. Modern designs may call for walls of only 9 in. thickness in many areas of the incinerator, while at the same time temperatures in excess of 2000°F must be resisted. Because of these requirements many modern incinerators have a suspended-wall or suspended-roof construction. Where temperatures must be maintained at the higher levels, insulation may be placed behind these suspended refractory walls. Where air cooling is employed, the insulation behind the refractory is often omitted. Suspended-wall construction has an additional advantage in that any deterioration of the refractory lining may be easily repaired; in older construction, entire sections of the walls or roof had to be replaced, since most of the chambers were designed as bearing walls.

The refractory materials are subjected to the highest degree of wear and destruction in the primary combustion zone. Within this zone, high temperatures, flame contact, slagging, spalling, thermal shock, mechanical wear from agitated tumbling wastes, and erosion from high-velocity gases are all at a maximum. The greatest concentration of refractory destruction is immediately above the grate line and in the charging area. In these areas of high destruction a refractory of a very dense material such as silicon carbide may be used. Lower-quality refractory materials may be used in other portions of the combustion chambers where less mechanical wear and fewer detrimental conditions exist. As mentioned previously, two mechanisms of wear of the refractories are spalling and slagging. Spalling generally results when internal thermal stresses are produced through differential expansion of the material—i.e., differential heating or cooling of the surface of the refractory relative to the interior portions of the lining, causing a differential expansion of the two zones with the result of the outer zone breaking away. Slagging, on the other hand, results from the buildup of a layer or deposit of flux composed of oxides of sodium, potassium, iron, calcium, and other elements on the surface of the refractory. The increased load on the latter due to the weight of this flux layer may then cause a structural failure within the refractory itself. A combined slagging and spalling failure may occur when a bonded slag layer and the underlying refractory experience differential expansion or contraction.

Several additional comments should be made concerning the design of the combustion chambers and the combustion process itself. First, consideration should be given to the use of water-tube walls in the incinerator furnaces. These walls consist of closely spaced steel tubes welded together to form a continuous wall and which contain water or steam. Such a water-tube wall affords the operator of the incinerator more precise control over temperatures and also helps to provide an airtight enclosure. Such furnaces have been used extensively in European incinerators, and in this country have found extensive use in the power industry for a number of years. Second, in any incinerator operation it is quite useful to provide facilities for the use of auxiliary fuels.

156 SOLID WASTE MANAGEMENT

Auxiliary fuels are useful in furnace warm-up, in initiating primary combustion when the charged solid wastes are wet or are of low heat value. Additionally, auxiliary fuels can lead to more complete secondary combustion and thus provide some additional odor and smoke control and will also give higher heat recovery from incinerators where such heat recovery is practiced. Auxiliary fuel in the United States is either gas or oil. Finally, some consideration should be given to the operation of starting combustion in the furnace. In particular, care should be exercised in the curing and preheating of furnaces where new refractory materials have been installed. To prevent any occurrence of damage in new refractory linings, a curing period of several days is often required. Many municipal incinerators have experienced difficulties with damage to refractory linings when operating personnel have not been aware of the requirement for proper curing of new linings.

9-3. WASTE HEAT RECOVERY

The combustion of solid wastes produces a significant amount of heat. Unless this heat is used in some way, its waste constitutes an added expenditure of recoverable resources. However, in the United States, at the present time only about a half-dozen incinerators are operated so that heat recovery is practiced. The advantages of waste heat recovery are several, particularly when as mentioned the heat is recovered through a water-tube wall furnace lining. First, the volume of gas which must be cleaned before it is exhausted is reduced; thus, the size of any air pollution control equipment required is also reduced. Second, the increase in the heat content of municipal solid wastes in recent years points toward a greater and greater potential for recoverable heat in incinerated solid wastes. However, even though these advantages exist, generally heat recovery is not practiced unless it can be shown to be economically advantageous in the form of income from the sale of heat or power. The difference between a conventional refractory-lined furnace and a water-tube wall furnace is significant in the amount of excess air required for operation. Common refractory combustion chambers usually require 150 to 200 percent excess air whether heat recovery is practiced or not; on the other hand, water-tube-wall combustion chambers usually require only 50 to 100 percent excess air. This reduction in excess air requirements of water-tube construction versus the conventional refractory type indicates the potential for reduction in the volume of flue gases from refractory furnaces. A consequent reduction in the amount and size of air pollution control devices required for effluent gas treatment would also be realized with the use of water-tube-wall combustion chambers.

In regard to the sale of waste heat, the mechanism for sale of the recovered heat is ordinarily through the generation of steam. The amount of steam produced in waste heat recovery systems has ranged from approximately 1 to 3½ lb of steam produced per pound of solid waste burned. The variation in steam production is mainly attributable to the variable consistency and characteristics of the incoming solid wastes. If a market for the steam produced can be secured, it can generally be sold for 25 to 50 cents per 1000 lb. On the basis of these prices, a recovery of from 50 cents to $3.50 per ton of solid waste burned could be achieved. At first glance, it appears that there are many advantages and few disadvantages connected with the recovery of waste heat from the incineration of municipal solid waste. Why then is waste heat recovery not practiced on a wider scale?

To date, in the United States, it has not been practiced because of several factors. First, there are the added costs of the heat recovery equipment and the extra maintenance required for the system. In addition to these costs there is an inherent uncertainty factor in any recovery operation because of the variability of the heat value of the solid wastes. A significant difficulty results in any attempt to match the supply of waste heat to the demand for heat. Therefore, in the design of any recovery system, provision must be made for generating some heat even when no solid wastes or when low-heat-value wastes are being burned. On the other hand, provision must also be made for dissipation of any excess heat generated. Excess heat dissipation will involve the construction and operation of condensers or some other means for heat dispersion. This equipment is quite costly in both capital expenditure and operating costs. The obvious uses for waste heat such as heating the incinerator plant, supplying hot water to the plant, generating in-plant electricity, and other special uses such as desalting seawater as practiced in the United States will usually absorb only a small fraction of the available heat from the incineration process. Thus, the most ideal means for utilization of recovered heat lies in the supply of steam to a power system having a demand for power at all times larger than the greatest amount of power produced at the incinerator. Under these operating conditions, no waste of generated steam will occur and any fluctuations in power generation at the incinerator plant will be made up from auxiliary sources within the power company installation. If a location near such a large utility is not possible, then some provision must be made for equalization of the steam output of the incinerator; i.e., the incinerator during periods of low-solid-waste incineration must be supplemented from an auxiliary power supply so that the generation of steam is at a constant value. If the generation of steam cannot be done as a reasonably constant value, little application in manufacturing, heating, or other areas can be made. Thus, except for those situations where an incinerator may be built adjacent to a large power plant, generation of recovered heat will entail the added costs of auxiliary fuel, auxiliary fuel burners, and additional manpower for the operation of all such systems.

Facing any proposed waste heat recovery system are several striking costs: the expenditures for additional equipment, additional manpower, and auxiliary systems to supplement the system. The advantages of waste heat recovery are not nearly so obvious. Therefore, in many instances the apparent disadvantages of added costs have discouraged planners and designers who failed to recognize the inherent advantages. The fact remains, however, that if the heat generated by solid waste incineration is not used, a significant waste of resources results. In addition, the resultant dispersion of waste heat may create a problem of thermal pollution. For these reasons, especially in an era when resource recovery and reuse are being emphasized, the planners or designers of any proposed incineration installation must consider recovery of waste heat.

9-4. INSTRUMENTATION AND CONTROL DEVICES

Since it has been emphasized that incineration is a controlled combustion process, it is obvious that devices must be incorporated to adequately monitor and control the combustion process. Generally the instrumentation associated with an incinerator will fall into one of two categories, either process control devices or protective monitors. The latter serve a dual purpose in providing protection both for the environment as well as for the operating equipment. Obviously, the instrumentation for an incinerator system which is required to achieve these objectives will be complex and involved. No extensive discussion of it will be given here; however, a general statement may be made. The instrumentation for incinerator operation must include monitors, gauges, and other instruments to measure and determine most of the following parameters: the weight of the incoming solid refuse; the weight of the outgoing residue from the furnaces; the characteristics of the outgoing gases in the incinerator stack; the flow rate of overfire and underfire air; temperatures in the combustion chamber, along gas passages, in the stack itself, and in any air pollution control device; gas pressures in these same locations; and other utility measurements, such as of electrical power and water consumption. In order to achieve an economical operation which is not injurious or offensive to the environment, and which accomplishes the basic objective of burning solid waste in the most economical way, instrumentation is absolutely necessary.

9-5. INCINERATOR EFFLUENTS

Perhaps the most significant difficulty encountered in the operation of municipal solid waste incinerators is the control and prevention of pollution of the environment. Improper design or inefficient operation of an incinerator may produce pollution of land, air, and water.

In the charging of the incinerator itself, solid wastes are dumped from trucks in a dumping area and stored temporarily in a storage pit. At this operation, dust and litter may be generated if housekeeping procedures are not adequately monitored. In addition, if litter is allowed to accumulate at this location, a resulting odor will generally be produced. Odor problems arise from the putrefaction of organic materials and generally are concentrated in the storage pit area. These problems of odor, litter, and dust can be especially significant in the operation of an incinerator, since they cause unpleasant working conditions and consequent inefficiencies and poor attitudes on the part of operating personnel. Thus, frequent cleaning and disinfection of the tipping areas and dumping areas are essential. Moreover, storage pit areas must be periodically sprayed with water to control dust and with pesticides and disinfectants to control vectors.

In addition to the problems associated with charging the incinerator, the material removed from it also can cause pollution problems. The residue from the incinerator amounts to approximately 15 to 25 percent of the original input solid wastes and includes such items as ashes, tin cans, rocks, glass, clinkers, and unburned organic materials. The volume of the residue from an incineration operation, in an uncompacted state, amounts to about 10 to 20 percent of the original volume of solid waste in a storage pit. Ultimate disposal of residue from an incinerator must be in a landfilling operation. As such, the incinerator residue must be treated in the same way as any other solid waste material going into a landfill. The residue may contain soluble constituents, both organic and inorganic, and therefore must be protected from any leaching action.

An additional product of the incineration process is flyash. The latter is that portion of the residue which is carried as solid particulate matter in the exit stream of combustion gases and includes very small particles of ash, cinder, mineral dust, charred paper, other partially burned materials, and soot. Flyash particles range in size from about 120 to less than 5 microns (μ) in equivalent diameter and the distribution of particle sizes within this range is commonly quite erratic. Most of the flyash resulting from incineration of municipal solid waste is inorganic and consists of the oxides of aluminum, calcium, iron, and silicon. When flyash is collected from the exit gas stream, it constitutes a by-product solid waste. The collected ash is difficult to handle in a dry state and therefore is quite commonly transported from the incinerator site in the form of a water slurry. The slurry containing ash has a rather high percentage of solids and is acidic, i.e., it has a rather low pH. Unless the flyash is handled in water, significant problems with wind scattering, and water leaching of soluble compounds from the ash can occur. Therefore any dry flyash must be stored at the incinerator site in closed containers, must be carefully removed to a landfill disposal site, and at that site, must be disposed of very carefully under standard sanitary landfill operating procedures. If the flyash is disposed of in a water slurry operation, the process water from the ash disposal operation may be treated with the process water which results from the overall incineration operation. Most of the water in an incinerator operation is used in quenching the residue. In addition, however, water may also be used to cool various parts of the equipment in the incinerator plant. The allowance of water used will vary with the design of the incinerator and on the basis of whether or not such water is recirculated. In one investigation the water requirements without recirculation for a 300 ton/day incinerator with two 150-ton continuous feed furnaces was about 2,000 gallons of water per ton of solid waste burned; of that 2,000 gallons approximately 90 percent was used in quenching and conveying refuse and residue. Recirculation of water could probably have cut this 2,000 gallon figure to approximately 500 to 600 gal/ton (Ref. 9-6).

Numerous studies have shown that incinerator process water contains a significant amount of inorganic and organic materials in solution or in suspension, and other contaminants. Several studies have also indicated the presence of bacteria in wastewater from an incinerator. Process waters from incinerators can therefore be considered at all times to be contaminated and should be considered as requiring treatment before disposal. Table 9-4 shows characteristics of wastewater from incinerators as determined in a number of studies. The easiest solution for the disposal of wastewaters from the incinerator lies in discharging them into a sanitary sewer. However, if such discharge is anticipated the effects of this added amount of wastewater on the sewer system should be anticipated and should be taken into account. If it is not possible for the wastewater from the process treatment to be discharged into a sanitary sewer, then the incinerator itself must be supplied with a treatment plant.

Of much greater significance than the generation of waste process waters in incineration, however, is the creation of air pollutants during the combustion operation. These pollutants consist of entrained particulate matter (such as the flyash briefly discussed in a previous section) and gaseous by-products. For the particulate matter, the important characteristics as far as its collection are concerned are the sizes of the particles, their specific gravity, chemical composition, electrical resistivity, and the quantities of particles produced. The particle size distribution and specific gravity of the particulate matter is of importance in the design of collection systems, since most particulate collectors are designed to collect particles in rather a small size range. In general, the larger, denser particles are easier to collect. Thus, coarse, high-density particles can be easily collected in devices based upon the inertial characteristics of the materials; settling chambers and cyclones are devices of this type. Smaller-size, lighter particles require more complex procedures for their collection. Such sophisticated collection devices include high-energy wet scrubbers, fabric filters, and electrostatic precipitators. As a rule of thumb, a dividing line of approximately 10 μ equivalent diameter gives an indication of the type of collection system required; if the greater portion of the particulate matter is less than 10 μ in

TABLE 9-4 Incinerator Wastewater Data

Characteristic	Plant 1 Max.	Plant 1 Min.	Plant 1 Avg.	Plant 2 Max.	Plant 2 Min.	Plant 2 Avg.
pH	11.6	8.5	10.4	11.7	6.0	10.5
Diss. solids, mg/l	9,005	597	3,116	7,897	1,341	4,283
Susp. solids, mg/l	2,680	40	671	1,274	7	372
Total solids % volatile	53.6	18.5	36.3	51.6	10.5	31.2
Hardness ($CaCO_3$) mg/l	1,574	216	752	1,370	112	889
Sulfate (SO_4) mg/l	430	110	242	780	115	371
Phosphate (PO_4) mg/l	55.0	0.0	23.3	212.5	1.0	23.5
Chloride (Cl) mg/l	3,650	50	627	2,420	76	763
Alkalinity ($CaCO_3$) mg/l	1,250	2.5	516	1,180	292	641
5-day BOD @ 20°C	–	–	–	–	–	–

Characteristic	Plant 5 Max.	Plant 5 Min.	Plant 5 Avg.	Plant 6 Max.	Plant 6 Min.	Plant 6 Avg.
pH	6.5	4.8	5.8	4.7	4.5	4.6
Diss. solids, mg/l	1,364	7,818	8,838	6,089	5,660	5,822
Susp. solids, mg/l	398	208	325	2,010	848	1,353
Total solids, % volatile	–	–	–	24.69	23.26	23.75
Hardness ($CaCO_3$) mg/l	2,780	2,440	2,632	3,780	3,100	3,437
Sulfate (SO_4) mg/l	1,350	1,125	1,250	862	625	725
Phosphate (PO_4) mg/l	15.0	11.5	13.0	76.2	32.2	51.5
Chloride (Cl) mg/l	3,821	3,077	3,543	2,404	2,155	2,297
Alkalinity ($CaCO_3$) mg/l	28	16	23	4	0	1.33
5-day BOD @ 20°C	13.5	6.2	8.8	–	–	–

Sources: Plants 1 and 2 USPHS *Report on the Municipal Solid Wastes Incinerator System of the District of Columbia,* 1967.
Plants 5 and 6 USPHS unpublished data (SW-11ts) (SW-12ts).

Plant 1 110TPD Residue Quench (Batch).
Plant 2 425TPD Residue Quench (Batch).
Plant 5 300TPD Cont. Feed-Flyash Effluent.
Plant 6 300TPD Cont. Feed-Flyash Effluent.

equivalent diameter, sophisticated collection techniques must be used. Table 9-5 and Fig. 9-9 show some of the properties of particles and their sizes in effluent gas streams from incinerators. However, no sweeping generalizations concerning particle size distribution and specific gravity of entrained particles should be made on the basis of the data in the table. The variability in characteristics of charged solid waste will produce a corresponding variability in the characteristics of the entrained particulates. As shown in Fig. 9-9 the amount of particulate

162 SOLID WASTE MANAGEMENT

TABLE 9-5 Properties of Particulate Effluents from Furnaces

	1 (250tpd)	Incinerator 2 (250tpd)	3 (120tpd)
Specific gravity	2.65	2.70	3.77
Bulk density (lb/ft^3)	–	30.87	9.4
Loss on ignition @ 750°C (%)	18.5	8.15	30.4
Size distribution			
<30μ	30.0	40.4	50.0
<20μ	27.5	34.6	45.0
<15μ	25.0	31.1	42.1
<10μ	23.0	26.8	38.1
< 8μ	21.0	24.8	36.3
< 6μ	19.0	22.3	33.7
< 4μ	16.0	19.2	30.0
< 2μ	13.5	14.6	23.5

matter emitted from refractory furnaces has been found to be dependent upon the quantity of underfire air utilized in promoting combustion. The quantity of flyash, for example, produced in the incinerator may vary from 10 lb/ton of waste burned to 50 or 60 lb/ton burned. Careful consideration must be given in the design of collection and control devices to the amount of underfire air anticipated for the combustion operation. Finally, another significant property of entrained particulate matter is the electrical resistivity of the particles themselves. The latter is of prime interest whenever electrostatic precipitators are considered as the mechanism for particulate collection. In general, collection devices for particulates with high resistivity values will have to be more complex, larger and more expensive than would the equipment for particulate with low resistivity values. If the resistivity of the material is very high, wet scrubbers or fabric filters may be more suitable than electrostatic precipitators. In order for the most suitable resistivity collection equipment to be selected, resistivity of the particles leaving in the gas stream must be determined. Since such resistivity has been found to depend to some extent on the temperature of the exit gas stream, adequate information on this must also be obtained. Figure 9-10 shows some data on three installations where the electrical resistivity was found to be a function of the temperature of the exit gas stream.

Figure 9-11 shows the products of combustion of solid wastes as a function of the amount of excess air supplied. In general over 99 percent of the effluent gases consist of carbon dioxide, nitrogen, oxygen, and water vapor; these materials are not generally considered air pollutants. The amounts of these are

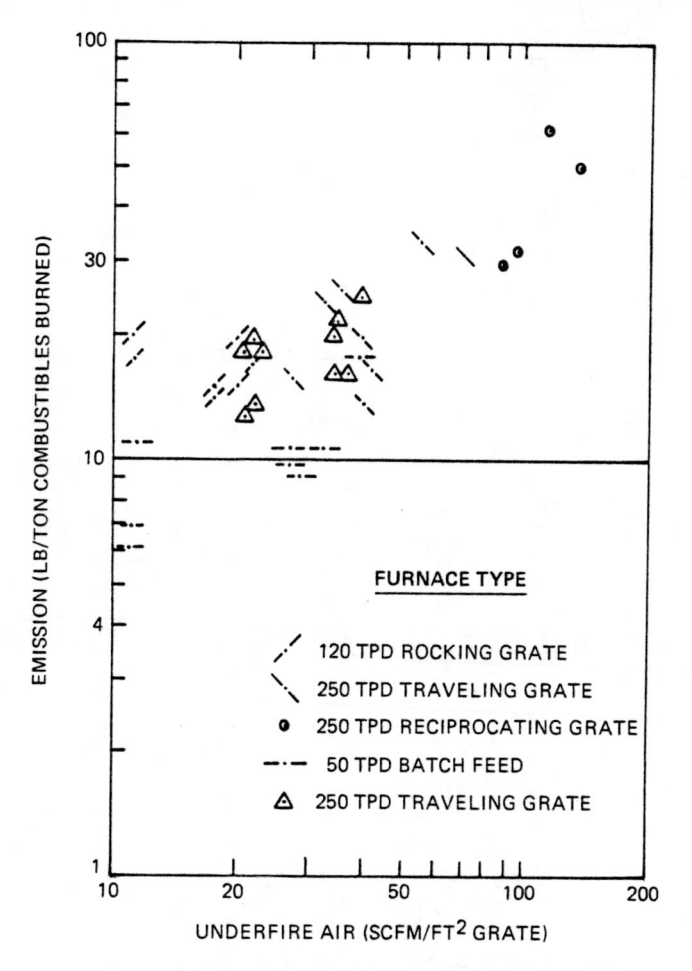

Fig. 9-9 Particulate emissions.

also dependent upon the proportion of excess air employed in the combustion process; thus, the use of water-tube wall incinerators would seem to be one solution to the problem of gaseous pollutant generation, because of the lower excess air requirements for water-tube wall designs. Compared to the other combustion processes that occur daily in the United States and which contribute significant amounts of gaseous pollutants to the atmosphere, the contribution of pollutants due to incineration of solid waste appears to be low. For example, the summation of nitrogen oxides and sulfur oxides generated in the combustion of such wastes is generally from 1 to 10 percent of the amounts generated per ton during

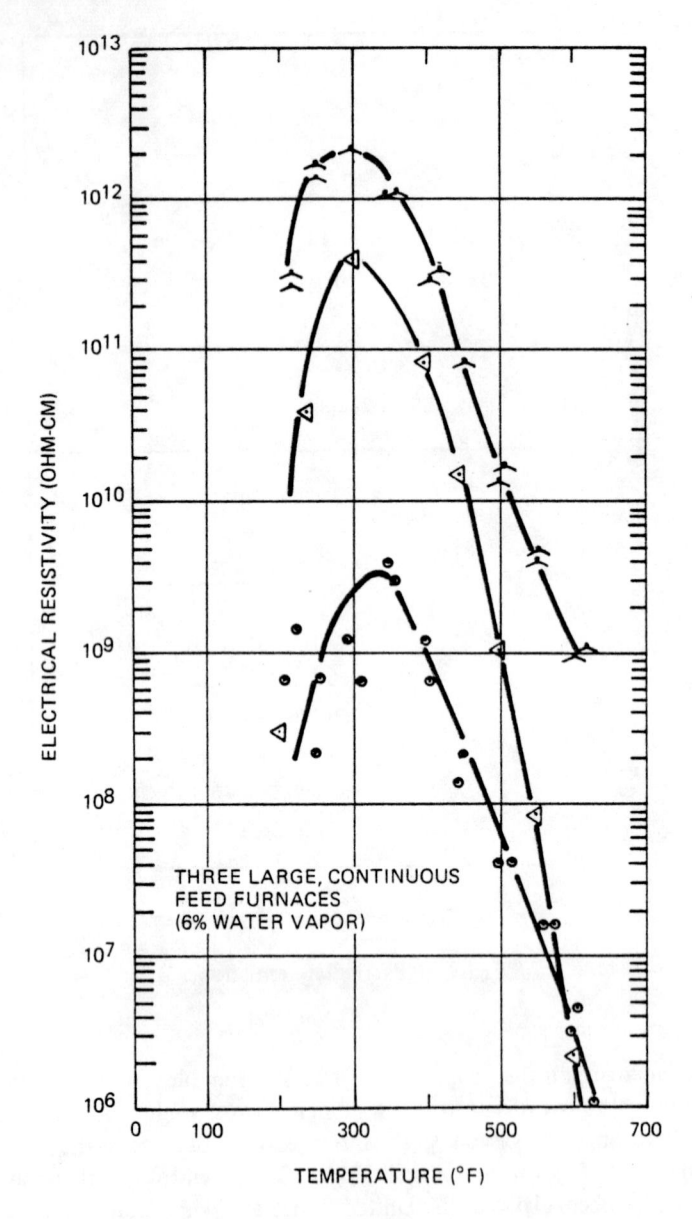

Fig. 9-10 Electrical resistivity of entrained particulate matter.

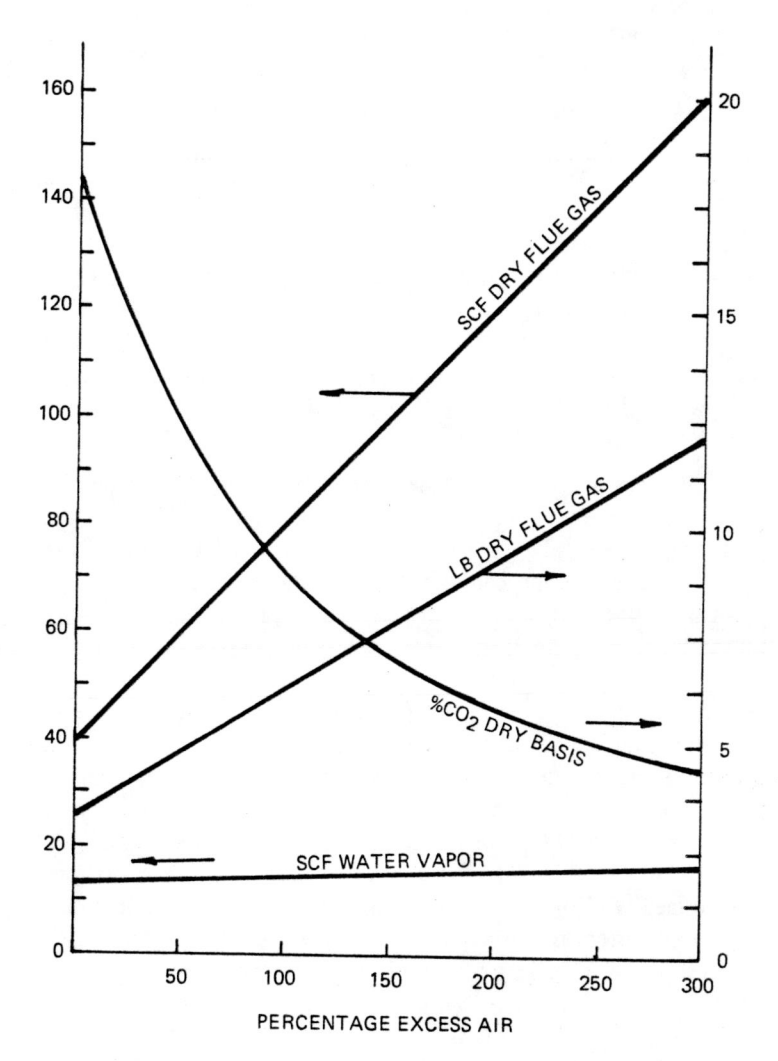

Fig. 9-11 Combustion products per pound of solid wastes.

the combustion of fossil fuels in power-generating plants. The common trace gases present in incinerator effluent streams are shown in Table 9-6.

In addition to generation of particulate matter and gaseous products during combustion, another undesirable aspect of solid waste incineration is the generation of water vapor plumes. If water is not employed in cleaning, collecting, or cooling exhaust gases the occurrence of such plumes from incinerator stacks will be limited and probably will occur only during periods of very low, outside

TABLE 9-6 Trace Gases in Incinerator Effluent

Gaseous Emissions (lb/ton)	Typical Municipal Refuse	Mostly Twigs and Branches (no garbage)
Aldehydes	23.6×10^{-4}	1.1
Sulfur oxides	–	1.9
Hydrocarbons (expressed as CH_4)	0.8	1.4
Organic acids (expressed as acetic acid)	–	0.6
CO	0.51	–
Nitrogen oxides	2.7	2.1
Ammonia	–	0.3

Sources: Ref. 9-9 and SRI, "Smog Problem in L.A. City."

ambient temperatures. However, where gases are cooled to acceptable temperatures by vaporization of water, the occurrence of such plumes is much more likely. Even so, generally the temperature of the effluent gases in the stack of the incinerator will probably be quite high and vapor plumes will only occur when outside temperatures are quite low and the relative humidity of the outside atmosphere quite high. When water methods are used in gas cooling and in particulate collection, however, the occurrence of water vapor plumes should be anticipated regularly. In this situation the exit gases are saturated with moisture and the absolute humidity of the exit effluent streams are quite high. In addition, the temperature of the effluent is low; consequently, dispersion of the gases is reduced, and condensate plumes occur under amost all atmospheric conditions. Condensate plumes in themselves are usually not harmful and their major detrimental effect is in causing poor public relations for the incinerator facility. However, in some instances there have been complaints of automobile corrosion as a result of condensate fallout, dangerous decrease in ground level visibility on surrounding terrain, and other rather particular complaints, which may or may not have been factual, but have produced undesirable end results in poor public relations for the facility.

9-6. EMISSION CONTROL

9-6.1 General

As an introduction to the topic of pollutant control, mention must be made of the criteria which have been established or will soon be established for levels of pollutant output. Any mention of specific levels for control of pollutants would be inappropriate, since air pollution control codes are currently being revised all over the country. However, some general statements can be made. As far as gaseous emissions are concerned, the principal types from incinerators (carbon dioxide, water vapor, nitrogen, and oxygen) are all normally present in the atmosphere and at this time are not considered pollutants. However, other trace gases produced during combustion in an incinerator, such as sulfur oxides, nitrogen oxides, carbon monoxide, hydrogen chloride and other hydrocarbon gases, are all considered to be noxious pollutants. General inspection of incineration practices and currently proposed criteria for emission of these noxious gases indicate that emission levels are such that very little concern need be expressed over the control of such pollutants from incinerators. The situation in regard to particulate emissions in the effluent gas stream from incinerators is somewhat different. In many localities throughout the country local ordinances are being adopted concerning the amount of suspended particulate matter which will be considered permissible in the atmosphere of that given community. Almost all of the regulatory agencies which have adopted air pollution control codes have required that these emission levels be stated with reference to a specific datum or a specific reference condition. The most commonly used reference conditions are the expression of particulate emissions as pounds of suspended solids per thousand pounds of dry flue gas corrected to 50 percent excess air—or, as a second reference, expression of particulate matter in grains per standard cubic foot of dry flue gas corrected to 12 percent carbon dioxide content. The standard cubic foot refers to the volume of a cubic foot of dry flue gas at 29.92 in. of mercury and 70°F as ambient conditions. For the incineration of municipal solid wastes, these two standard reference conditions are approximately equivalent. Figure 9-12 provides a means of correcting any given reading to these basic conditions of 50 percent excess air or 12 percent carbon dioxide. In addition to the limitation of quantities of particulate matter in suspension in the atmosphere, many air pollution control codes are based upon a visual emission level. The emissions from a combustion operation must not reduce visibility in a particular area nor produce opacity there beyond a certain level. The basis for the opacity determination is the Ringelmann number. This number is determined by a visual observation of the stack plume; the observer compares the opacity of the stack plume with a series of reference grids of black lines on white. When the reference grids are properly placed they appear as shades of gray to the observer, and the proper shade of gray is selected to match the shade of the stack plume. Basic difficulty arises in the application of any visible emission level code, since most plant design is based upon the idea that a quantitative reduction in the weight of particulate matter per unit volume of escaping flue gases must be obtained. The correlation between the quantitative values of pounds of particulate matter per unit volume of flue gases and the visibility criteria is rather complex. In general, some relation must be made between the "visibility limit" of the gas and its entrained particulate matter and a definite quantitative value for pounds of the latter per unit volume of

168   SOLID WASTE MANAGEMENT

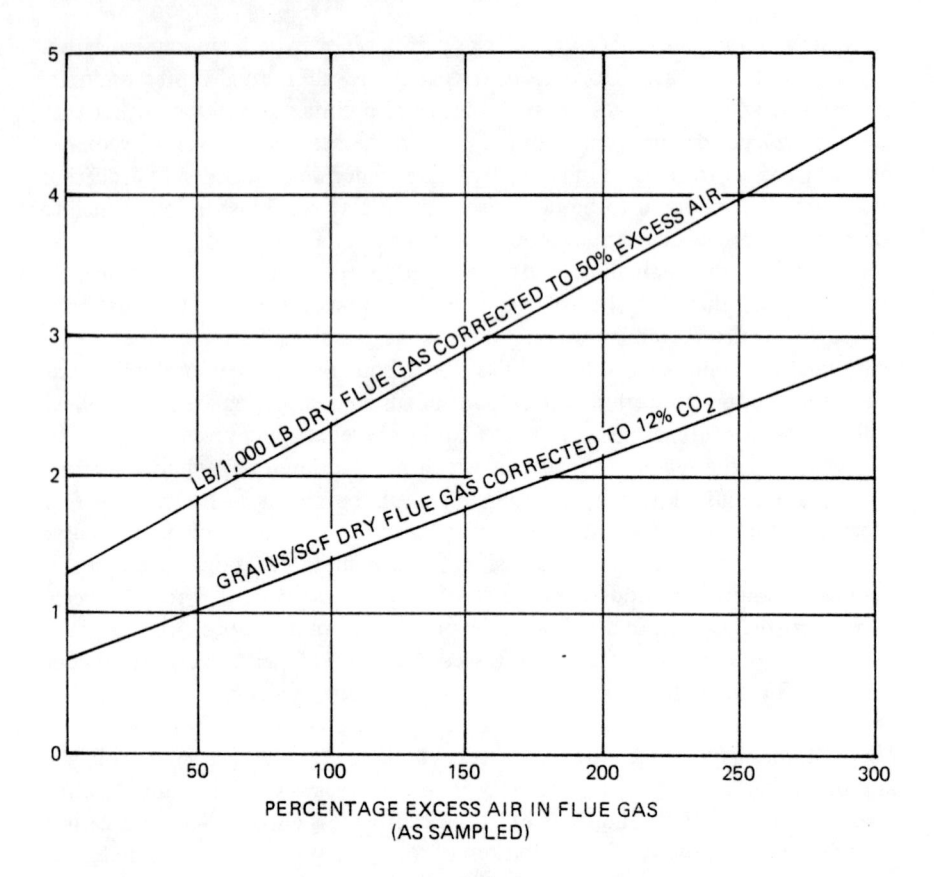

Fig. 9-12   Dust concentration equivalents.

escaping gases. In general, quantitative figures proposed for certain levels of visibility are as follows: an entrainment of from 0.01 to 0.02 grains per cubic foot of effluent gas will produce a gaseous effluent which is optically clear (Ringelmann number less than 1). Experience with coal-fired steam generators indicates that collector efficiency in excess of 99 percent weight will be required to achieve optically clear effluent gases. The similarity of particle size distribution from fossil fuel plants and that from incinerators would seem to indicate that similar efficiencies for incinerator collection devices and coal-fired plant types must be attained to obtain comparable Ringelmann numbers.

To furnish an estimate of the emission levels currently permitted for combustion effluents in certain codes, Table 9-7 is included to show the present (1973) requirements of the federal government (primary and secondary standards), the state of Kentucky, and Jefferson County, Kentucky. In forecasting future trends, it is apparent that even stricter regulatory codes will soon be enforced.

TABLE 9-7 Air Quality Standards

Pollutant	Federal (1973) primary	(1973) secondary	State of Kentucky	Jefferson County, Kentucky
Particulate	$\mu g/m^3$			
Annual geom. mean	75	60	65	65
24 hr[a]	–	–	180	150
24 hr[b]	260	150	220	180
SO_2	ppm			
Annual geom. mean	0.03	0.023	0.02	0.015
max one month	–	–	0.05	0.030
24 hr[b]	0.14	0.10	0.08	0.10
2 hr avg.	–	–	0.15	0.30
1 hr avg.	–	–	0.20	0.40
CO	ppb			
Any 8 hr period	8.7	8.7	8.0	8.0
max 1 hr	13.1	13.1	30	20
Total oxidant	ppm			
max 24 hr	–	–	0.02	0.02
max 1 hr	0.064	0.064	0.05	0.03
Total HC	$\mu g/m^3$			
any 3 hr period	125	125	100	100
NO_2	$\mu g/m^3$			
Annual arith. mean	100	100	–	–
24 hr avg.	250	250	–	–

[a]Not to be exceeded 1% of the time.
[b]Not to be exceeded more than once per year.

9-6.2 Methods of Control

In discussing methods of control, some separate discussion of each type of pollutant must be given—i.e., gaseous emissions, suspended particulate emissions, and odors from combustion. The gaseous emissions from combustion of municipal solid waste consist primarily of materials which are not considered air pollutants; on the other hand, both nitrogen oxides and sulfur oxides are produced during incineration, and these gases are certainly air pollutants. However, the amounts of these gases produced per ton of solid waste burned is quite small compared to the amounts produced in the combustion of a ton of fossil fuel. Solid wastes are considered to be relatively clean fuels in that they have low sulfur content, generally on the order of about 0.1 to 0.2 percent by weight. Most fossil fuels such as coal and fuel oils in use today have sulfur contents which range from 1 to 3 percent. The amount of sulfur in a fossil fuel is thus from 5 to 30 times the amount in solid waste. In addition, most of the sulfur in the solid waste is retained in the ash portion of the residue rather than being discharged from the combustion chamber as an oxide in the effluent gases. Thus, few difficulties are anticipated in the control of sulfur oxide emissions. In a like manner, little difficulty is anticipated in the control of nitrogen oxide emissions to regulated levels, since the emissions per ton of fossil fuel are over 10 times greater than the emissions of nitrogen oxides per ton incinerated. A further pollutant which may be produced in increasing amounts is hydrogen chloride gas, which results from the incineration of plastics such as polyvinyl chloride. Since PVC packaging materials are appearing in ever greater amounts in the solid waste stream, it appears reasonable to expect an increase in the emissions of hydrogen chloride in stack gases. However, it also appears that hydrogen chloride will be relatively easy to control, since the gas is highly soluble in water and therefore can be removed by wet scrubbers.

In the control of particulate emissions, a number of different devices now have been used and will continue to be used. The oldest form of particulate collector is a settling chamber having either a wet or a dry bottom. This type of collector was in general the only type of device used in incinerators in the United States until the last 20 years. Since the efficiency of the settling chamber in collecting particulate matter is only about 33 percent at best, the use of this type of chamber as the sole particulate collection device in an incinerator has been virtually discontinued in this country.

Cyclones and multiple cyclones have been used in about 20 percent of all the incinerators built in this country within the last 15 years. Average collection efficiencies for this type of apparatus have been established at between 60 and 65 percent with a maximum efficiency attainable in the range of 70 to 80 percent. For this type of device the gauge pressure drop in the stack is in the range of 2½ to 4 in. of water. Installation costs for cyclone collectors are in the range of 15 to 25 cents/cfm of treated gas.

A more common type of particulate collector is the wetted baffle spray collection device. This system, consisting of vertical baffle screens that are wetted by flushing sprays or overflow weirs, has been installed on over half of the incinerators built in the last 15 years. The efficiency for this type of device has been measured at from 10 to 50 percent, depending upon the number of screens used in the collector. Installation costs for these apparatus are from 2 to 4 cents/cu ft of gas treated per minute.

A variation of the wetted baffle spray system is the wet scrubber. Wet scrubbers have been installed in approximately 20 percent of the incinerators built during the last 15 years. These scrubbers differ from the baffle collectors in operation, since collection of the particulate matter is accomplished through direct contact of the particles with the water itself rather than with a wetted screen surface. Particulate matter collides with water droplets and the larger droplet-particulate particle is easily collected in some type of inertial collector. The droplets are formed by atomizing the liquid into the gas stream or by

INCINERATION   171

separating coarse water droplets into smaller droplets with high-velocity gas jets. The efficiencies of wet scrubbers have been measured at from 94 to 96 percent with accompanying pressure drops in the range of 5 to 7 in. of water. Water requirements for wet scrubbers range from 5 to 15 gal per 1,000 cu ft of gas treated per minute. A disadvantage of the use of wet scrubbers is that the effluent gases will be saturated with water vapor which can produce a water vapor plume from the stack of the incinerator. Additionally, wet scrubber systems require corrosion-proof construction. The installation cost for these collectors is consequently somewhat high; costs for the basic device itself (without auxiliary costs such as foundations, etc.) is from $0.25 to $1.25/cfm of gas treated.

Another method for collection of particulate matter is with the use of electrostatic precipitators (Fig. 9-13); such devices operate by electrically charging the suspended particulate matter and then causing it to be deposited

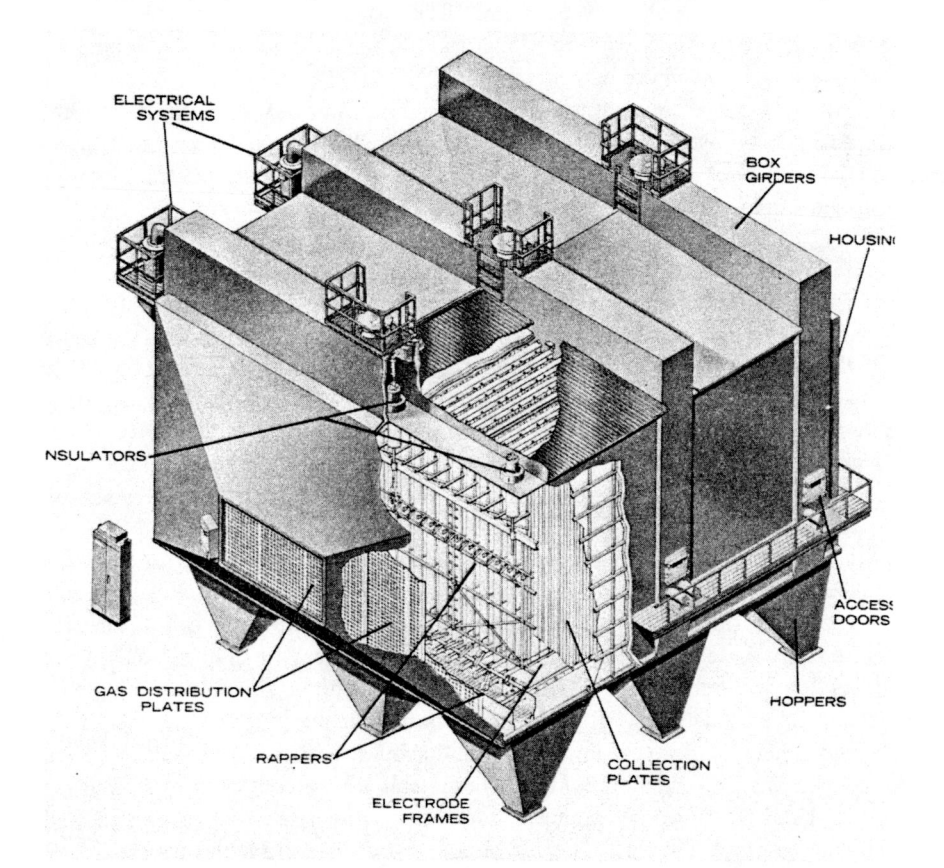

Fig. 9-13 Electrostatic precipitator (*courtesy* Wheelabrator-Fray, Inc.).

on the surface of an oppositely charged electrode. The particulate matter is removed from the collection electrode by either vibrating the collecting surface or washing the electrode with water. Several factors determine the efficiency of an electrostatic precipitator: the electrical properties of the particulates; moisture content of the effluent gases; and temperature of the latter. Gas temperature is very important (as indicated previously in Fig. 9-10). Precipitators operate most efficiently in a temperature range from about 450 to 520°F. The efficiencies reported for electrostatic precipitators are from 96 to 99.6 percent with accompanying gauge pressure drops of less than 0.5 in. of water. The electric power requirements for precipitators in operation have been measured from 200 to 400 W per 1,000 cu ft of gas treated per minute. The advantages of electrostatic precipitators are several: the precipitator has a high efficiency at low operating costs; it operates at low pressure drop and low electric power input; and it achieves high-collection efficiency with no use of water and therefore creates no water pollution problems. The installation cost for electrostatic precipitators (basic cost with no auxiliaries) is from $0.85 to $1.50 per cubic foot of gas treated per minute.

Another type of collection device is the fabric filter; however, these have not generally been used in incinerator operations in the United States. Such filters operate in much the same manner as does a vacuum cleaner. A dust cake is formed on the filter, and further filtering is done predominantly by the cake itself; high efficiencies are therefore obtainable (as much as 99.9 percent or more). Many different types of fabric filters exist and the particular application of a fabric filter to an incinerator will depend upon the temperature, moisture content, and chemical characteristics of the effluent gas stream. The resistance of the fabric in the filter to high temperatures and corrosive chemicals obviously will be of primary importance. A disadvantage of the filter is the high operating pressure drop which occurs with efficient collection; the drop being from 4 to 7 in. of water. The initial cost for fabric filters is from $0.75 to $1.50 per cubic foot of gases treated per minute, were they to be installed in an American incinerator.

In summary, settling chambers and wetted baffles appear to be inadequate as collection devices for solid waste incinerators. Cyclone collectors may be able to produce effluents which meet intermediate codes, but in the near future it is quite likely these will be inadequate as collection devices. In all likelihood, future collection requirements will have to be met through the use of direct impaction wet scrubbers, bag fabric filters, or electrostatic precipitators.

Two other topics are of importance and should receive some consideration in any discussion of emission control. These items are the control of odors and the control of water vapor plumes. Odors may effectively be controlled and eliminated from effluent gases by retaining the gases from the primary combustion chamber in a secondary chamber at sufficiently high temperatures long

Fig. 9-14 Typical incinerator charge (*courtesy* Wheelabrator-Fray, Inc.).

enough for all of the hydrocarbons to be reduced to carbon dioxide and water. A 0.5-sec retention time at a temperature of 1500°F will usually be sufficient to eliminate most odors. Generally, water vapor plumes will be caused when wet scrubbers are used as the collection device for control of particulate emission. If the optimum solution to the collection problems seems to be high-efficiency wet scrubbers, then some other means of eliminating the vapor plume must be employed. One such scheme is the removal of heat from the hot furnace gases before they are introduced into the wet scrubber and the reintroduction of this heat after the gases exit from the wet scrubber. This scheme would reduce the relative and absolute humidities of the stack effluent and would insure adequate dispersion with little likelihood of creating a vapor plume. On the other hand, the effluent gases which are saturated with water vapor by the wet scrubber may be processed in a dehumidifier to lower the humidity and eliminate the problem of plumes. The dehumidifying operation may be carried out by cooling the flue gases below the scrubber outlet temperatures. By means of such a cooling technique, the scrubbing waters may be recovered and recycled and the operating horsepower required for the high-pressure-drop fans in the scrubbers may be reduced. Dehumidification schemes of this type have been applied in industrial applications but have not to date been used on incinerator installation.

9-7. SPECIAL WASTES

Occasionally in the solid waste stream entering an incinerator, bulky items such as mattresses, automobile or truck tires, tree stumps, logs, crates, boxes, or discarded furniture may appear. Other, noncombustible items such as major appliances (stoves, refrigerators, water heaters) may also be present. These items are all collectively designated "bulky solid wastes." Such wastes cannot be disposed of in the normal municipal incinerator, since they do not burn sufficiently in the normal detention time; in addition they might damage the charging and removal mechanisms or be simply too large to enter the incinerator. Disposal of bulky wastes has been handled in the past by deposition in a sanitary landfill or by open burning. If haul distances are long, landfilling of bulky wastes is uneconomical. Open burning of any sort of solid waste is inexcusable and usually illegal. If no efforts are made to collect and dispose of bulky wastes, as has been done in some communities, open burning and dumping on an illicit basis will soon be practiced.

Two approaches to this problem have been taken in the past: special incinerators and/or size reduction. Several communities in the United States today are using special incinerators to burn oversized items. These are generally rectangular refractory-lined furnaces. They are charged by the batch method and the materials are retained within until combustion is complete. Auxiliary fuel is also used in most of these installations in order to initiate ignition and to complete combustion. On the other hand, size reduction of bulky wastes has not generally been successful in the United States. A variety of equipment including hammer mills, flail mills, and chipping devices has been used. Some mills and some chippers are successful with certain types of materials, but generally large items containing metal frames or metal skeletons are not reducible in these apparatus. Furthermore, hazards to operating personnel are inherent in the operation of this heavy equipment, which is extremely noisy and dusty. It appears likely, therefore, that special incinerators offer more promise for the disposal of bulky wastes.

In addition to bulky items which are difficult to incinerate, other special wastes include potentially hazardous kinds. Included in this category are those which are highly flammable or explosive, toxic chemicals, and radioactive materials. If such hazardous materials are generated at an industrial source, some special arrangement between the industry and the incinerator operating agency should be made for handling them. When solid wastes of a hazardous nature are generated in domestic areas or in residences, it is virtually impossible to prevent them from entering the solid waste stream. In the event that such wastes do enter this stream if detected, they must be removed by the operating personnel at the incinerator; they may be fed into the incoming waste stream in small quantities if no general hazard is caused. Radioactive materials should never be handled in any incinerator.

Other special wastes which must be dealt with in a different manner at a municipal incinerator include those which are highly objectionable and unpleasant in appearance or dangerous to health, i.e., hospital wastes, carcasses from slaughter houses or from streets, etc.

The best solution to the disposal of hospital wastes is pathological incineration. In the case of slaughterhouses, butcher shops, or other plants which generate highly putrescible organic materials, the best solution is special drying equipment and incinerators. Some provision should be made in a municipal incinerator, however, for animals which are collected from city streets and suburban highways and thoroughfares. Some incinerators are equipped with a special hearth in the secondary combustion zone where dead animals can be placed until the carcasses are reduced by the hot gases and flames.

Finally, another special material which should be considered for incineration is sewage sludge. In many communities in the United States, such sludge is mixed with domestic and municipal refuse and the mixture is then incinerated. Advantages of this method derive from the fact that the excess heat generated when the solid wastes are burned is available for use in drying the partially dewatered sludge, which then burns much more readily. Other advantages stem from the use of a single facility instead of separate incinerators—one for the sewage sludge and the conventional incinerator for the municipal solid wastes. However, incineration of combined materials is still in an experimental stage in many aspects. Therefore, careful consideration must be given to all aspects of the problem before such a method of combined incineration is adopted.

9-8. SUMMARY

It should be apparent that the subject of incineration of municipal solid waste is a rather complex topic. Many variable parameters enter into the picture. The variability of the solid wastes themselves account for much of the uncertainty in design and planning of any incineration operation. Little can be done to standardize or make uniform the refuse coming into an incinerator. Certainly if reclamation and recycling operations are carried out as proposed by many solid waste management experts, these operations will change the character of the solid waste charged into the municipal incinerator. In the event that such changes do occur, innovations in the design and operation of the incinerator must of necessity result.

In addition to these considerations, added emphasis will be given in the future to the recovery of excess heat from the incineration of wastes. Advances in the state of knowledge of waste processing, and the realization on the part of the general public of the tremendous quantities of materials currently being wasted, will lead to greater and greater emphasis upon reuse and recovery of every portion of the solid waste stream. Obviously, some portions of this stream can be reused most efficiently by recovering their heat values. Thus, recovery of waste heat should receive added emphasis in any design or plan for a proposed incinerator. In addition to the observations made in the foregoing sections, some discussion should be devoted to some of the new developments in incineration of solid waste. These new developments, which include high-temperature incineration, partial oxidation of various fractions of the solid waste stream, and pyrolysis of various types of wastes, are now essentially in developmental stages. However, some discussion is devoted to each of them in Chapter 11.

REFERENCES

Basic information, design characteristics and cost data for incinerators are given in an excellent text on the subject edited by Richard C. Corey, *Principles and Practices of Incineration*, Wiley-Interscience, New York, 1969.

Additional References

9-1. Kaiser, E. R., "The Incineration of Bulky Refuse," *Proc.*, Natl. Incin. Conference, ASME, New York, 1966.

9-2. ———, "The Incineration of Bulky Refuse, II," *Proc.*, Natl. Incin. Conference, ASME, New York, 1968.

9-3. ———, "Refuse Reduction Processes," *Proc.*, Surgeon General's Conference on Solid Waste Management for Metropolitan Washington, D.C., USPHS Publ. No. 1729, Govt. Printing Office, 1967.

9-4. Amer. Public Works Assoc., *Municipal Refuse Disposal*, 2d ed., Public Admin. Service, Chicago, 1966.

9-5. Jens, W., and Rehm, F. R., "Municipal Incineration and Air Pollution Control," *Proc.*, Natl. Incin. Conference, ASME, New York, 1966.

9-6. Matusky, F. E., and Hampton, R. K., "Incinerator Waste Water," *Proc.*, Natl. Incin. Conference, ASME, New York, 1968.

9-7. Rogus, C. A., "Weigh Refuse Electronically," *The American City*, Vol. 72, No. 4 (April 1957), pp. 128–130.

9-8. Cohan, L. J., and Fernandes, J. H., "Potential Energy Conversion Aspects of Refuse," paper presented at ASME Annual Meeting, Pittsburgh, 1967; ASME Paper No. 67WA/PID-6.

9-9. Stenburg, R. L., Hangebrauck, R. P., von Lehmden, D. J., and Rose, A. H., Jr., "Field Evaluation of Combustion Air Effects on Atmospheric Emissions From Municipal Incinerators," *J. Air Pollution Control Assoc.*, Vol. 12, No. 2 (Feb. 1962), pp. 83–89.

9-10. Pearl, D. R., "A Review of the State of the Art of Modern Municipal Incineration System Equipment", Pt. 4, V. 4, USPHS Publ. No. 1886, Govt. Printing Office, 1969.

9-11. Schoenberger, R. J., and Purdom, P. W., "Classification of Incinerator Residue," *Proc.*, Natl. Incin. Conference, ASME, New York, 1968.

9-12. Kaiser, E. R., "Refuse Composition and Flue-Gas Analyses from Municipal Incinerators," *Proc.*, Natl. Incin. Conference, ASME, New York, 1964.

9-13. Walker, A. B., and Schmitz, F. W., "Characteristics of Furnace Emissions from Large, Mechanically-Stoked Municipal Incinerators," *Proc.*, Natl. Incin. Conference, ASME, New York, 1966.

9-14. Stephenson, J. W., and Cafiero, A. S., "Municipal Incinerator Design Practices and Trends", *ibid.*

9-15. Conner, W. D., and Hodkinson, J. R., "Optical Properties and Visual Effects of Smoke-Stack Plumes," USPHS Publ. No. 999-AP-30, Cincinnati, 1967.

9-16. Mandelbaum, H., "Incinerators Can Meet Tougher Standards," *The American City*, Vol. 82, No. 8 (Aug. 1967), pp. 97–98.

9-17. Fernandes, J. H., "Incinerator Air Pollution Control Equipment," Pt. 5, V. 4, USPHS Publ. No. 1886 Govt. Printing Office, 1969.

9-18. Shuster, William W., "Partial Oxidation of Solid Organic Wastes," USPHS Publ. No. 2133, Govt. Printing Office, 1970.

10

Sanitary landfill

10-1. INTRODUCTION

Perhaps the oldest method of waste disposal practiced by man is that of landfilling. The midden heaps of ancient man now furnish archeologists and paleohistorians extensive information about early man and his life-style. Until rather recently there has been no real modification in the practice of disposal by open dumping except for the addition of fire. Fire to medieval man symbolized purification and therefore was used in burning over refuse dumps outside castle and city walls during the Middle Ages. With greater understanding of the combustion process, modern man has come to the realization that burning of an open dump simply trades one form of pollution for another, trading land pollution for air pollution. Consequently, during the last 30 to 40 years the practice of solid waste disposal in a landfill has been modified to minimize environmental pollution and insult. The modified practice is now referred to as sanitary landfilling. Sanitary landfilling is defined by the American Society of Civil Engineers as "A method of disposing of refuse on land without creating nuisances or hazards to public health or safety, by utilizing the principles of engineering to confine the refuse to the smallest practical volume, and to cover it with a layer of earth at the conclusion of each day's operation, or at such more frequent intervals as may be necessary." Thus, above all else, the sanitary landfill is an engineered construction. Four basic steps in that construction should be considered: (1) the deposition of solid waste in a prepared section of a site in such a way that the working face has a minimum area; (2) the spreading and compaction of the waste in thin layers; (3) the covering of the waste with a layer of compacted cover soil either daily or more frequently if required; (4) the final cover of the entire construction with a compacted earth layer 2 to 3 ft thick.

10-1.1 Preliminary Planning

As with other engineered constructions, most of the difficulties associated with a successful practice of sanitary landfilling may be resolved with adequate preliminary planning and design, and competent supervision during operation. The daily operations of a sanitary landfill should not be considered merely the daily activities of a business venture, but should be considered analogous to the construction phase of an engineered project such as a dam or high-rise structure. With this consideration in mind, competent supervision of that operational phase will follow. Before the operational phase, however, the first consideration in developing a sanitary landfill is adequate preliminary planning.

Preliminary planning for a proposed landfill operation should include several considerations: (1) an adequate investigation of the types and amounts of refuse that are generated in the area to be serviced by the proposed landfill must be conducted; (2) the overall economics of the landfilling operation as compared to other methods of waste disposal such as incineration or composting must be critically examined; (3) the overall availability of land for a sanitary landfill operation must receive thorough consideration; and (4) the factors of public acceptance of a sanitary landfilling operation should be considered over and above all of the other planning considerations previously mentioned. The acceptance by the public of a sanitary landfilling operation will in the long run depend upon such factors as competent supervision of the operation, its overall economy, and pleasing appearance of the site. However, a preliminary educational campaign to secure public acceptance of the proposed facility will do much to preclude later social problems. If the populace to be served by the proposed landfill are made aware of the benefits gained in landfilling—low cost of disposal; no pollution of air, land, or water; and future use of the site—there will be little likelihood of irate citizens seeking injunctions to stop construction and operation of such a landfill. While the public relations program is being conducted to educate the public to the benefits of a proposed landfill, other preliminary planning measures should be undertaken. Included in such preliminary planning of a like facility are the engagement of engineers to complete a satisfactory design of the operation, selection of a site with proper consideration for both environmental and social factors, development of an operational plan whereby the actual construction and operation of the fill and the decision as to its future use are assured, and development of a financing plan so that the operation can be undertaken. Each of these topics will be considered in detail.

10-1.2 Basic Data for Landfill

As was discussed in a previous chapter, basic data on the types, characteristics, and relative amounts of the particular wastes which are to be disposed of must be obtained for any operation such as sanitary landfilling, incineration, or composting. Types and amounts of wastes and the point of origin for each type must be determined. Basic data on the types, characteristics, and amounts of waste generated in each subsection of a collection area must be gathered and analyzed so that the landfilling operation may be suitably designed to accommodate the relative proportions of materials in the waste stream. For example, if a large portion of industrial wastes of an inert nature are anticipated, adequate provision in the landfill operation may be made for use of this inert material, possibly as a road sub-base or possibly in special fill areas where structures may be located in the future use of the landfill area. Ample consideration should be given to the changes which are projected for the community to be served by the landfill; for example, the latter should be designed to have, if possible, a minimum 30-year life and the changes in community waste generation practices for that 30 years should be anticipated. Using this basic data the community planner then should engage an engineering staff to design a landfill after selecting a site for the operation. Site selection is one of the most important actions in obtaining a satisfactory sanitary landfill. The considerations in selecting a site for the fill are given in the following sections. A primary consideration is the economic feasibility of sanitary landfilling in the study area. For this reason a detailed description of landfill economics is presented next.

10-2. ECONOMICS OF SANITARY LANDFILLS

10-2.1 General

In considering various methods of waste disposal, a planning agency or design engineer must compare the costs of sanitary landfilling with the costs of other methods of disposal, such as incineration or composting. The costs for landfilling consist of capital ones such as the investment for land; the purchase of equipment, and the construction expenses of supporting facilities such as temporary roads; and the operating costs associated with the day-to-day operation of the landfill. Of the initial costs, the purchase price for available land is generally the outstanding item. Land costs are quite variable from locality to locality around the country and will also vary from point to point within a given metropolitan area. Of course, reflected in the land cost will be certain other amounts necessary in order to render the site usable for a sanitary landfill. A low purchase price for land may reflect the fact that extensive and costly site improvements are necessary in order to render the site available for use as a fill or as the site of any other constructed facility.

There are some disadvantages in the purchase of land by a municipality, the chief one being that the land is removed from the taxable land bank and therefore revenues accruing to the municipality are reduced. Furthermore, the amounts of capital investment which the municipality or community government may undertake may be limited by law, and the capital outlay associated with land for a sanitary landfill may be beyond the financing capabilities of some communities. In such cases, it may be advantageous for the disposing agency to lease or rent land. Of course, in such a situation the eventual resale value of the property is lost to the disposing agency and accrues to the leasing or renting individual or corporation. The balance of benefits and costs associated with either purchase or lease of land must be determined for the particular locality in which the landfill is to be constructed.

10-2.2 Development Costs

In addition to the price of land there will be certain particular costs associated with the development of the site for use as a landfill. Such amounts include the required costs of a preliminary survey on the solid wastes generated by the community; sums for the investigation of the site with respect to topography, hydrology, geology, and soil characteristics; fees for retention of engineering consultants and designers; and the actual fees for the completed designs, plans, and specifications for operation of the fill. Another major category of costs in landfill development will be those associated with the physical development of the site, such as clearing and grubbing, landscaping, installing drainage, building access roads, installing utilities such as electricity, water, and telephones, and building fences and adequate signs for the operation. All of these items should be allowed for in a cost estimate for the facility. Additionally, some physical plant construction will be necessary for administration buildings, buildings for the storage and maintenance of hauling and placement equipment, for personnel, and to house scales for the weighing of incoming refuse. These structures are all part of the initial investment associated with a landfill. Finally, the equipment which is to be used in the fill will constitute a rather large cost item and may be either financed as an outright purchase or in a rental or lease arrangement much in the same manner as the purchase or rental of the land for the fill.

10-2.3 Operating Costs

Operating costs for a sanitary landfill will cover such items as labor, operating and maintenance of equipment, purchase and hauling of cover material if necessary, administrative services, and various miscellaneous charges for maintaining facilities or paying for such services as utilities, insurance charges, etc. General labor costs will account for approximately half the total operating expense of a sanitary landfill. An additional third of the total cost will be accounted for in equipment expenditures. The latter will encompass such items as gas and oil, and will also include maintenance and repair costs.

In general, the operating expenses for sanitary landfills will vary with the size of the operation, since a certain proportion are in the category of initial costs and therefore are relatively less for a larger operation (for example, roads to the site, fencing, etc.). Also many small operations are operated on a part-time basis and therefore are rather inefficient. Collected data on sanitary landfills show that the operating costs of a small operation (less than about 160 tons/day) will vary between $1.25 per ton and approximately $5.00 per ton. On the other hand a larger operation (more than 300 tons per day) will have operating costs in the range of $0.75 to $2.00/ton, with the lower costs associated with the very large operations. Figure 10-1 shows the range in costs per ton for operating sanitary landfills as a function of their operating capacity.

If an economic comparison between sanitary landfill and other alternate methods of disposal such as incineration or composting is to be made, then the entire cost of each operation must be estimated. Included in the total cost of sanitary landfilling must be the initial costs previously discussed, the operating costs mentioned, and also the total cost of hauling the refuse to the landfill site. Ideally, hauling costs are kept to a minimum by locating the fill at the weighted center of the waste collection area. However, this is not always possible and increased hauling costs for fills located at the perimeter of a waste generation area may preclude the use of such fills. In the event that a landfill site is located at a significant distance from the refuse generating area, the operation of a transfer station may provide needed economy, rendering the landfilling operation competitive as to total cost with incineration and composting. Of course, the total cost for the incineration operation must also be considered; such costs include the disposal figures for the noncombustible residue which is left after incineration. For composting operations, the total cost must include all the expenses associated with non-compostable material disposal and also the marketing costs associated with the sale of the final product. In many instances,

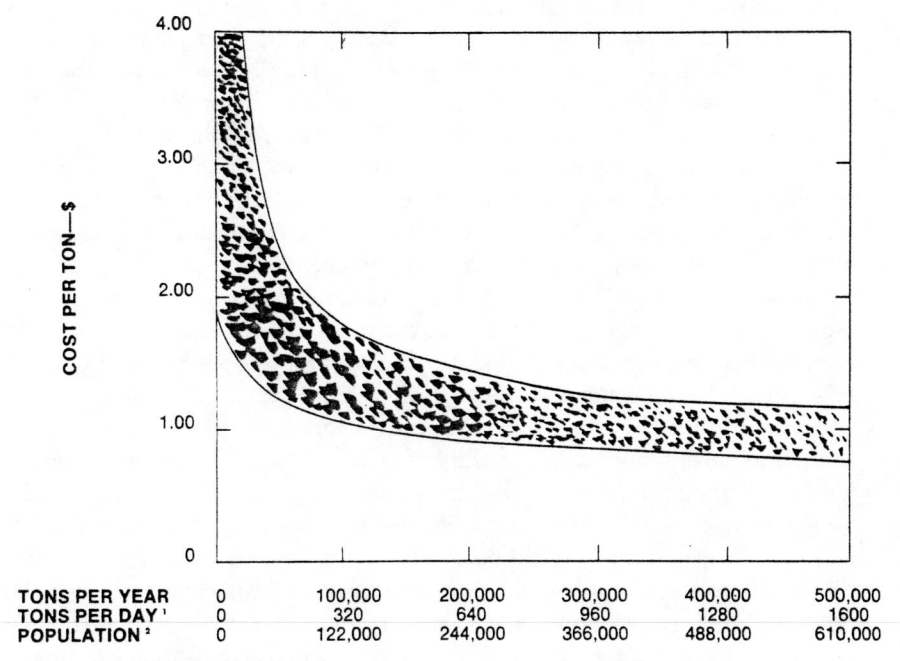

TONS PER YEAR	0	100,000	200,000	300,000	400,000	500,000
TONS PER DAY [1]	0	320	640	960	1280	1600
POPULATION [2]	0	122,000	244,000	366,000	488,000	610,000

[1] Based on 6-day work week.
[2] Based on national average of 4.5 lbs per person per calendar day.

Fig. 10-1 Operating costs for sanitary landfills (*source: Sanitary Landfill Facts*, USPHS, 1968).

the location of an incinerator in the center of a waste generation area will make the incineration disposal method more economical than sanitary landfilling. However, the sanitary landfill also has the advantage that a portion of generally unusable land is converted to useful acreage and may be employed as the site of a recreational facility or the site of very light structures. All of the advantages and disadvantages mentioned above should be considered when comparing the cost for landfills, incinerators, and composting plants.

10-4. SITE SELECTION FOR SANITARY LANDFILLS

10-4.1 General

An initial factor in any site selection activity for a sanitary landfill is the determination of what land, and which sites, are available for use. In this connection, some attention should be paid to the zoning of various areas of a community. Land which has been zoned for commercial or industrial development is much more likely to become available for sanitary landfill operations than that which has traditionally been zoned for residential use. Likewise, land which is still in the developmental stages and has not been surrounded by previous developments will be much more amenable to a landfill operation than land which is surrounded by developed areas, particularly residential ones. Near many large metropolitan areas there are suburban belts which contain sites suitable for landfilling operation. However, suburban residents generally are resentful and enthusiastically resist any attempts to bring urban refuse to the suburban area for disposal. Adequate planning, which should include a public relations program and competent leadership by elected officials, may overcome such suburban resistance, especially when an overall regional plan for waste disposal is drawn up so that both the metropolitan and suburban communities may dispose of wastes in the same landfill. With a comprehensive regional plan, the overall economies achieved with large-volume operations will make costs to both suburban and metropolitan consumers much lower, and the quality of collection and disposal service should improve. If the social and political considerations for location of a landfill have been given adequate attention and a number of sites are deemed acceptable, the technical characteristics of individual sites should be examined. The technical considerations for each site may be grouped as to environmental and economic factors. In this discussion of site selection, predominant emphasis is placed upon environmental considerations. The overall economics of a landfill operation, including the influence of site selection, were discussed previously.

10-4.2 Basic Considerations

The selection of a site for a sanitary landfill which will not be offensive and injurious to the surrounding environment will be based upon several different considerations. Before enumerating these, it is pertinent to ask a question concerning the basic philosophy of the sanitary landfill. Is the landfill a refuse cache, a place where materials are to be stored for possible later use, or is the landfill a particular form of waste treatment plant? Many persons, including some sanitary engineers who design fills, consider the sanitary landfill as primarily a storage area where reusable materials, because of the present unfavorable economics of separation, must be temporarily stored with degradable nonreusable materials. If the landfill is viewed from this standpoint, then the site for it should be selected so that little contact between the environment and the contained, temporarily stored materials occurs. For example, if the intention in burying wasted cans or other metals is that at a future date they will be returned to the manufacturing cycle, then the landfill will have to be designed so that the least amount of moisture infiltrates the cover soil and leads to oxidation of the metals within. If on the other hand, the sanitary landfill is considered to be a specialized form of waste treatment plant, then different criteria will apply for site selection and operation. For example, in such a case, considerable amounts of moisture will be necessary for complete and rapid degradation of the degradable wastes within the fill and the latter should be located in an area where rainfall will be sufficient. In addition the materials which are leached from the landfill by percolating waters will of necessity need to be collected and perhaps reinserted in the fill to accelerate the degradation process. In general, at the present time, the criteria for adequate sanitary landfilling are such that the contacts between the environment and the contained wastes are kept at a minimum. When such philosophy is adopted for the overall operation of the landfill, primary consideration must be given to avoiding the generation and dispersion of contaminants from the fill, and also avoiding the placement of a fill so that the surrounding environment is visibly marred.

10-4.3 Siting to Minimize Difficulties

A landfilling operation may be designed so that the aesthetic impairment is minimal. For example, surrounding the entire filling area with an embankment of earth which has been covered with grass and shrub plantings will effectively isolate the working face from its environs. Such measures may preclude any detrimental appearance of the landfilling operation. However, it should be recognized that certain facets of the operation may lead to poor appearance of the area or to nuisances for area residents. For example, the travel of collection and transfer trucks through the area where the landfill is located may create problems of litter, noise, or odor. Obviously, consideration must be given to minimizing such problems. The selection of routes for the trucks bringing refuse into the landfill site should be such that contact with residential areas is minimized. In addition, all trucks which are transferring refuse into the landfilling area should be adequately covered to prevent blowing paper and the generation of odors and flies. At present the problem of noise from transfer and collection trucks is not as amenable to solution: however, any general solution to the noise pollution problem will include a solution for landfill refuse trucks also. Of more direct concern to the landfill planner is the possibility that contaminants generated within the fill during decomposition of the contained refuse will enter the surrounding environment and prove hazardous or noisome to the residents of the area.

The introduction of contaminants from a landfill into the surrounding environment can occur only if those contaminants are transported in some way from within the landfill. One of the major sources of transport for contaminants is water which seeps or percolates through the fill and enters either the groundwater system or the surface system. Another source of transportation may arise from the contaminant itself; i.e., gases which are generated within a landfill as a result of degradation of the organic materials therein obviously may move in response to pressure and concentration gradients.

The most important single consideration with respect to contamination is the introduction of water into the landfill and the emergence of that water carrying contaminant materials. The water content of the fill material will also have much to do with the progress of degradation in the landfill. Thus considerable attention should be given to the possibilities that water will be introduced into a fill and will migrate through it carrying out of the deposited refuse contaminants which may be in suspension in the water or may be dissolved therein. Additionally, the gases which are created during decomposition of the degradable refuse within a landfill may migrate laterally or vertically to enter the atmosphere or possibly to collect in dangerous concentrations near structures or in depressions. Thus, consideration must be given to the topography of a site, the climate of the area in which the landfill is proposed, the hydrology of the region both with respect to surface and groundwaters, and the overall geology of the site including the characteristics of both soils and rocks there.

10-4.4 Physical Factors in Siting

In general, any topographic land form may be used for the site of a landfill. However, the variations in topography from site to site will cause special difficulties and will necessitate special design considerations for the successful operation of a fill. For example, flat or gently rolling terrain may be most suitable for the landfilling operation itself, but may also receive high priority as the future location of recreational areas or industrial sites. Likewise, low-lying flatlands may be periodically flooded and may therefore not be suitable for landfilling operations. On the other hand, severely eroded topography, such as is found in many depressional areas where canyons and ravines are the dominant features, may be quite suitable for landfilling from the point of view that a considerable volume of refuse may be deposited in each acre of land because of the depression therein. However, the origin of these depressions must not be forgotten—erosion from flowing water has created them. Therefore, any landfilling operation contemplated for depressional areas such as ravines or canyons should take full account of the condition that any flowing water must be intercepted before it reaches the landfill. Adequate surface drainage features must also be designed so that erosional waters cannot interfere with the operation of the fill. On the other hand, many man-made depressions or other man-made features may function quite well as landfill sites. For example, abandoned strip mines may furnish an ideal location for a sanitary landfill, since the surrounding terrain is certainly not amenable to residential zoning. The problem of water seeping up into refuse location ordinarily will not be met in such strip mining areas. Very often, where coal seams have been exposed in a strip mining operation, the soil which underlies the coal beds may be an underclay or a shale formation, which in either case amounts to an almost impermeable layer quite suitable for excluding groundwaters and keeping the fill dry. In addition, if porous strata are encountered, an artificial barrier may be created between the fill and the porous bed by placing fine-grained soil blankets over the strata. If these expedients are employed, a suitable site may be attained and an area once blighted by man's activities may be reclaimed and put back into the category of a useful parcel of land. Other sites which may be amenable to such treatment include abandoned clay pits, sand and gravel quarries, and limestone quarries. Some special consideration should be given to the migration of groundwaters and generated gases whenever pervious formations such as gravel strata are encountered in a man-made excavation.

Special consideration must be given to any proposal for sanitary landfill operations in tidal areas or marshlands near the sea coast. Such operations should never be undertaken. Dumping of refuse in a saturated high-water environment such as marshland is totally undesirable from an engineering point of view. Moreover, the use of swamps or marshland as landfill sites totally ignores the very productive nature of these areas. The ecological considerations of the overall use of wetlands, whether marsh, swamp, or tidal, as landfill areas make obvious the fact that the gains in ability to dispose of refuse are almost always outweighed by the losses to man in the reduction of wildlife and scenic beauty occurring when such wet lands are altered.

The next factor to be considered in the selection of a landfill site is the overall climate of the area—the amount and time of occurrence of rainfall, the velocity and direction of the prevailing wind, and the overall temperature-time relationship for the site. Of special importance in many localities is the wind situation at any particular site. Quite often a windy site may be very dusty if rainfall is not sufficient to control the dust. Additionally a windy site may be covered with blowing paper. To prevent the occurrence of the latter, litter fences and other expedients may be used; of course, any extra expense associated with such measures must be taken into account. In general, excessively windy sites should be avoided if at all possible. When consideration is given to the amount and intensity of rainfall at a particular site, the relationships of the rainfall and the topography of the site should be considered. If the topography at a site is such that steep grades must be climbed by collection equipment or if the site is covered with a soil that is easily eroded, then the occurrence of high-intensity rainfalls will be very detrimental. If the soil is easily eroded, then the site must be flat or gently rolling in order for the intense rainfall not to have an adverse erosional effect on the project. A well-drained non-erodable soil will do much to reduce the problems associated with high-intensity rainfalls. On the other hand, in an area where rainfall is minimal, the problems of dust and blowing paper (where the site is windy) must be considered.

The primary consideration as to temperature in landfill design is the amount of time during the year when it is below freezing. If examination of records in a given area show that during the winter months the temperature remains below freezing for a considerable time, then adequate cold weather operational measures must be employed. For example, cover material will be hard to obtain if frost has penetrated the upper few feet of the ground in an area where freezing temperatures are common during the winter months. In such a situation cover materials may be stockpiled during milder weather to be available for use during frozen ground operations.

Obviously, when consideration is given to the interrelationships between rainfall and topography, the hydrological aspects of a particular site are examined. Other important hydrologic aspects of landfill sites, in addition to surface runoff characteristics and rainfall intensities, are the location of the groundwater reservoir and the surface drainage pattern. Water passing over a fill can lead to infiltration, leaching of contaminants from the refuse, and pollution of surface and groundwater systems; obviously then the surface water drainage pattern at a particular site is of extreme importance. Surface water flow is important also from the point of view of daily operations of both collection and landfilling equipment. Finally, surface waters will be responsible for erosion of the land surface and may remove cover material *in situ* or may remove such material from completed refuse cells. Hydrologic investigations of the proposed landfill site should include determination of the frequency, duration, and intensity of storms during the period of recorded history of the area, and should also include records of the water losses to infiltration in the soil and evaporation and transpiration in the atmosphere. As a general principle, a sanitary landfill site should be designed so that surface waters do not invade the site from upland areas around the fill. In this way, the only source of problem water will be the rainfall which actually falls on the surface of the landfill itself. The amount of water which percolates down through the cover material in a landfill will depend upon the characteristics of the cover soil and also the hydraulic configurations of the area. For example, the hydraulic configuration of the area, including the grading of the surface, will determine the amount of time that surface water remains on the site; the longer the retention time of surface water on a site, the higher the amount of infiltration through the surface layer. A competent hydrologist supplied with information on the amount and intensity of rainfall, the infiltration and evapo-transpiration characteristics of the soil cover, and the water storage characteristics of the soil cover can predict the quantities of water which will percolate into a landfill and reach the deposited refuse. The hydrologist must also, in order to adequately survey the proposed site, determine the location and characteristics of groundwater. If the proposed landfilling operation involves excavation below the permanent groundwater table, then flowing groundwaters may also, in addition to infiltrating surface waters, come into contact with deposited refuse and remove contaminatory materials from it. If contact between refuse and flowing water is anticipated on the basis of the hydrological analysis, then remedial measures must be employed to prevent the generation of a leachate rich in contaminants. For example, surface waters should be intercepted before they reach the landfill site, and rainwaters falling on the site itself should be rapidly and efficiently removed. Removal of falling rainwaters may be easily achieved by covering the site with impermeable materials such as clay soils and grading the surface areas to produce rapid runoff. If groundwater/refuse contact is anticipated, then revision of the design plans for the landfill must be made to include either elevation of the refuse above the groundwater table, or the placement of an impermeable barrier between groundwater and deposited wastes. On the other hand, if the placement of the landfill above the groundwater table makes the particular site economically unattractive, it may be possible to make provisions for collection and treatment of the leachate which is generated from the landfill. Such collection and treatment of the leachate will be an adequate solution to the water/wastes contact problem if no contaminatory materials are introduced as a result into either the surface water system or the groundwater system.

In summary, adequate hydrological investigations of a site will include gathering of data on surface drainage (rainfall characteristics, infiltration rates, evapotranspiration rates, storage capacity) and data on groundwater characteristics and quality (groundwater table location, direction and flow rate of groundwater, location of groundwater wells, water quality in aquifers). In some instances the problems of contamination of ground or surface waters from landfill leachate may be superfluous, since the quality of some aquifers and some surface bodies of water is already quite poor. However, the potential for purification of surface waters is high, and therefore no contamination of surface waters from landfill leachates should be tolerated. On the other hand, the quality of water in some groundwater systems is such that contaminants from a landfill do little to further impair its quality. Additionally, water flow in groundwater systems is generally much slower than in surface systems and the chances of purification of the former therefore is much lower than for the latter. On the basis of the foregoing reasoning some persons have drawn the conclusion that, if no immediate danger to residents of the surrounding area exists, some pollution of already polluted groundwaters may be tolerated. Extreme caution should be used in implementing such reasoning because growing demands for groundwater, and improved sanitation techniques (conversion of septic tank service to sewer service), may do much in future years to gradually purify presently polluted groundwater systems, while the possibility of purifying a landfill site which is polluting groundwater reservoirs is quite low. Therefore, it appears that only in extreme exceptions to general situations would it be permissible for a landfill to be designed on the assumption that any further pollution of groundwaters is acceptable.

Since the flow of water through a site is of extreme importance to the overall success of the landfilling operation, it appears important to consider the flow characteristics of water in various types of earth materials. Additionally, the structural integrity of earthen materials is of importance in providing support for the proposed landfill construction. Therefore it is reasonable to give considerable attention to the geological aspects of any proposed landfill site, including the flow characteristics (permeability) of the soils and rocks present, and their structural characteristics. To be adequate, a geological investigation of a proposed site should include procurement of data concerning the depth and types of soil present at the site, the depth to and the characteristics of the types of rock which are present, as well as other significant properties of the underlying bedrock (general rock type, discontinuities within the strata—bedding planes, joints, faults—and microstructures in the rock). This data may be obtained from various sources, such as federal, state, or local agencies; for example, the U.S. Geological Survey, state geological surveys, the Soil Conservation Service, the U.S. Department of Agriculture, the U.S. Army Corps of Engineers, and local offices or university departments concerned with soil science, soil engineering, or geology. Some generalized statements may be made concerning the geology and soil characteristics of sites as they pertain to the successful operation of a landfill.

Engineers separate earth materials into two broad categories of "soil" and "rock" on the basis that rocks are those natural mineral aggregates which are not easily separated into their individual constituents by mild mechanical action; soils obviously are mineral aggregates which are easily separated by mild mechanical action, such as stirring in water, into their individual grains. For the different types of rocks, the geological classifications are, generally: sedimentary; metamorphic; and igneous. Igneous rocks can be considered primary in that they are formed during the cooling and solidification of molten magma. Erosional forces remove individual grains from these igneous rocks and transport them to other locations; and there they form sediments. If the sediments are acted upon by other natural forces, so that with time the individual sediment grains again become bonded together, the resultant rock is called a sedimentary rock. Metamorphic rocks are both igneous and sedimentary types which have been acted upon by natural forces and have been subjected to intense heat and/or intense pressure, so that the basic rock fabric has been changed. In general, the primary rocks, the igneous formations, have low permeability with respect to the fabric of the rock itself. However, bodies of igneous rock contain discontinuities such as joints and faults, and the overall mass permeability of the igneous formation may reach a significant value. Joints in igneous rock may result from contraction on cooling, expansion and contraction during subsequent movement, and tectonism. Most igneous rocks contain some joints and, therefore, at least preliminary consideration and investigation must be made of the permeability of an igneous formation; these rocks should not be considered impermeable unless proved so.

The statements made concerning igneous rocks may be repeated, in essence, for metamorphic rocks. Many of the latter have low fabric permeability but may have significant secondary permeability due to the presence of joints, faults, or other discontinuities. Again, as in the case of igneous types, the permeability of metamorphic rock should not be simply discounted but should be investigated. The permeability obviously will govern the potential for groundwater flow through the site and also the movement of leachate from a landfill. The third category of rocks, sedimentary, displays a wide range in permeabilities.

In discussing sedimentary rocks it is pertinent to first discuss sediments in general. Sediments are usually classified according to their average grain size—for example, as gravels, sands, silts, clays, etc. In general, gravels consist of sediments with grain sizes such that the grains are visible and generally greater than about 2 mm in diameter. Sands contain visible particles and are generally considered to consist of particles from a maximum diameter size of 2 mm down to the limit of visible particles, about 1/10 mm in diameter. Silts are those sediments which contain particles below the visible range in size but which are not in the colloidal size range. Generally this size range is considered to be from about 1/10 mm to about 0.005 mm. The soils which have particle sizes below 0.005 mm in equivalent diameter are generally classified as clays. Permeability, because of a number of factors, generally varies almost inversely with grain size in soils. Thus gravels are much more permeable than sands, while sands are much more permeable than silts and clays. Natural soils consist of various proportions of gravel, sand, silt, and clay particles. Thus a material which is predominantly gravel but which contains fine clay-size particles in its pores may be as impermeable as a material which contains essentially all clay-size particles. If a sediment, whether it be gravel, sand, silt, or clay is lithified and made into rock, then the character of the resultant sedimentary rock will reflect the properties of the original sediment. For example, a conglomerate created from gravel, or a sandstone created from a clean sand, will be much more permeable (fabric permeability) than will a shale created from clay-size particles. In general, then, the permeability of the rock fabric will vary with the particle size of the original constituent. Additionally secondary pores created by movement and distortion within a rock mass (joints, faults, etc.) also lend permeability to the mass and so should be considered. In general, the most pervious rocks and the formations which constitute the best aquifers are sedimentary formations of sandstones and limestones. Sandstones generally have a porous fabric and may also contain many joints. Limestones generally are jointed and also, in many cases, contain large solution cavities where percolating waters have enlarged existing joints. On the other hand, siltstones and shales have low fabric permeability and ordinarily do not contain closely spaced extensive joints or other fractures.

In considering the generation of leachate, the movement of groundwater through a fill, or the movement of leachate out of a fill into a groundwater system, it is necessary to investigate the permeability of any underlying rock formation. Obviously, the permeabilities of igneous, metamorphic, and fine-grained sedimentary rocks many times will be negligible. However, the permeabilities of porous or highly fractured sediments are likely to be extremely high. In all cases, a thorough geologic investigation of the subsurface materials at and around a proposed landfill site should be conducted.

In addition to the rock strata which are present at a particular site, the engineer designing a landfill must also be concerned with the soil cover which is present. This soil will be used, in many cases, as the cover material during the construction of the landfill; or if the depth to bedrock is great, the landfill must be founded upon the existing soil at the site. For these reasons a comprehensive investigation of soil properties and characteristics is mandatory for any proposed landfill operation.

Cover material must function to control the infiltration of water into a landfill, but it must also function to prevent the exit and entrance of disease vectors such as flies, rats, and other vermin. Additionally, cover soils must act to control movements of gases produced during degradation of the contained refuse. Of course, the cover soil on a sanitary landfill acts to control blowing paper and other litter and finally serves the additional function that when the landfill is completed it provides a basis for the growth of vegetation. The cover soil, which acts as a barrier between various refuse cells in a sanitary landfill, will prevent the spread of any combustion which is initiated in an individual cell. If all of the uses of soil as a cover material are considered a conflict as to requirements becomes apparent. For example, many soils which are suitable for the control of disease vectors are not also suitable for the control of gas movement nor would they be suitable for the control of moisture movement. For example, gravel layers may effectively block the ingress of rodents into a fill, but would not prevent the movement of water or gases. (In addition gravel layers may not prevent the movement of small vectors such as fly larvae.) On the other hand, soils at the other end of the size range, such as silts or clays, are highly cohesive and have exceedingly small pores. In a saturated state, silts and clays have low permeabilities to both gas and water, and effectively prevent the movement of vectors. However, a clay soil is subject to shrinkage on drying, and a cover layer of clay may shrink and break open in cracks when it dries during summer months. Additionally, clay soils are very difficult to handle, place, and compact and do not furnish suitable traction for hauling and placement equipment. The establishment of a vegetative cover on a highly plastic clay soil would also be quite difficult. When all of the needs for cover soil are considered it appears that the best type of soil for cover purposes in a sanitary landfill is a well-graded type, a mixture of both fine and coarse constituents. The coarse constituents, such as sands and gravels, will lead to high strength and easier compaction in the finished product and the inclusion of the fines in the cover soil will produce a material which can be rendered impervious to the flow of water, the movement of gas, and the movement of vermin and disease vectors.

Investigations in connection with any proposed sanitary landfill should include a soil sampling program. Samples of the soil should be obtained from the existing surface and, if feasible, down to and including (by means of coring operations) underlying bedrock. An inventory of the particular soil types at the proposed site should then be made. In the inventory the characteristics of each of the particular soil layers which are present should be listed. Then, all possible uses of the available materials may be considered. For example, very plastic clay soil present at the site might be removed and stockpiled for use in a compacted barrier to prevent the movement of water or gas. Sandy and gravelly soils could be removed and stockpiled for use as subgrades in temporary road construction or, because they do not retain water and are therefore not subject to freezing problems, they could be stockpiled for use as cover material during winter operations. If the soils present at the site, taken individually, do not in themselves constitute good cover materials, it is possible that they might be blended in some way to produce a suitable cover material. It is often economical to transport cover material to a landfill site if the available materials are unsuitable. Since much of the operation of the landfill will consist of excavating, hauling, spreading, and compacting cover soil, ample consideration should be given to the workability and handling characteristics of the available soils. Generally, it is feasible to mix various types of coarse-grained soils such as sand and gravels, but practical mixing of fine-grained ones such as clays with coarser materials is not feasible because of the poor handling characteristics of the cohesive, sticky, fine-grained soils. Certain soils will have no use in a proposed landfill. For example, peats or highly organic materials have little value in a landfilling operation. The only possible use of a peat material would be as a very thin final cover layer to provide organic matter to enrich the upper layer of soil on which vegetation is to be established.

10-4.5 Summary on Site Selection

In summary, the factors to be considered in selecting a site for a proposed landfill include the location, economics, topography, climate, hydrology, geology, and the characteristics of the available soils which are present at the site. In essence, the site should be selected so that any contamination of the environment through waterborne pollutants or through the generation of effluent gases is eliminated.

10-5. SANITARY LANDFILL DESIGN

10-5.1 General

The design of a sanitary landfill operation must include: the planning and specification of the appropriate fill method to use at the particular site chosen, the organization of the fill to minimize pollution problems, such as the generation of leachate and the movement of gases, and also the improvement of the site so that the fill operation can be easily performed (temporary roads, installation of housing and facilities for workers, installation of scales, installation of utilities, etc.)

10-5.2 Methods

The designer of a sanitary landfill has as one of his responsibilities the task of selecting the most appropriate method of filling the site chosen by the planner or the developer for the proposed landfill. Several methods of filling of refuse are used at the present time and no one is the most advantageous for all sites. Three general methods will be discussed and each of these has particular advantages with varying degrees of suitability for different sites. It is the designer's responsibility to select the method of filling which is most economical and technically feasible for the site proposed. To do this he must give consideration to the types and amounts of refuse which are to be filled and to the type and quantity of equipment which is available or which is likely to be available during the construction of the fill.

In general, the three types which are available for landfill operations consist of the trench method, the area method, and a rather standard modification of these two. In some cases, both the area method and the trench method will be used at the same site and in other cases a modification of both will be used.

In all landfilling methods, certain practices are common. For example, the

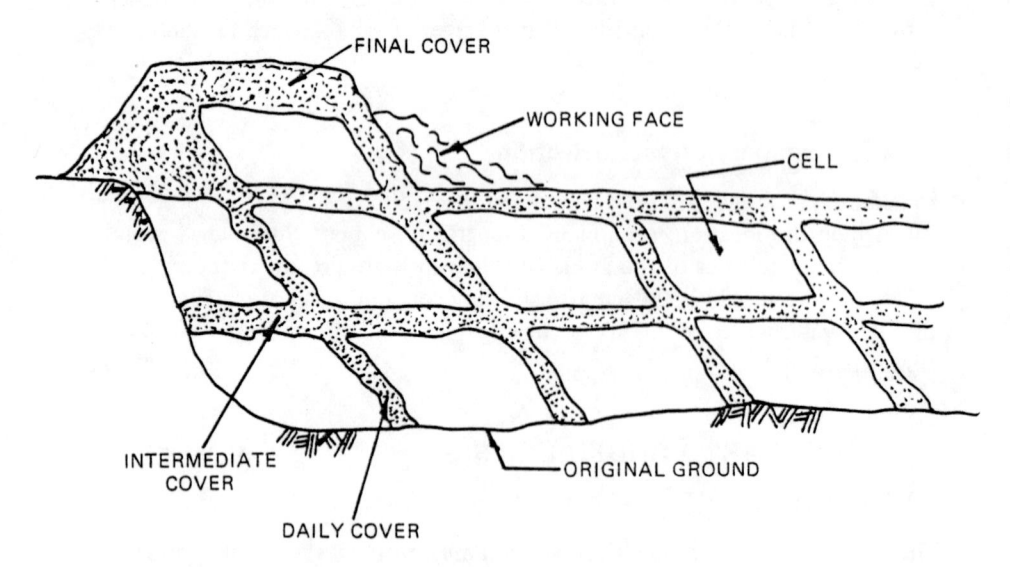

Fig. 10-2 Typical landfill operation.

refuse is always delivered to the site, spread and compacted in thin layers within a small area. The structure which is created as a result of this layering of refuse is referred to as a cell. Figure 10-2 shows the typical cell construction for a sanitary landfill. At the end of the working day the compacted wastes are covered with a layer of soil which is spread uniformly over the solid waste and then compacted to a rather high density. In some instances, the soil cover is placed over the refuse more frequently than at the end of each working day. The compacted refuse and the containing soil layers constitute a refuse cell. When a number of cells have been constructed and the elevation of the composite construction has reached the final elevation desired, a 3-4 ft thick layer of cover soil is placed and compacted over the top of all of the refuse cells (this layer is usually placed in smaller lifts of 12 in. or less thickness and compacted to a final thickness of 3-4 ft).

10-5.3 Densities

The landfill designer is generally called upon to predict the usable life of the fill site and therefore he must calculate on the basis of a given quantity of refuse delivered daily to the site how many days capacity may be obtained there. Obviously, of primary interest to the designer is the compacted density of solid waste within the landfill cells. In order for the landfill to be economically competitive with other means of disposal the compacted density of the refuse must be 800 lb/cu yd or more. In some areas of a fill, the density will be much higher than 800 lb/cu yd; where inorganic materials such as construction or demolition wastes or industrial wastes are concentrated, higher densities are attained. In contrast, in areas where very bulky wastes such as paper wrappings or plastics are concentrated, the density is likely to be considerably lower than 800 lb/cu yd. It is particularly difficult to attain high densities in compaction of materials, such as plastics and rubbers, which have natural resiliency. Where it is necessary to dispose of such resilient materials, it is often advantageous to mix layers of soil into the refuse cell; in other words, a thin layer of refuse may be covered with a thin layer of soil, and the weight of the cover soil will act somewhat to prevent rebound of the compacted resilient materials.

10-5.4 Cell Design

The overall dimensions of a refuse cell will vary with the characteristics of the cover soil, the characteristics of the refuse, the availability of land, and the topography of the site. In general, the most economical operation will be obtained when the requirements for cover material are kept at a minimum but yet are maintained at sufficient amounts to adequately cover the exposed compacted refuse. A preliminary estimate of cover material requirements can be made by a rule of thumb that one part of cover soil will be required for every four parts of refuse. However, in the design of a landfill, with attention given to the characteristics of the particular cover soil, a more accurate estimate of requirements can be made. The thickness of the compacted cover which will be required in a landfill will be dependent upon such variables as the permeability characteristics of the cover soil, and the topography and climatic conditions at the site. These relationships were discussed in the previous section. Usually the most economical operation will result if the layers of refuse are spread thinly (about 2 ft thick) and thoroughly compacted, and the dimensions of the refuse cell maintained such that the length and width of the cell are approximately equal. Adequate compaction of the refuse may be easily obtained if the material is spread in layers and worked from the bottom of a working slope up to its top, with the slope maintained at an angle of approximately 30 to 40 deg with the horizontal. The working face should also be maintained wide enough so that a long waiting line of collection trucks is not created. However, the working face should not be so wide that blowing paper or problems with vector entrance and exit from the refuse occur. In general, problems with litter and vectors will be minimized as the area of the working face is reduced. Therefore, a balance between ease of operation and maintenance against vectors and litter must be obtained. The depth of the refuse cell will govern the total amount of cover material required for the landfilling operation. Therefore, if land acquisition is difficult, or if suitable cover material is not available and must be purchased and hauled to the disposal site, deep lifts of up to 30 ft may be advantageous in reducing the total requirement for cover soil. However, in localities where such soil is ample, the overall operation of the landfill will benefit from the restriction of cell depth to approximately 6 to 8 ft. With smaller depth of refuse as recommended, the problems of disease vectors, litter, and water infiltration into the fill will be minimized.

The particular method of cell construction to be used at a given site will vary with its characteristics—topography, cover soil, rainfall, etc. In general, the trench method will be used wherever a deep layer of cover soil is available and where deep excavations are possible (low groundwater table). Figure 10-3 shows the general principles of the trench method of sanitary landfilling. In this method a trench is excavated in the ground and solid wastes are placed therein. The cover material for the refuse cells is obtained from the trench itself and therefore a minimum amount of hauling of cover material results. Since a deep excavation is made, the trench method will be unsuitable where the groundwater table is at or near the surface. Where the soil at the landfill site consists of highly cohesive materials which will stand on vertical faces, the trench is very economical; since the soil will stand the refuse cells can be constructed close together with only a narrow wall of *in situ* soil between cells. If the soil at a landfill site is a clean granular type that does not possess cohesion, the sidewalls of a trench will require considerable slope and the refuse cells must be constructed at considerable

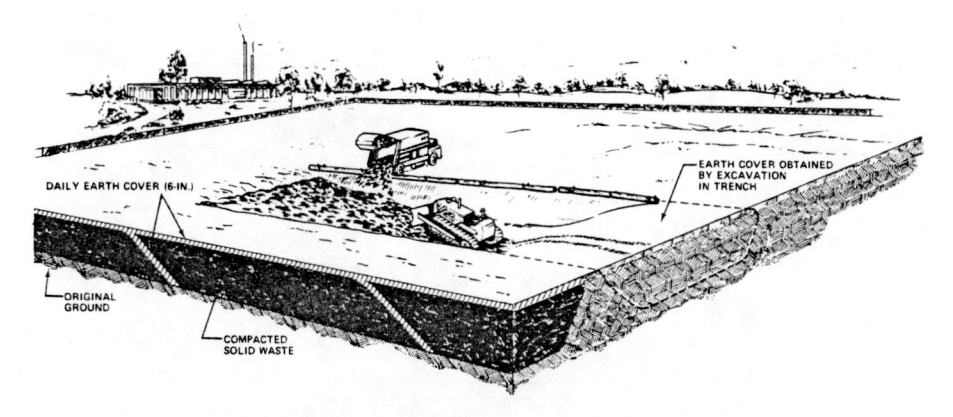

Fig. 10-3 Trench method of sanitary landfill (*courtesy* Office of Solid Waste Management Program, EPA).

distances apart. Therefore, if the cover soil at a particular site is such that immediately or during the time that the working face is exposed it will collapse or slump, the trench method of landfilling may not be suitable for that site.

An advantage accruing to use of the trench method is that the surplus soils excavated from the trenches may be stockpiled and used in a later area fill operation over the top of the completed trench-fill ones. Another advantage is that the refuse may be placed in an excavation out of the wind, so that even at a windy site trenches excavated transversely to the prevailing wind direction will greatly reduce the amount of blowing paper and will thus reduce the cost of litter control at the site. If the trench method is used, some provision should be made for removing surface drainage waters which collect in the trench at the low end.

In areas where the groundwater table is near the soil surface, it is not possible to create deep excavations for the placement of refuse. In such situations, and also over areas which have been previously filled by the trench method, the area method of landfilling can be used. Figure 10-4 shows the area landfilling method. Essentially, in the area method, the refuse is spread and compacted on an existing ground surface and cover soil is then spread and compacted over the waste to form a completed cell. The primary differences between the trench and area fill methods are the excavation in the trench method and the filling on the original ground surface in the area method, and the differences in the equipment required for both: In the trench approach, the equipment commonly required will be bulldozers or front-end loaders, equipment capable of excavation and of spreading and compacting the refuse. Since the cover soil is obtained from the excavation, hauling equipment for the soil is not required in the trench method. In the area method of landfilling, in addition to the bulldozers or

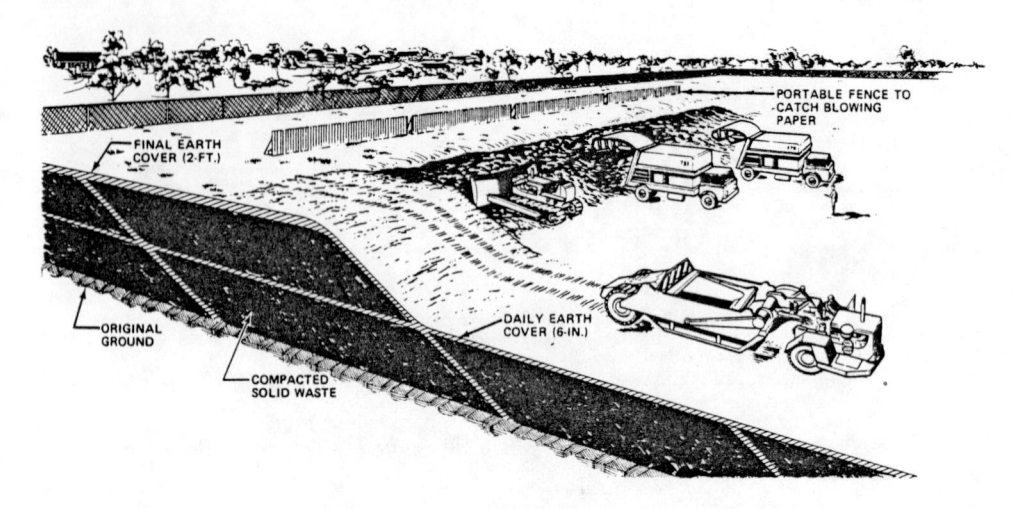

Fig. 10-4 Area method of sanitary landfill (*courtesy* Office of Solid Waste Management Program, EPA).

loaders which are required for spreading and compacting the refuse, hauling equipment for cover soil will be required. Scrapers and other forms of hauling equipment will generally be used at area method landfills. In general, the area method is favored where groundwater is near the surface, or where an existing depression is to be filled and cover material is at a minimum.

There are many possible combinations of the two basic landfilling methods. The particular combination of methods or single method used at a given site will depend upon the characteristics of the latter. In many localities, a variation of the two basic methods has been used with considerable success; this variation has been called the "progressive slope" or "ramp" method. Here, the refuse is spread and compacted on a slope and then covered with soil which has been obtained immediately in front of the working face. Figure 10-5 shows the ramp method of landfilling. The advantages of this method include the fact that a small excavation is made in front of the working face and cover soil is obtained there for the wastes just deposited. This is generally more economical than the area method, since it requires no hauling of cover soil and a certain amount of excavation below the original ground surface is accomplished.

10-5.5 Prevention of Pollution

Whatever method is used must insure adequate prevention of groundwater pollution, surface water pollution, or deleterious concentration of waste gases. As

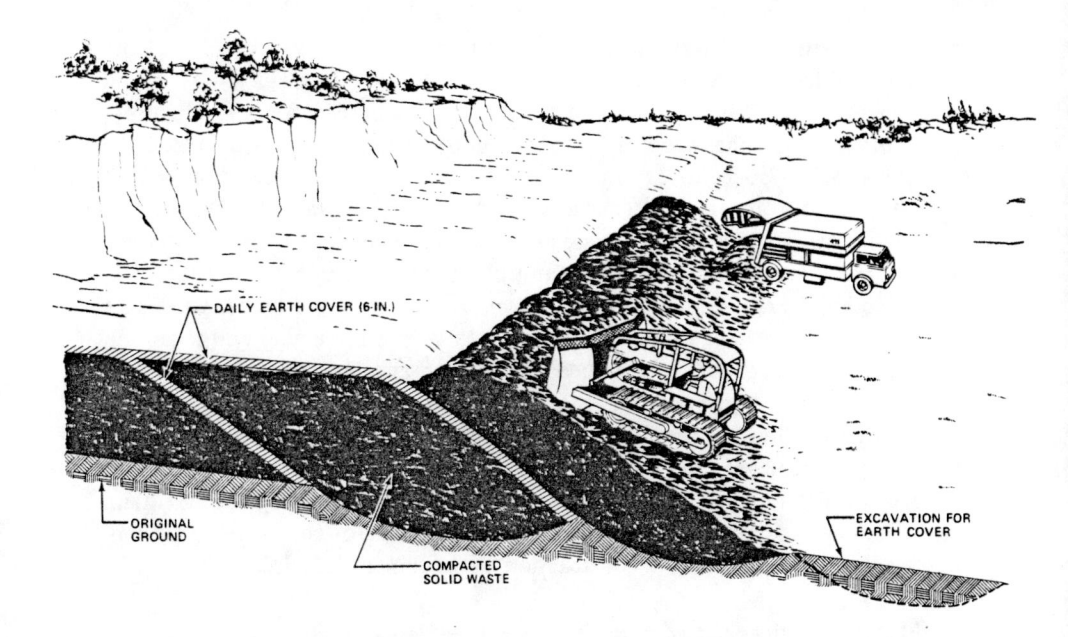

Fig. 10-5 Ramp method of sanitary landfill (*courtesy* Office of Solid Waste Management Program, EPA).

mentioned previously, the control of surface and groundwaters in a landfilling operation will depend in large part upon the overall design philosophy of the planning agency. In other words, is the landfill a refuse treatment plant wherein the degradation of waste should be accelerated as much as possible? Or is the sanitary landfill to be protected from the influence of water, so that the degradation and decomposition of materials will proceed at a minimum rate? Rapid decomposition may appear attractive, since rapid stabilization of the landfill will permit the use of the site at an earlier date than would be associated with slow decomposition of the refuse fill. However, rapid decomposition will be facilitated only with the introduction of significant quantities of water into the fill in order that the biological degradation processes are accelerated. The introduction of significant quantities of water into the fill may also generate significant amounts of leachate which may require collection and treatment in order to prevent pollution of groundwater resources. At the present time, the balance of advantages and disadvantages associated with both philosophies (treatment plant versus refuse cache) seems to favor that of minimizing decomposition rates by preventing contact of water and filled refuse. The designer of the sanitary landfill therefore must concentrate on preventing contact between contained refuse and both surface and groundwaters.

10-5.5.1 Surface Water Pollution. The prevention of surface water pollution may easily be accomplished by two methods: (1) interception of all surface drainage entering the landfill site and rerouting of those waters around the landfill; and (2) the collection and treatment of all surface waters which have passed through the site. The collection and treatment of polluted waters is inherently more difficult than the rerouting of non-polluted waters, so, in most cases, the designer should concentrate on rerouting surface waters around a landfill site so that no contact between surface drainage and contained refuse occurs. The rerouting facilities in most cases should be designed on the basis of a 50-year storm-pattern design and if the landfill is located in an area subject to possible flooding, drainage or flood protection provisions should be designed with respect to a 50-year occurrence. In addition to waters impinging on the landfill site from other parts of a watershed, rainfall striking the surface of a fill area may also produce leachate from refuse cells. Therefore, provisions for drainage of the waters falling on the fill area itself must be made. Consequently, the surface of the completed cells should be graded and sufficiently compacted so that surface retention times are minimal and surface runoff occurs at a maximum rate. Temporary drainage facilities such as half-sections of corrugated pipe may be advantageous to use during construction of the sanitary landfill.

10-5.5.2 Groundwater Pollution. Several research investigations have illustrated the fact that groundwater mounds may be created in sanitary landfills. In general such a mound is simply a rise in the water table associated with increased recharge at the point of table rise; the rise in the table usually is due to increased amounts of infiltration through exposed refuse as opposed to the original cover soil, but also may be partially due to the restriction of groundwater flow by the creation of impermeable barriers around a landfill site. The formation of a water table mound beneath a landfill is a good indication that pollution of the groundwater system probably is occurring. If not, it is likely to occur in the immediate future as the groundwater rises into contact with the deposited refuse.

As mentioned previously, the generation of leachate in a landfill may be prevented in large measure by the interception of rainfall and surface waters before such waters can infiltrate the refuse cells. If such interception is impossible because of extremely high rainfall intensities, or for some other reason, the prevention of groundwater pollution may be attained by the construction of a barrier between the contained refuse and the groundwater system. In general, the methods of prevention of such pollution will be either through the use of an impermeable barrier or of a groundwater lowering system. Groundwater elevations may be lowered in permeable strata such as gravel or sand layers or sandstone strata by means of drains, ditches, or pumping wells. Several considerations must be taken into account when any groundwater lowering system is constructed: (1) the system must be fail-safe so that any failure of a power source for the pumping system may be accommodated by the use of an alternate power source—

any rise in the groundwater level which would produce groundwater/leachate contact must be avoided in all cases; (2) design of a lowering system will require comprehensive data on the characteristics of the groundwater system and the permeability characteristics of the soils at the site—such data-gathering can be quite expensive; and (3) the situations in which a low-maintenance gravity flow system can be installed are not very common. Because of the preceding reasons, it is generally more advantageous to control leachate and prevent groundwater pollution through the construction of some sort of impermeable barrier. Barriers of asphaltic or bituminous materials have been suggested for use in sanitary landfills, but the settlement of refuse after decomposition may lead to cracking or rupturing of asphaltic liners and hence their value is questionable. Impermeable liners may be constructed of a synthetic material such as polyethylene, polyvinyl chloride, or rubber. In general polyvinyl chloride is more suitable for use than polyethylene since it is also highly impervious to the movement of gases as well as liquids. If such a synthetic material is to be used, extreme care will be required in the construction of the liner so that no puncturing of the thin liner film results. A bedding course of sand beneath the liner will be required, and a further course of very carefully placed sand over the liner will also be necessary. In most situations, a cheaper and more effective liner may be obtained by compacting a layer of clay soil to replace the impervious synthetic liner. The use of natural soils has the additional advantage that many soils act as natural biological filters to remove bacteria and other contaminants from leachate which might flow through such soils. Previous research has demonstrated that a significant thickness (approximately 5 ft) of compacted soil will act as a rather efficient filter for removal of readily decomposed organic material and also as a filter to remove pathogenic materials such as coliform bacteria (see Ref. 10-1 on Design). In any case, the construction of a compacted clay liner will produce a very slow flow of materials from the landfill to any groundwater reservoir beneath the compacted layer of soil. Since the flow is kept at such a low rate, the materials leached from the filled refuse cells may easily be collected at the surface of the impermeable clay layer and disposed of separately. Since the lining of a landfill site creates a large reservoir for the rainfall which falls on the area, some provision must also be made for the collection and disposal of leachate produced by that rainfall. Table 10-1 shows the characteristics of several samples of collected leachate. The characteristics of this leachate are such that it would require considerable treatment before release to surface waters. Facilities similar to those used for the treatment of liquid industrial wastes would be required for the treatment of collected leachate from a sanitary landfill. Biological and chemical treatment should be considered in addition to simple primary treatment by settling. Disinfection of the leachate waters should be considered to destroy any pathogenic materials therein.

In any case, sufficient provision must be made so that the groundwater quality

TABLE 10-1 Composition of Leachate From Municipal Solid Waste[a]

		Study 1	
Component	Mean	Low	High
pH	6.25	6.0	6.5
Hardness, $CaCO_3$	6983	890	7600
Alkalinity, $CaCO_3$	8045	730	9500
Ca	2234	240	2330
Mg	378	64	410
Na	1507	85	1700
K	1373	28	1700
Fe, total	163	6.5	220
Ferrous	8.7	—	—
Chloride	2215	96	2350
Sulfate	614	84	730
Phosphate	12.8	0.3	29
Organic N	309	2.4	465
NH_4-N	269	0.22	480
BOD	24,642	21,700	30,300

		Study 2	
Component	Mean	Low	High
pH	5.4	3.7	8.5
Hardness, $CaCO_3$	3070	200	5500
Na	614	127	3800
Fe, total	635	0.12	1640
Chloride	299	47	2340
Sulfate	152	20	375
Phosphate	2.9	2.0	130
Organic N	47.4	8.0	482
NH_4-N	22.2	2.1	177
COD	17,251	809	50,715
Zn	35.2	0.03	129
Ni	0.27	0.15	0.81
Suspended solids	349	13	26,500

[a]Average composition, mg/l of first 1.3 l of leachate per cubic foot of a compacted, representative municipal solid waste.

Source: Study 1. *Investigation of Leaching of a Sanitary Landfill,* Publication No. 10, California State Water Pollution Control Board, Sacramento (1954). Study 2. Summary Report, *Pollution of Subsurface Water by Sanitary Landfill,* USPHS Grant No. 5-RO1-UI-00516, Drexel Inst. of Technology (Feb. 1969).

TABLE 10-2 Groundwater Quality Near a Landfill

Parameter	Ambient (mg/l)	Landfill (mg/l)	Monitor Well[a] (mg/l)
Total dissolved solids	636	6712	1506
pH	7.2	6.7	7.3
COD	20	1863	71
Total hardness	570	4960	820
Sodium	30	806	316
Chloride	18	1710	248

[a]Monitoring well located downstream, approximately 150 ft from the landfill, at a depth of 11 ft in sandy clayey silt.
Source: Ref. 10-12.

in and around a landfill may be periodically measured and the status of any groundwater pollution may be evaluated. Observation wells and surface sampling stations should be installed for periodic sampling of the water quality in both surface and subsurface flow systems. Table 10-2 presents data gathered in this way.

In the future, it may become possible to recirculate collected leachate into a sanitary landfill in much the same manner as activated sludge is recirculated in sewage treatment plants. In such a facility the sanitary landfill would function primarily as a waste treatment plant.

10-5.5.3. Gases. In addition to the control of leachate production and its introduction into either surface or groundwaters, a sanitary landfill designer must account for the production and movement of gases in the fill. During the early days after deposition of wastes in a fill, the degradation of the wastes proceeds by aerobic decomposition. If the water content of the wastes in the fill is below about 40 percent decomposition will proceed at a slow rate. However, if the water content is between 40 and 80 percent rapid aerobic decomposition will occur and water and carbon dioxide will be produced by the acting microorganisms. If the landfill is subjected to considerable rainfall, the moisture of the solid wastes may increase to the point where the waste decomposes anaerobically and in addition to the gases previously mentioned methane will be produced during anaerobic decay (see Table 10-3). This may apply to small areas of the fill as well. The microbial decomposition of solid waste can also produce odorous gases which will be detrimental to the vegetative layer designed for the final cover soil in a landfill if oxygen is excluded from the vegetable root zone. Decay may also present hazards of explosion if methane is present in the proper concentrations. Generally methane presents an explosion hazard if it is present in a concentration of from 5 to 15 percent; at lower concentrations the quantity is insuffi-

TABLE 10-3 Landfill Gases

Time Interval Since Start of Cell Completion (Months)	Average % by Volume		
	N_2	CO_2	CH_4
0–3	52	88	5
3–6	3.8	76	21
6–12	0.4	65	29
12–18	1.1	52	40
18–24	0.4	53	47
24–30	0.2	52	48
30–36	1.3	46	51
36–42	0.9	50	47
42–48	0.4	51	48

Source: Merz, R. C., and R. Stone, *Special Studies of a Sanitary Landfill*, Final Summary Report and Third Progress Report, HEW Grant No. UI 00518-08 (1969).

cient for combustion and at higher concentrations insufficient amounts of oxygen are present to support combustion. Therefore only between concentrations of 5 and about 15 percent (methane in air) is there an explosion hazard. The landfill designer must be cognizant of the detrimental effects of the gases produced and must design the fill so that the movement and dissipation of these gases is controlled. In connection with the control of gas movement, the properties of existing soils and rocks at the landfill site are of extreme importance. Any saturated formation will present an effective barrier to the movement of gases. In this regard, clay soils in a saturated state are extremely effective in preventing the movement of gases; while coarse materials such as sand and gravels when saturated present effective barriers to gas movement, the ease of dewatering of such materials creates a potential hazard of such movement should the water table temporarily drop in a given location where gases are intended to be retained by a gravel or sand layer in a saturated state. In many cases, the natural environment will produce effective saturated soil or rock barriers to the movement of gases away from a sanitary landfill. In other instances, the presence of an unsaturated pervious stratum of rock or soil may provide easy exit for generated gases which then migrate laterally along the stratum away from the landfill and may collect in dangerous concentrations beneath structures or in natural gas traps. In situations where a possible gas hazard exists, the landfill designer must seek to prevent gas movement. The methods of prevention or control of such movement may employ either a permeable drain stratum or an impermeable barrier.

In the use of a permeable stratum as a collector for generated gases in a landfill, the most commonly used material for the pervious stratum is clean dry gravel. Figure 10-6 shows some commonly used gravel vents and gravel-filled trench collectors for landfill gases. Ideally, the collector trench should extend at

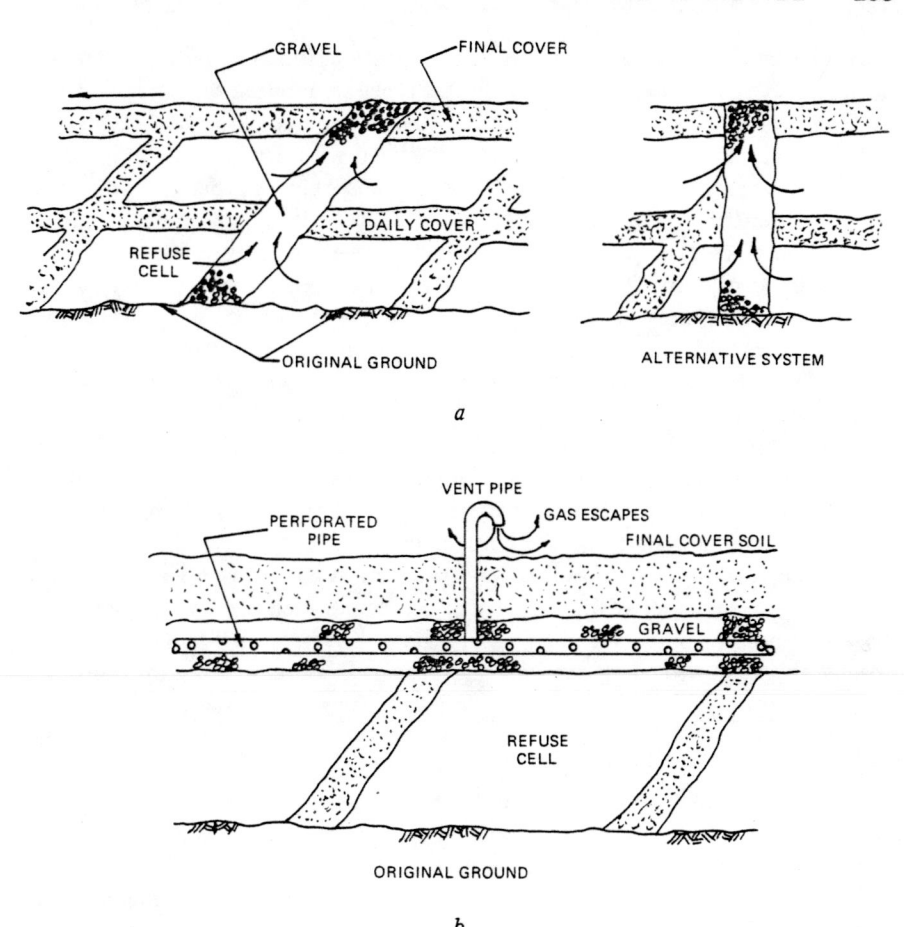

Fig. 10-6 (a) Gravel vent; (b) pipe vent.

least as deep as the fill and quite commonly somewhat deeper so that all generated gases are intercepted and any lateral flow of gas is prevented. The overall width of the trench and the gradation of the gravel therein should be specified so that the open trench is not clogged by infiltrating finer materials from the adjacent soil. The trench, to be pervious to gases, must remain dry and therefore should be dewatered either naturally or through mechanical means. Natural drainage is preferred to mechanical because of the high reliability of the former as opposed to the lower reliability and high cost of any mechanical drainage. The surface of the gravel trench should remain bare and should not be planted, since a topsoil cover will provide a barrier to the movement of gases. In other instances pipe vents as shown in Figure 10-6 have been used to collect and dis-

206 SOLID WASTE MANAGEMENT

perse the gases generated in fills. In this connection, the fill is generally covered with an impermeable layer and the collector pipes are inserted through that layer and then are spread over the top surface of the fill generating the gases. Collector pipes are placed in trenches which are backfilled with dry gravel, and are connected to a vertical riser pipe vented to the atmosphere. In general, if generated gases are vented no further problems should be anticipated. However, where production rates of gases are extreme or where the possibility of their concentration in nearby structures exists, an automatic burn-off system may be installed in a riser pipe so that collected methane is periodically burned and reduced to carbon dioxide and water vapor, which is then vented to the atmosphere. Nevertheless, this method is costly and should only be employed where absolutely necessary.

In addition to the collection of gases by use of a pervious layer or a pervious wall of earth materials, the movement of gases can be controlled by the creation of impermeable barriers. Where the barrier blocks the egress of gases from a landfill, a venting system may be constructed to collect the blocked gases and carry them through the topsoil cover to the atmosphere. As noted, the most commonly used material for such barriers is compacted clay. Clays normally have high water contents in their natural state and therefore *in situ* are impermeable to gas flow. During handling and placement of these clays, the moisture content will not reduce to the point where the clay soil becomes pervious to gas flow. Therefore such soils are quite suitable for use in construction of gas barriers. Care should be exercised to insure that any constructed barrier is continuous over the entire surface area of the fill cells, and no contact between solid wastes and pervious soil or rock should be allowed. Figure 10-7 shows such barriers. In addition, where clay soils are compacted to form an impervious membrane, care should be taken to insure that the clays remain moist after placement and are not allowed to dry out, since upon drying they will exhibit shrinkage cracks. A thickness of clay from 2 to 3 ft has proven effective in controlling gas movement. Certainly thicknesses of impermeable barriers of less

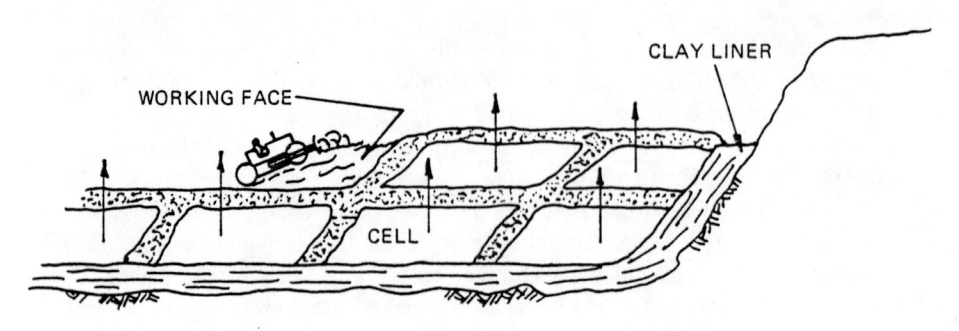

Fig. 10-7 Clay liner scheme.

than 2 ft are undesirable because the placement of a layer of less than that thickness by common earthwork methods may lead to gaps in the continuous membrane.

10-5.6 Improvements

In addition to the design considerations previously presented in connection with water movement, gas movement, and landfill method, the landfill designer must also be concerned with improving the *in situ* site to render it useful as a sanitary landfill. Included in site improvement work are some of the operations commonly used to improve any construction site: clearing and grubbing of brush, construction of windbreaks, construction of temporary slopes. In addition to the customary considerations of site improvement, rather special site improvements must be made in connection with sanitary landfilling. For example, clearing and grubbing should proceed in increments so that erosion and cover soil removal are minimized. Some consideration should be given to construction of an embankment around the entire site which, when planted with grass and shrubs, would effectively isolate the landfilling operation from nearby residences or commercial establishments. The public relations benefits associated with such construction are enormous. Permanent roads should be constructed from the primary road system to the landfilling site, since a large volume of traffic will use these routes. It is desirable to maintain grades at less than about 7 percent for uphill ways and less than about 10 percent for downhill ones to minimize wear on hauling equipment. It is important to remember that a large number of ton-miles of travel will take place over the roadways associated with the landfill, and initial expenditures to improve those roadways should produce significant economies in hauling costs. In addition to permanent roadways some temporary roads must be constructed from the end of the permanent access to the working face of the landfill.

Utilities must be furnished to the landfill site both for personnel convenience and also for administrative operations such as monitoring and weighing of incoming refuse. Therefore, electrical, sanitary sewer, and water services should be provided. Additional physical plant will be required to house special structures such as records offices for large landfill operations and also structures will be needed to house scales. Scale requirements for sanitary landfill operation are generally the same as the requirements for operation at an incinerator. Automated electronic scales and recorders are expensive as to initial investment but furnish economy in the long-run operation of large landfills. The capacity of scales for most landfill should be approximately 30 tons and the size of the scale platforms should be sufficient to handle the largest hauling vehicle anticipated for use at the site. A platform with dimensions of at least 10 by 34 ft will generally service most collection vehicles, while a 50-ft platform will suffice for transfer vehicles. As in incineration operations, the accuracy of the weighing scales should be plus-or-minus 1 percent. An additional special consideration for sanitary landfills is the provision for peripheral and litter fencing. For good operation of a sanitary landfill the entire site must have a limited-access character. General, unauthorized dumping by the public must be avoided and therefore peripheral fencing, and in many cases peripheral lighting, must be provided. In addition to peripheral fencing litter fencing should be provided to prevent the occurrence of blowing paper and other light wastes. The advantages accruing to the prevention of blowing litter far outweigh the initial cost associated with litter fencing. The small wooden fencing used in many northern states as highway snow fences (to prevent the occurrence of drifts) is quite suitable to prevent blowing litter in a landfill.

10-5.7 Site Operation Plan

Finally, the designer has the responsibility to evolve an operating plan for the construction and operation of the landfill. The operating plan should include designation of operating hours, weighing practices, traffic routing, and general waste-handling procedures. Provisions must be made for the specification of cover material (source, physical characteristics, compaction requirements, moisture control); provision must also be made for periodic maintenance of equipment, facilities such as scales, and systems such as leachate collection types. An important consideration in operating a sanitary landfill is the provision of an operating procedure for wet-weather conditions. With such conditions adequate provision of materials for stabilizing temporary roads, provision, where necessary, for temporary storage of refuse until the abatement of precipitation, and regulations for temporary covering of the working face should be made. Other "weather" requirements include the stockpiling of cover material for use during freezing-temperature conditions, and the provision of watering facilities for easing any dust problem and adjusting the moisture content during dry weather. Finally, the landfill designer must include in the operating plan provisions for such emergencies as fires within the compacted refuse of the fill. An additional consideration which should receive more study in the future will be the provision of salvage facilities at the landfill site. At the present time allowing scavengers and other would-be salvage operators into a landfill site is an extremely poor practice. Detrimental placement conditions and personnel safety hazards result from the practice of letting such persons have access to the fill. However, in future as more emphasis is placed upon reuse and recycling of waste materials, provision may be made for the organized and controlled salvage of such collected materials. These salvaging operations of necessity will involve the construction of a processing plant and the installation of separation and sorting equipment.

10-6. EQUIPMENT FOR SANITARY LANDFILLING

10-6.1 General

In general the equipment required for a sanitary landfill falls into two categories: that required for excavation and placement of cover material and solid wastes, and that required for supporting operations. The placement and compaction of refuse requires less energy than heavier and denser material such as soil. Therefore equipment requirements for spreading and compacting refuse are somewhat less than for spreading and compacting an equal volume of soil. Additionally, refuse is not easily compacted by vibratory means, so the primary means of compaction is direct compressive stress beneath a compactor foot or wheel. Supportive activities for a sanitary landfill would include the construction and maintenance of roads, fire protection services, and the control of special problems such as dust or blowing litter. For large landfilling operations, special equipment is often purchased to fulfill the supporting functions. The equipment used in a sanitary landfill may be broadly classified as either crawler or rubber-tired types, with the understanding that certain kinds of equipment may be rather specialized and not fit into either of these categories. Generally, rubber-tired machines are somewhat faster in operation than crawler equipment but do not excavate as well. While in many instances rubber-tired equipment will compact earth more efficiently than crawler-mounted machines (because of the low contact pressures associated with the wide-track crawlers), the rubber-tired types used in sanitary landfilling often do not provide essentially better compaction of refuse. Moreover, rubber-tired vehicles must be equipped with special tires which resist puncture by any sharp constituents of the refuse. Crawler machines are advantageous for use in compacting landfill materials, since they are ideal for excavating and moving over unstable surfaces. Therefore, for any given installation, the individual designer must establish the most advantageous balance for that particular site, with the particular refuse, and the particular costs of the various types of equipment, to decide whether to purchase and use crawler-mounted or rubber-tired equipment. Figures 10-8 and 10-9 show typical equipment items.

10-6.2 Special Equipment

In addition to the traditional rubber-tired and crawler-mounted types of equipment used in earthwork contracting, specialized forms of landfill compactors have been developed by several manufacturing concerns in the United States. The primary advantage of these compactors is the large-toothed steel wheels with which they are equipped. In several studies of the relative abilities of special compactors with load concentrators on their wheels operating in comparison to the conventional rubber-tired or crawler-mounted equipment, it has

Fig. 10-8 Crawler tractor for spreading refuse and earth cover (*courtesy* John Deere Co.).

Fig. 10-9 Rubber-tired tractor (*courtesy* International Harvester Co.)

become apparent that greater densities are obtained (as much as 10 percent greater) with the special steel-wheel load-concentrator compactor. However, these compactors have disadvantages in that there is a lack of traction obtained with the wheels on steep surfaces or when the compactor is being used for excavating cover material. Therefore, it appears feasible to use special compactors only in conjunction with other types of equipment such as rubber-tired or crawler-mounted excavators. Figure 10-10 shows such a compactor.

Where large landfills are constructed, with requirements for haul distances in excess of 1,000 ft, it becomes economically attractive to employ large scrapers for the transfer of cover material. In very large operations, the necessity of excavating and covering bigger areas with soil may make the use of draglines attractive. Draglines are most commonly employed in large landfilling operations where cover material is obtained from a borrow pit or where a significant amount of excavation is necessary (as in the trench method). Figures 10-11 and 10-12 show these apparatus.

a

Fig. 10-10 (*a*) Tractor-type spreader with special cleated wheels for fill compaction (*courtesy* John Deere Co.); (*b*) Trashmaster cleated wheel tractor spreading and compacting refuse (*courtesy* Rex Chainbelt, Inc.).

b

Fig. 10-10 Continued

At sanitary landfills, special hazards to the equipment exist because of the nature of the material being worked. Engine screens and radiators often are fouled by blowing paper or extraneous wire. Materials from the refuse pile may also find their way into the engines of working equipment. Finally, tracked vehicles may be rendered inoperable by the wedging of large solid constituents between the tracks and roller bearings in the crawler power train. In order to minimize such hazards to the operating equipment, several types of accessories are available for use with these machines. These include engine screen and radiator guards, reversible fans to blow paper off radiators, under-chassis guards to protect engines, and special protecting covers for hydraulic lines.

10-6.3 Equipment Difficulties

The principal difficulty in obtaining and specifying equipment for landfilling operations is in suiting the adopted vehicle to the landfill requirements for small sites where only a single machine will be purchased, or in cases where it is not feasible to purchase even a single standard machine. At very small sites (handling less than about 10 tons/day), other types of equipment such as farm tractors equipped with bulldozer blades may be used to spread the solid waste. However, a farm tractor is unsuitable for the efficient compaction of solid waste. Therefore, a larger volume of landfill will be required for the poorly compacted wastes.

a

Fig. 10-11 (*a*) Typical scraper for earth haulage (*courtesy* of WABCO); (*b*) cutting trench for landfill site. Earth removed will be used for cover of previously filled site (*courtesy* John Deere Co.).

214 SOLID WASTE MANAGEMENT

b

Fig. 10-11 Continued

Additional costs for the extra volume associated with poorly compacted wastes should be compared with the additional costs of purchasing and operating a larger piece of equipment capable of a higher degree of waste compaction. Table 10-4 lists typical equipment requirements for landfills.

10-6.4 Equipment Costs

The costs associated with owning and operating equipment for a sanitary landfill are significant. The capital costs for equipment are exceeded possibly by only those of buying the land for the fill. However, at present the market for sanitary fill equipment is competitive and many manufacturers will be vying for the equipment contract. In approximate cost, the initial investment associated with a medium-size crawler bulldozer is $30,000 to $35,000. This type of equipment will have a useful life of approximately 5 years or 10,000 working hours. In contrast, more specialized equipment, such as scrapers or draglines, has a much higher cost. A dragline will have a cost approaching $100,000 and a modern scraper will also approach this figure.

The cost of operating the landfill equipment will consist of both direct and indirect costs. The principal item in direct operating cost is fuel, which constitutes about 90 percent of direct operation expenses for landfilling equipment. This cost may be approximated on a national average as $3/hr.* The indirect costs associated with equipment operation consist principally of maintenance and repair work. The parts and labor required for maintenance and repairs may be estimated per hour of operation by dividing the total cost of the expected

a

Fig. 10-12 (*a*) and (*b*) Draglines preparing sites in wet soil areas (*courtesy* Koehring).

Sanitary Landfill Guidelines, USPHS, 1970.

b

Fig. 10-12 Continued

maintenance and repairs for a piece of equipment by the expected total operating hours associated with that equipment (about 10,000 hours).

10-7. USE OF THE COMPLETED SANITARY LANDFILL

10-7.1 General

Any projection for future use of a sanitary landfill site must be made with full recognition of the characteristics of the deposited material. For example, decomposition of the organic material in a fill may require a time period of from 2 to 20 years depending upon climatic conditions. In addition, the amount of decomposition will vary with the amount of organic material present. Therefore, the density of the final material in a fill will vary with the composition of the originally placed material, the climatic and topographic conditions at the site, and

Table 10-4 Average Equipment Requirements

Population	Daily Tonnage	No.	Type	Size (lb)	Accessory[a]
0 to 15,000	0 to 46	1	tractor crawler or rubber-tired	10,000 to 30,000	dozer blade landfill blade front-end loader (1- to 2-yd)
15,000 to 50,000	46 to 155	1	tractor crawler or rubber-tired	30,000 to 60,000	dozer blade landfill blade front-end loader (2- to 4-yd) multipurpose bucket
		*	scraper dragline water truck		
50,000 to 100,000	155 to 310	1 to 2	tractor crawler or rubber-tired	30,000 or more	dozer blade landfill blade front-end loader (2- to 5-yd) multipurpose bucket
		*	scraper dragline water truck		
100,000 or more	310 or more	2 or more	tractor crawler or rubber-tired	45,000 or more	dozer blade landfill blade front-end loader multipurpose bucket
		*	scraper dragline steel-wheel compactor road grader water truck		

[a]Optional, dependent on individual need.
Source: *Sanitary Landfill Facts*, USPHS, HEW, Publ. No. 1792.

the adequacy of the landfill design. With decomposition of the contained refuse, settlement of the landfill surface must occur. Recorded settlements of fill materials have varied from approximately 2 to 40 percent. A rule of thumb for estimating this is that fill settlements are likely to be at least 20 percent of the initial height of the fill.

When considering the completed landfill site as the possible location of a future structure, the problems with continued settlement of the site are only a few of the difficulties which must be met. The bearing capacity of the completed landfill structure will be quite variable from point to point within the area and in fact may not be relied upon to support any significant structural loading. Consequently, even very light structures founded upon a fill must be supported by some sort of foundation which transmits the load of the structure through the completed refuse cells down to and into the original soil or rock stratum underlying the site. Difficulties in constructing such a throughgoing foundation should be anticipated. For example, the gases generated during decomposition of the refuse in the fill cause hazards of gas collection in any constructed facilities and also will cause hazards of gas concentration during construction of foundation elements. Ample provision in the design of anticipated structures should be made for the prevention of gas permeation and collection. In addition, the leachate produced in a decomposing landfill will be excessively corrosive and therefore will attack any structural materials placed in contact with the refuse material. Hence any foundation elements such as piles or piers which are inserted through the refuse must be designed to withstand the corrosive effects of both gases and leachate.

10-7.2 Present Uses of Completed Sites

Because of the difficulties associated with construction of facilities on completed sanitary landfills, the primary uses for such fills to date in this country have been as green space, recreational areas, and locations for agriculture. The most common use of completed sanitary landfills is as green areas, particularly where such areas may be located close to metropolitan centers. A certain amount of maintenance will be required in order to keep the quality of the site constant as a valuable green area. Such maintenance includes periodic grading of the cover material so that ponding of water and erosion of cover material is prevented. In addition to these considerations, a special design should be formulated for the final cover material when a green area use is anticipated. This is necessary since, if the cover material is thin, only shallow-rooted vegetation will flourish. In addition, the accumulation of gases in the root zone of the cover vegetation may interfere with the normal growth of the plants and might in effect kill all vegetation. In order to overcome these obstacles a moist, fertile cover soil of sufficient thickness must be used in conjunction with a gas venting system. Choice of the vegetation to be used as cover in the area will depend upon the climate of the landfill site, the soil used as final cover material, and the allowable depth of the root system above the top of the final refuse cell. Any type of vegetation which requires irrigation should obviously be disregarded for planting on a completed fill surface.

Similar considerations must be kept in mind when a completed landfill site is to be used for agricultural purposes. Here too, gases generated in the decomposing refuse may interfere with the growth of plants in the overlying soil unless venting for these gases is provided. Deep-rooted crops should not be planted over a completed landfill site, since the root system may encounter decomposing solid wastes which are toxic to the plants. For example, a 9-ft-thick final cover would be required to prevent the penetration of wheat root systems into an underlying refuse layer, whereas a 6-ft-thick final cover layer would probably suffice for the prevention of contact between corn roots and refuse. In any event, if agricultural activities are established over the final refuse cell, the ultimate cover material layer should be sufficiently thick so that cultivation does not interfere with the bottom two feet of the final cover.

An additional popular use for completed landfill sites is for recreational facilities. Around the country in various localities completed landfills have been used as the site for ski slopes, toboggan runs, coasting hills, golf courses, baseball diamonds, football fields, parks, outdoor amphitheaters, and other recreational areas. The problems associated with use of the completed sanitary landfill as a recreation site are similar to those associated with the use of the site as a green area. Problems with the ponding of water or the erosion of cover material, as a result of settlement or poor final grades on the cover material, may occur. In addition, periodic maintenance is required for gas and leachate collection and monitoring facilities, temporary roadways, and other facilities which might be adversely affected by the continuing settlement associated with further decomposition of the waste in the landfill. Small structures associated with recreational use of the area may be constructed on sanitary landfills using the top layer of cover material as the bearing surface for the structures, provided that the overall loading of the bearing surface is very light. For example, sanitary facilities, equipment storage sheds, bench storage areas, and concession stands may be constructed on the upper surface of the final cover material in a completed landfill used for recreational purposes.

10-7.3 Structures on Completed Landfills

As mentioned previously, the construction of structures on completed sanitary landfill entails many difficulties, but in some localities a completed site will be strategically located and therefore the possibilities of building residences or manufacturing facilities on a completed fill may be attractive. If it is considered desirable and necessary to locate such facilities on the completed landfill, the design of the proposed structure must include careful consideration of the special features associated with the fill. These include continuing settlement and a low and erratically varying bearing capacity of the upper cover layer. The engineer engaged to plan and design the substructure for such a facility must be aware of these problems and must consider ways in which to exclude noisome gases and corrosive liquids from the foundation of the structure. Additionally, during construction, workmen may be faced with the hazards of explosion of methane or the hazard of hydrogen sulfide, a toxic gas produced during anaerobic decay of the landfill materials. The superstructure of a proposed facility must also be designed to accommodate the settlements associated with continuing decomposition of refuse. Differential movements between different points in a structure should be anticipated.

If it is known in advance that a particular landfill site, upon completion, will be used as the location of future structures, then some consideration should be given to segregating the deposition of waste during the operation of the sanitary landfill. In this way, inert, relatively dense materials can be placed in islands within the general site to furnish a stable nonsettling construction site for the proposed structures. Each island may be constructed of the soils found at the site or they may be constructed from selected solid fill materials such as construction and demolition wastes which are brought to the sanitary landfill for disposal. At other sites, when foundation engineers are faced with highly compressible soils, an expedient frequently followed is to excavate the compressible materials and backfill with nonsettling soils. Such a tactic employed at a landfill site would be expensive, unpleasant, and possibly hazardous because of the pollutants and contaminants associated with the decomposing wastes.

10-7.4 Inspection

Every completed landfill should be inspected thoroughly by the governmental agency responsible for its operation. Then for any future reference, a completed plan of the site with a detailed description should be recorded with the proper authority within the regional area where the site is located.

10-8. SUMMARY

To summarize this discussion of the use of sanitary landfill as a means for disposal of solid wastes the following statements may be made:
1. Sanitary landfilling is a complete disposal operation, whereas alternate disposal methods such as incineration and composting produce materials which require further treatment of non-compostable residue, noncombustible residue, finished compost, process waters, flue gases, etc.
2. If land is available within short distances of a waste-generation area, sanitary landfill will usually be the most economical disposal method, will have lower capital investments than other methods (particularly if land is leased), and can be put into operation almost immediately.
3. A sanitary landfill is capable of receiving variable amounts of many different types of wastes without the difficulties associated with variable loading of process operations such as composting or incineration.

4. In highly urbanized areas land suitable for landfill sites may not be available within economic haul distances and the alternative location of a landfill in an unsuitable area (such as a residential district) may as a result generate public animosity toward the landfill operation.
5. Because refuse decomposes in a sanitary landfill, constant vigilance is required to prevent environmental pollution during construction. Additionally, because of surface settlement, gas generation, and leachate production during decomposition of refuse in a completed fill, special design and construction practices must be followed in the utilization of the completed landfill site. Explosive concentrations of methane and toxic concentrations of hydrogen sulfide may be encountered during construction and during future use of a constructed facility. Generation of leachate will present a continuing potential problem of pollution of surface and groundwaters.
6. Land with little potential for use may be converted through a landfilling operation to a "green-space" area, a recreational facility, a farm site for limited agricultural activities, or a construction site for specially designed structures.

REFERENCES

General

10-1. American Public Works Assoc., *Municipal Refuse Disposal*, 2d ed., Chicago, Public Admin. Service, 1966.

10-2. Committee on Sanitary Engineering Research, "Survey of Sanitary Landfill Practices: Thirtieth Progress Report," *J. San. Engrg. Div.*, ASCE, Vol. 87, No. SA4 (July 1961), pp. 65–84.

10-3. Committee on Sanitary Landfill Practice of the Sanitary Engineering Div., "Sanitary Landfill," ASCE Manual of Engrg. Practice No. 39, 1959.

Sanitary Landfill Site Selection

10-4. California State Water Pollution Control Board, "Investigation of Leaching of a Sanitary Landfill," Publ. No. 10, Sacramento, 1954.

10-5. Qasim, S. R., "Chemical Characteristics of Seepage Water from Simulated Landfills," Ph.D. Dissertation, West Virginia University, Morgantown, 1965.

10-6. Anderson, J. R., and Dornbush, J. M., "Influence of Sanitary Landfill on Ground Water," Am. Water Works Assoc., Vol. 59, No. 4 (April, 1967), pp. 457–470.

10-7. Cook, H. A., Cromwell, D. L., and Wilson, J. A., "Microorganisms in

Household Refuse and Seepage Water from Sanitary Landfills," *Proc.*, West Virginia Acad. of Sci., Vol. 39, No. 107 (1967).

10-8. Anderson, D. R., "Gas Generation and Movement in Landfills," *Proc.*, Natl. Ind. Solid Waste Management Conference, University of Houston, Houston, 1969.

10-9. Remson, I., Fungaroli, A. S., and Alonzo, A. W., "Water Movement in an Unsaturated Sanitary Landfill," *Proc.*, ASCE (*J. San. Engrg. Div.*) Vol. 94, No. SA2, pp. 307–317, April, 1968.

10-10. Black, R. S., and Barnes, A. M., "Effect of Earth Cover on Fly Emergence from Sanitary Landfills," *Public Works*, Vol. 89, No. 2 (Feb. 1958), pp. 91–94.

Sanitary Landfill Design

10-11. California State Water Quality Control Board, "Waste Water Reclamation in Relation to Ground Water Pollution," Publ. No. 24, Sacramento, 1953.

10-12. Hughes, G. M., Landon, R. A., and Farvolden, R. N., "Hydrogeology of Solid Waste Disposal Sites in Northeastern Illinois," Interim Report, USPHS, HEW, Cincinnati, 1969.

10-13. *Pollution of Subsurface Water by Sanitary Landfill; a Summary on a Solid Waste Demonstration Grant Project*, USPHS, HEW, Cincinnati, 1968.

10-14. Anderson, D. R., Bishop, W. D., and Ludwig, H. F., "Percolation of Citrus Wastes Through Soil," *Proc.*, 21st Ind. Waste Conference, Purdue University, Lafayette, Ind., 1966.

11
Recent developments

11-1. GENERAL

During the past twenty years and particularly during the last decade many new methods, techniques, apparatus, and products have been developed for use in solid waste management. To describe in detail and evaluate all of these developments would require an amount of space equivalent to several books such as this one. Therefore, in this chapter only a few examples of particular developments which appear to offer significant promise will be presented. Some of the methods have been mentioned in Chapter 6; examples of model or pilot operations will now be discussed.

Included in this discussion are such items as (1) the Melt-Zit high-temperature incinerator; (2) power generation through incineration, CPU 400; (3) pyrolysis of organic wastes; (4) pyrolysis of used rubber tires; (5) new landfill stabilization methods; and (6) geo-ecological studies of dangers associated with landfill disposal of hazardous wastes.

11-2. HIGH-TEMPERATURE INCINERATION
11-2.1. General

A new development in the incinerator art has been the so-called "high-temperature incinerator" as typified by the Melt-Zit model of the American Design and Development Corporation. In this newly developed incinerator the noncombustible fractions of refuse are melted in a bed of high-temperature coke and are then drained from the furnace area as molten slag and iron. All of the organic material in the refuse is removed in this manner and the residue remaining after incineration is completely sterile, is inert, and possesses a high density. Figure 11-1 shows the features of the Melt-Zit pilot plant furnace built at Whitman, Massachusetts, as a demonstration model for individuals and communities considering high-temperature incineration. The components of the incinerator are the refractory-lined furnace (with option of several different diameters); the refuse charging conveyor; the coke elevator, and the limestone and air charging mechanisms. The limestone, coke, and refuse are all distributed in a random fashion by being dropped into the furnace from a conveyor; each constituent is charged separately into the furnace as required by furnace performance. Air is supplied by means of blowers, with the air entering at two levels by means of air ports spaced around the furnace (see Fig. 11-1). Air at the lower level is fed into the coke bed and produces the high temperatures required for the melting of metal and slag from the refuse. Air forced into the furnace above the coke bed burns the incoming refuse. The construction of the furnace is such that it induces a slight negative pressure at the input chute and as a result additional air is drawn through the charging opening. The upward rush of hot gases causes partial entrainment of the solids which are dumped from the conveyor. Therefore, the refuse is burned partly in suspension within the shaft of the furnace, and partly on the surface of the lower coke bed. Slag and metal resulting from the melting of the noncombustible portions of the refuse are drained from the bottom of the incinerator through a taphole. The taphole is located at the level of a sand hearth which underlies the coke bed. The drained slag and metal in a molten state flow out of the furnace into a water reservoir where it immediately congeals into a black granulate frit, and iron pellets. The furnace is put into operation in stages, as follows: (1) a sand bed is laid at the bottom of the furnace and is sloped toward the slag spout; (2) coke is laid over the sand hearth to a depth of 3-4 in.; (3) the coke is ignited by means of small propane torches which are inserted into the furnace through the lower air ports; (4) the coke is gradually brought to operating temperature by means of forced air supplied through the air ports; (5) congealed slag from a previous run is charged into the furnace to establish a flow of slag from the spout as an indication that the coke bed is at a sufficiently high operating temperature; (6) after the proper temperature has been developed in the coke bed, air is forced into the upper air ports over the

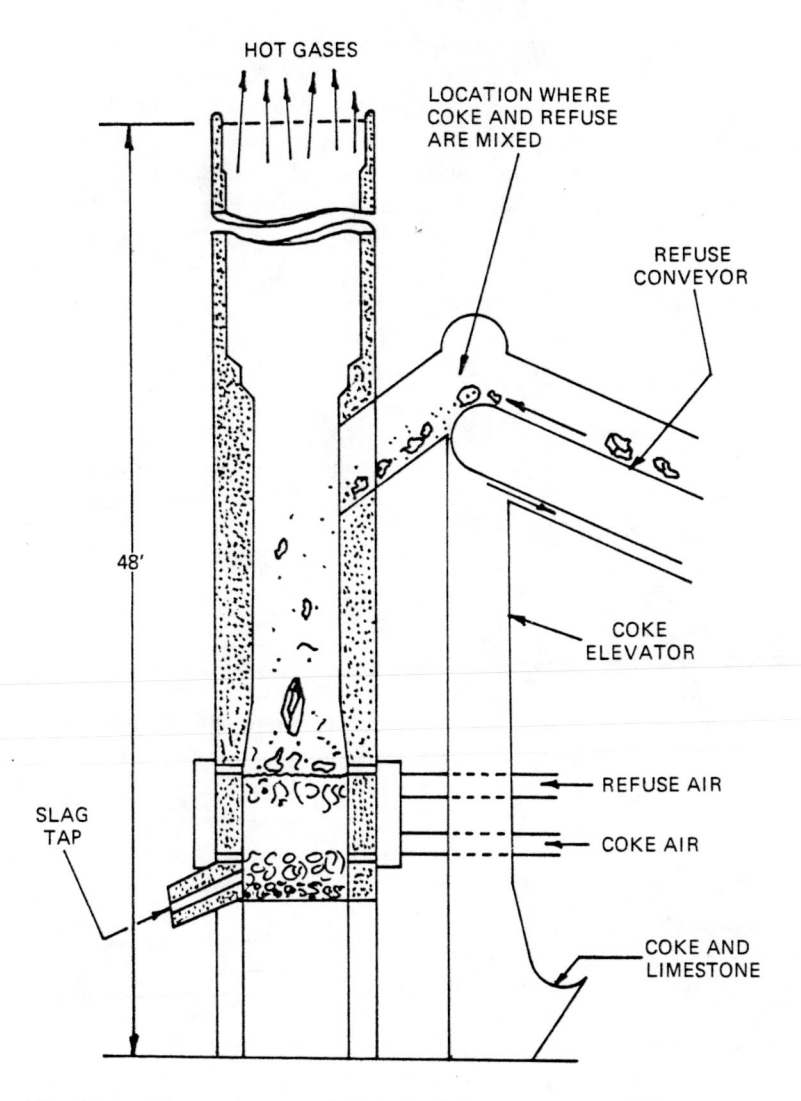

Fig. 11-1 Schematic view of Melt-Zit high-temperature incinerator.

coke bed and refuse combustion is begun; (7) the refuse charging is begun in small intermittent batches until the process has been well established, after which the rate of feed becomes continuous.

Coke is charged periodically into the furnace to maintain the desired thickness of 3-4 ft of the coke bed. The refuse feeding ordinarily is interrupted for approximately 5 minutes every half-hour, at which time the coke bed is charged and heated. When operating in the steady state, ignition and combustion of the refuse is aided by the high temperature of the refractory lining of the shaft. The flaming refuse in turn reheats the lining. When coke is charged into the underlying bed, the top of the coke bed ceases to radiate a large amount of heat, but radiation from the refractory lining reestablishes ignition promptly and the temperature of the coke bed is soon restored. The refuse layer is maintained on the coke at a thickness of only a few inches so that the flow of gases from the bed is not unduly interrupted. At the end of the incineration run, the bottom of the furnace is opened and the remaining coke, slag, and molten metal are dropped into a refuse container.

11-2.2. Evaluation of the Melt-Zit Process

In a critical evaluation of the Melt-Zit high-temperature incinerator by the USPHS (see Ref. 11-1), several process problem areas were outlined, including:

1. A low refuse firing rate. During optimum test operation, the refuse firing rate was approximately 1.6 tons/hr.
2. A relatively high consumption of coke. At least one ton of coke was required for every 6–8 tons of refuse incinerated.
3. An inability to drain molten slag and iron in order to maintain continuity of operation.
4. A consumption of limestone in the amount of 1 ton for each 38 tons of refuse. This consumption rate was necessary to promote fluidity of the slag.
5. Flyash, including combustibles and other ashes in the amounts of approximately 288 lb/hr (or more) were emitted from the furnace area itself; this amounted to approximately 10 percent of the refuse charged.
6. A high labor requirement was indicated to manage the furnace properly; at least three men were required per shift to tend the furnace.
7. A high consumption of oxygen and iron pipe lances were required to maintain the slag flow. In addition, the refractory lining in the furnace was attacked by the high temperatures in the slag ingredients.
8. The electric power consumption of the pilot plant was not indicative of the power consumption to be expected in full-scale operation.
9. The burning rate of the refuse was limited by the melting rate of the inert material and/or by the loss of solids as carry-over from the furnace.

Several distinct process advantages were also noted in this USPHS critique including:

1. A molten residue was received from the incinerator free of putrescible and combustible materials; thus, the residue could be deposited in landfill without cover, would not attract vermin or rodents, and would be odorless and sterile. The slag is in the form of a friable, black, glassy granulate if it is quenched in water. If it is cooled more slowly, the slag becomes a friable solid.
2. A high yield of iron was obtained from the refuse. This may be produced as

flattened or spherical drops if the molten slag is allowed to run into a water tank. During test runs about 20 percent of the residual material consisted of iron in this form and it could be easily separated magnetically from the remainder of the residue.

Based upon initial tests, burning rates for several sizes of incinerators would be as follows:

Furnace Diameter (ft)	Top Diameter (ft)	Tons of Refuse (per hr)
3	5	1.60
4	6.7	2.84
5	8.3	4.45
6	10.0	6.40
7	11.7	8.72

The Melt-Zit shaft furnace as it stands by itself is only the primary combustion chamber for a total incineration plant. A secondary chamber would be required for complete combustion of the carry-over solids which move rapidly up in the hot gas stream and which must be prevented from exiting from the incinerator. A dust collector would also be required after the secondary chamber in order to reduce the particulate load of the exhaust gases.

The technical advantages of high-temperature incineration are (1) a residue free of putrescible matter; (2) maximum density in subsequent landfill; (3) minimum pollution of groundwater and streams; (4) possible salvage or use of the residue; and (5) minimal use of water. A significant problem in high-temperature incineration is the high viscosity of glass and other silica-rich inert materials. Prior removal of glass as cullet or separate collection of glass would alleviate this problem. Other methods include: (1) preheating the coke-bed air so that a higher coke-bed temperature is attained which would reduce the glass slag viscosity, or (2) more effective mixing of limestone with glass which would also lower the slag viscosity. Preheating of the coke air would be helpful also in reducing the amount of coke used, since ample excess heat is available in the operation of this unit. A coke rate as low as one-twentieth of the refuse may be feasible in practice. However, any decrease in coke consumption will be accompanied by higher proportions of metals, inerts, and limestone in the coke bed, and consequently lower temperatures there.

11-2.3. Economics

The economics of this system may be compared with those of conventional municipal incinerators. A high-temperature incineration plant probably requires the same land area, buildings, roadways, handling equipment, secondary combustion chamber, and other appurtenances as would a conventional incinerator.

In addition to these facilities, other units for receiving, storing, and conveying coke and limestone would be required. The maximum size currently projected for high-temperature incinerators is approximately 360 tons/day, which is roughly equivalent to conventional rectangular furnaces of 500 sq ft of grate area. Because of its tall cylindrical construction and similar volumetric combustion rate, the high-temperature furnace will probably have a refractory wall area and cost comparable to conventional furnaces. The cost of the stoker and hydraulic feed drive would be saved in the high-temperature apparatus. Only small costs would be incurred in a granulator and conveyor for the frit, which is recovered from the quenching tank. If continuous casting operations for large pieces of slag and metal were undertaken the additional expense of such apparatus could be met through sale of the end products. An additional savings for the high-temperature incinerator would be the much smaller water requirements as opposed to conventional incinerators.

Capital costs for municipal incinerators are currently in the range of $7500 to $10,000 per ton/day of capacity, if the incinerator is assumed to operate on a 24-hr schedule. Present estimates for costs of high-temperature incinerators indicate that the cost for a comparable plant for the same site would be 90-100 percent of that of a conventional incinerator.

In respect to operating expenses, little savings are anticipated in operating a high-temperature incinerator. In addition, receiving, storing, and conveying coke and limestone would require additional manpower over and above the requirements for the conventional apparatus. Finally, the high-temperature incinerator incurs the cost of coke and limestone, which are not common to other incinerators. These added costs must be supported and justified through reduced costs of residue disposal or sales of the final end product.

11-2.4 Other Significant Factors

Disposal of the final residue of this high-temperature process is much easier than disposal of conventional incinerator residue, since there is a complete absence of combustible and putrescible material; thus, the residue could be deposited anywhere without a daily cover of soil. Reuse or resale of the other residual matter from the high-temperature incinerator, the metal pellets and the slag and frit, at present is a highly questionable venture. Some profit may be realized from sale of the metal but the resale market for this is uncertain. The frit produced during high-temperature incineration is friable and crushes readily underfoot. Its strength is much lower than that of sand or stone and the material could not be used in structural functions. In comparison with conventional incinerators, however, the Melt-Zit process is very economical as to water usage. The residue quench water absorbs approximately 300,000 Btus/ton of refuse initially inserted into the furnace; if residue heat were converted by allowing the quench water to boil, the water consumption would be only 36–48 gal of water/

ton of incoming refuse. Even if boiling were prevented, the water consumption would be only slightly more. Water use in the cleaning of flue gases would be slightly more than that for a conventional plant of the same tonnage because of the additional heat from the coke. However, the runoff water from the flue gas scrubbers could also be used for the quenching operation. Little or no process water would be discharged to sewers or streams. For comparison, the total water consumption for a conventional incinerator would be approximately 2,000 gal/ton.

As a result of preliminary tests and evaluations, the pilot-plant unit was modified in 1969 to use gas or oil as auxiliary fuel to replace coke. Additionally, the primary combustion zone was enlarged and a steam injection system was installed to replace the original air pollution control equipment.

11-3. POWER GENERATION BY INCINERATION OF SOLID WASTE, CPU-400

11-3.1. General

In recent years two factors have combined to make the possibility of electric power generation through incineration of solid waste a proposition attractive to many people. These factors are (1) the increasing demand for power as a result of increased automation and use of air conditioning appliances, and (2) the increasingly high paper content of solid wastes, which gives the waste materials a high caloric value and therefore causes them to have great potential as an energy source. One of the most promising experimental energy recovery programs to date has been that conducted by the Combustion Power Company, Inc., in which a subscale model was developed to study the technological and economic feasibility of waste heat utilization through gas turbine generation of electrical power. The turboelectric generator-powered plant which will be developed in this program is called the Combustion Power Unit-400 (CPU-400) and is designed to process 400 tons of refuse per day and produce up to 15,000 kw of electric power from that refuse. Although the major claim for the CPU-400 is that it is a pollution-free disposal method, the potential for power production in the plant cannot be overlooked. Initial cost estimates, including a reduction in expenses through the sale of produced power, indicate that the CPU-400 may operate at costs below comparative ones for standard incinerators and may actually be competitive with landfilling operations (see Ref. 11-2).

11-3.2 CPU-400 Operation

The basic configuration of the CPU-400 pilot plant is shown in Fig. 11-2. Collection trucks dump wastes in the enclosed receiving area, shown in the upper right of the system schematic. The solid wastes are pushed from the receiving

230 SOLID WASTE MANAGEMENT

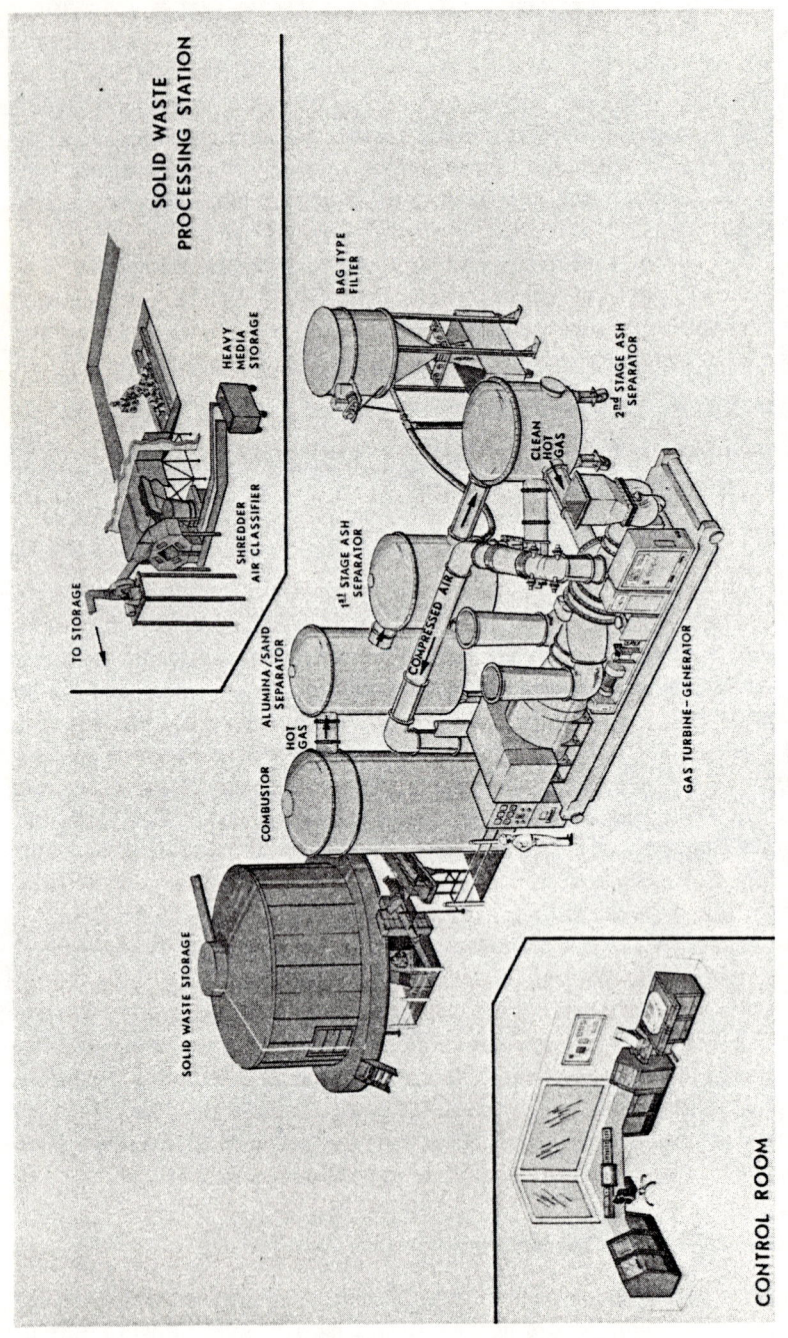

Fig. 11-2 CPU-400 pilot plant (*courtesy* Office of Solid Waste Management).

area by front-end loaders onto conveyors which feed material shredders. After shredding, the processed solid wastes are air-classified; heavy materials such as metals and glass are removed to temporary storage. The heavy media may be further separated by magnetic and optical separators into secondary fractions for recycling. The lighter materials travel from the air classifier to the storage carousel. This unit is designed to supply a constant stream of combustibles to the combustor. The shredded wastes enter the fluidized-bed combustor through high-pressure air locks. In the combustor, inert sand-sized particles are held in suspension as the fluidized bed by air flowing up through the combustor from the compressor of the gas turbine. The airflow velocity is maintained at the proper value to just support the particles without blowing them out of the bed. The fluidized bed heats the incoming refuse to ignition temperature and holds the wastes in place sufficiently long for complete combustion of all burnable waste constituents. Because of this complete combustion, the production of air pollutants such as carbon monoxide and unburned hydrocarbons is practically eliminated. However, particulate matter is entrained in the exhaust gas stream from the fluidized-bed combustor. This particulate matter is a serious air pollutant and also poses an erosion threat to the gas turbine. Consequently, a series of collector devices (alumina/sand separator, ash separators, bag filter) are used to remove first any of the inert bed material and then any particulates originating from burning of the refuse. The cleaned hot gases then enter the turbine-generator unit. Several turbine models are commercially available from a number of manufacturers in size ranges suitable for the CPU-400. In the pilot plant constructed at Menlo Park, California, the gases pass from the turbine unit through exhaust vents and are wasted to the atmosphere. In full-scale operations these gases could be passed through a secondary steam-generation heat-recovery unit for more complete utilization of the heat value of the combusted wastes. This waste heat could be used to preheat the incoming solid wastes, to desalinize seawater, or to dry and incinerate sewage sludge.

The principal advantages claimed for the CPU-400, as mentioned previously, are the recovery of the heat value of solid wastes through generation of electricity, steam, etc., and the pollution-free disposal of waste materials. The fluid bed insures complete combustion and eliminates unburned gases such as carbon monoxide and complex hydrocarbons. The low temperature of the bed (1500 to 1650°F) precludes the formation of nitrous oxides. Furthermore, the inert components of the fluid bed may be selected to control the formation of sulfur oxides and hydrogen halides, including HCl. Finally, particulate matter is collected, as discussed above.

11-3.3 Economics

Tables 11-1 and 11-2 show *the manufacturer's claims* for the estimated performance economics of the CPU-400. The costs and income breakdown shown

TABLE 11-1

CPU-400—Electricity Production Only

BASIS: 3 power modules burning 110 ton/day each 24 hr/day, 7 day/wk, each generating 2250 kW of electricity; "as received" solid waste includes solid waste burned plus 15% air classifier fallout plus 5% water loss on shredding

"As received" solids waste = (110 ton/day) (3)/0.80 = 390 ton/day @85% utilization (310 day/yr) = 121,000 ton/yr

COSTS

Capital costs = $6 million
Annual capital cost @6%, 20 yr = $522,000
Unit capital cost @121,000 ton/yr = $4.30/ton

Operating cost
  Labor—16 men (4 shifts) @$13,250/yr = $212,000
    Payroll extras @25% = 53,000
  Maintenance = 252,000
  Utilities (900 kW @$0.010/kWh) = 76,700
  Total annual operating cost = $593,700
  Unit operating cost @121,000 ton/yr = $4.90/ton

Total cost = $9.20/ton

INCOME

Electrical: 5.02×10^7 kWh @$0.007/kWh = $352,000/yr
Unit income @121,000 ton/yr = $2.91/ton

Net disposal cost = $6.29/ton

Source: Combustion Power Company

---

in these tables indicate that the use of the CPU-400 plant could produce significant savings in the operation of waste disposal. Field tests on one-tenth scale models began in 1972. An air classification system has been installed on the scale models currently being tested. It is anticipated that income from the sale of the recovered noncombustible items (shown in Table 11-2) will more than offset the initial capital and operating costs associated with the air-classification system.

In summary, it appears that the generation of electrical power by means of controlled solid waste incineration is technically feasible and potentially economically attractive. The future of the Combustion Power Unit-400 looks bright, providing that the final full-scale equipment performs as well as has model equipment. *If the equipment performs according to the manufacturer's claims*, this type of operation should be very economical and possibly even profitable for many municipal waste disposal agencies.

TABLE 11-2

Material Recovery System (without shredding costs)

BASIS: Cost of shredding *not* included, i.e., the MRS system is added onto a system that includes shredding

COSTS

Capital cost = $600,000
Annual capital cost @6%, 20 yr life = $52,000
Operating costs

Labor—4 men (1 per shift) @$10,000/yr =	$40,000	
Payroll extras @25% =	10,000	
Maintenance =	20,000	
Total cost =	$122,200	
Total unit cost @121,000 ton/yr		= $1.01/ton
Total unit cost @186,000 ton/yr		= $0.66/ton

INCOME

Ferrous @5% of "as received" solid waste @ $15/ton =	$0.75
Aluminum @0.5% "as received" solid waste @ $200/ton =	1.00
Glass, sand, etc., @15% "as received" solid waste @ $3.00/ton =	0.45
Other metals @0.25% "as received" solid waste @ $300/ton =	0.75
Total income =	$2.95/ton
Net *income* (186,000 ton/yr.) = $2.95 - 0.66 =	$2.29/ton
(121,000 ton/yr) = $2.95 - 1.01 =	$1.94/ton

For ferrous recovery only assume:
costs = $0.25/ton
income = $0.75/ton
net income = $0.50/ton of "as received" solid waste

Source: Combustion Power Company

11-4. PYROLYSIS OF SOLID WASTES

11-4.1 General

Pyrolysis has been mentioned in Chapter 6 in conjunction with resource recovery. Two examples of pyrolysis operations will be described in the next two sections (see Refs. 11-3 and 11-4).

11-4.2 Pyrolysis of Organic Wastes

Considerable study has been devoted to the idea of utilizing the organic content of some solid wastes by pyrolyzing the high-molecular-weight organic compounds in the solid wastes into simpler compounds which can be utilized in industrial applications. Major waste components such as leaves, dried sewage sludge, and paper have been studied. The conversion of the organic molecules in the solid wastes is accomplished essentially by a process of partial combustion. For example, in one series of tests finely ground wastes were inserted into an air/nitrogen stream in a fluidized bed (see Ref. 11-3). The air/nitrogen streams were controlled to contain less oxygen than that required for complete combustion of all of the organic materials present. The gaseous products of the partial combustion reaction can be condensed and collected in a series of traps maintained at progressively lower temperatures. Tests have been performed using temperatures ranging from 250 to 1000°C and with air/nitrogen mixtures varying from 100 percent nitrogen to 0 percent nitrogen. The combustion products obtained in this operation included water, acetic acid, formic acid, formaldehyde, methanol, acetone, toluene, and many other hydrocarbon derivatives. Gases produced included methane, carbon dioxide, carbon monoxide, propylene, ethylene ammonia, ammonium carbonate, and hydrogen. In general, the tests wherein high nitrogen-to-air ratios were used favored the formation of hydrocarbons. On the other hand, where low nitrogen-to-air ratios were maintained, more highly oxygenated compounds were formed. Many of the same products were formed in all of the tests made, but the amounts of each product varied with the ratio.

11-4.3 Test Results

The conclusions drawn by the investigators in this particular testing program were to the effect that a considerable potential exists for the conversion of organics contained in solid waste to lower-molecular-weight compounds which have significant dollar values. In the tests, paper, dried sewage sludge, and dried leaves were all processed effectively. Four broad classes of partial oxidation products were obtained: (1) tars, (2) an aqueous mixture, (3) organic materials, and (4) a mixture of gases. In the words of the investigators, "Tars represented a relatively small portion of the total products formed, and decreased with increasing temperature. Products contained in the tars were not identified. The aqueous fraction was largely made up of water plus water-soluble organics. The organic fraction contained a complex mixture of at least 35 materials including dissolved gases. The gaseous products were materials not condensed at the dry ice-acetone bath temperatures."[1]

A summary of the products identified in each test where various materials were partially oxidized is shown in the table on the next page.

11-4.4 Pyrolysis of Used Tires

Rubber tires present a very significant solid waste problem. For example, about 215 million rubber tires were produced and shipped from rubber processing

---

[1] "Partial Oxidation of Solid Organic Wastes," final report to USPHS on research grant EC-00263, Cincinnati, 1970.

TABLE 11-3 Pyrolysis of Organic Wastes

Material Used	Paper	Sewage Sludge	Leaves
Products	Water	Water	Water
	Acetic acid	Methanol	CO_2
	Formic acid	Acetone	CO
	Formaldehyde	Toluene	Formaldehyde
	Methanol	A-nitrile	Methanol
	Methane	Propane	Acetone
	CO_2	Methane	Acetaldehyde
	CO	CO_2	Methyl acetate
	Propylene	CO	Ethyl vinyl ether
	Ethylene	Ethylene	
	Hydrogen	Ammonia	
		Ammonium Carbonate	

Source: Ref. 11-3.

plants in the United States in 1968. Over 5½ billion lb of rubber were used in those tires, most of which were eventually discarded in scrap heaps as worn out (see Ref. 11-4). This very large quantity of hydrocarbon-rich waste has been neglected for many years by the rubber processing industries. However, in recent times some projected recovery of hydrocarbons from used tires has been attempted, by use of a pyrolysis operation, by the Firestone Tire and Rubber Company. The initial attempts at developing a pyrolytic technique consisted of laboratory experimentation during which the feasibility of this destructive distillation technique was demonstrated (Ref. 11-4).

11-4.5 Test Results

In conjunction with the U.S. Bureau of Mines Coal Research Center at Pittsburgh, Pennsylvania, Firestone developed a pilot plant at the Research Center.

The test apparatus located there consisted of a hermetically sealed retort heated to the carbonization temperature of rubber in an electric furnace. The volatile effluents from the retort were passed into a series of condensors to selectively separate heavy process oils into crude fractions. Lighter components then passed through an electrostatic precipitator to remove any particulate matter entrained during carbonization, and through scrubbers where basic and acidic components from the volatile hydrocarbons and gases were removed. The gases were dried and a "dry ice"-acetone trap was used to collect the light oils. Noncondensable gases were passed through a metering system and retained in a holding tank where some of the gases were analyzed. The remainder of the noncondensable gases were burned.

An assortment of raw materials were tested. Shredded whole passenger tires, containing both rubber and fabric, were examined as well as ground rubber containing no fabric. In addition, samples containing bead wire were tested. Yields of liquid, solid, and gaseous products are shown in Table 11-4. Each test, which

TABLE 11-4 Pyrolysis of Used Tires

Test	Temp. °C	Residue[a]	Yields Heavy[a] Iol	Light[a] Oil	Gas[a]	Total[a]
pt-1	500	42.0	45.2	4.2	5.0	96.7
pt-3	900	52.3	14.0	6.5	20.8	97.3
pt-4	500–564	41.8	41.3	4.3	5.5	97.4
pt-2	500	44.8	45.6	1.7	3.2	95.6
pt-5	500	40.3	44.6	3.3	4.8	98.0
pt-6	500–900	42.1	42.7	5.5	7.2	100.0
tt-3	500	36.5	48.7	4.3	5.0	95.7

[a]Weight percent
Source: Ref. 11-4

required from 7 to 14 hr, was conducted on approximately 100 lb of raw material. Examination of the test results indicated several trends: (1) increases in carbonization temperature produced more solid residue, more gas, but less oil; (2) inclusion of beads in the charge had little effect on the outcome (the wire was stripped clean of all organic material and was easily separated from the residue); (3) using fabric-free rubber changed the product distribution very little; (4) tests pt-5 and pt-6 were conducted using much larger pieces of whole tires— 3 in. squares in test pt-5 and one-fourth passenger tires in pt-6—the only noticeable effect being a substantial increase in the rate of carbonization, probably because of better heat transfer in the larger pieces; (5) when shredded truck tires were carbonized, a significant decrease in the yields of solid residue was noted, and the yield of heavy oil increased, possibly because of the higher natural rubber content in truck tires which produced more volatile products. Table 11-5 presents some of the properties of the solid residues. An interesting point is that most of the sulfur was nonvolatile and remained in the residue during carbonization—the sulfur content of the residues being approximately double the sulfur content of the materials carbonized. Table 11-6 presents some of the data obtained from analysis of the heavy oils produced. The technical feasibility of pyrolysis of scrap tires has been demonstrated. The products so created can be expected to find application in a number of areas, since roughly 45 percent of the products are solid carbonaceous residues. However, process economics remain uncertain.

TABLE 11-5 Solid Residue

Test	% Fixed Carbon	% Ash	% Sulfur	Heating Value BTU/16
pt-1	86.3	9.6	2.0	13,470
pt-3	90.5	8.3	1.7	13,500
pt-4	87.7	9.2	1.9	13,490
pt-2	74.8	10.2	2.3	13,720
pt-5	83.1	12.7	2.0	13,020
pt-6	85.4	13.6	2.5	12,640
tt-3	80.1	16.5	3.2	12,250

Source: Ref. 11-4

TABLE 11-6 Heavy Oils

Test	% Residue Distill. @350°C, 14.7 psia	Volume; %; Neutral Oil Olefins	Aromatics	Paraffins and Nephthenes
pt-1	39.6	15.5	51.5	33.0
pt-3	30.1	13.0	84.8	2.2
pt-4	35.5	17.5	51.6	30.9
pt-2	39.8	16.0	54.2	29.8
pt-5	30.4	19.0	50.6	30.4
pt-6	35.2	12.0	56.8	31.2
tt-3	27.6	25.0	51.4	23.6

Source: Ref. 11-4

11-5. LANDFILL STABILIZATION TECHNIQUES

11-5.1 General

One of the principal difficulties in adapting the method of sanitary landfilling to refuse disposal is the fact that the landfill site may not easily be used until the solid wastes have decomposed within the refuse cells, the production of gases has ceased, and the settlement of the fill surface has stabilized. In order to reduce the difficulties associated with the long decomposition times in sanitary landfills, several techniques have been proposed for acceleration of the biodegradation of solid wastes. These techniques include incineration of the wastes within fill cells, forced aeration of the refuse within the cells to promote rapid aerobic degradation, and the recycling of leachate produced by abundant watering of the landfill surface to create, in effect, a solid waste treatment plant (see Refs. 11-5, 11-6, 11-7, 11-8).

11-5.2 Tests of Landfill Incineration

In a series of controlled landfilling operations in the state of California, the Ralph Stone Company, under contract to the Office of Solid Waste Management Programs, has tested the technical feasibility of placing refuse in conventional sanitary landfill cells, and then initiating drying of the refuse by means of a heating element placed within them, further heating the refuse to the ignition point, and allowing a contained combustion to take place. In a series of preliminary tests the operation required a period of more than one week for the drying and combustion of the refuse in a typical cell. Drying of the refuse to the ignition point required several days, and after ignition the primary combustion of most of the refuse required an additional period of several days. Finally, large, bulky slow-burning items such as sizable pieces of wood or logs burned for one or two days after the major portion of the refuse had been consumed. On the basis of preliminary tests, the investigating personnel concluded that such means of stabilization of a landfill is technically feasible. The economics of such stabilization have not been proven to date.

11-5.3 Forced Aeration

In an accompanying investigation, also performed by the Ralph Stone Company, the concept of forced aeration of landfill cells by means of a buried piping system was field-tested. In this method, excess air was supplied to the refuse within a landfill cell and the waste gases produced during decomposition were removed so that essentially aerobic decomposition took place within the fill. The maintenance of aerobic degradation within the fill cells accelerated the decomposition of the refuse and did not produce the explosive or toxic gases associated with anaerobic decay. As in the case of subsurface incineration, the investigating personnel concluded that aerobic degradation of wastes within a landfill by means of forced aeration is technically feasible.

11-5.4 Recycling Leachate

A third innovation in solid waste disposal in landfills has been proposed recently. This method involves the recirculation of leached fluids through the decomposing refuse cells in the fill. Here, the biodegradation of wastes within a fill would be accelerated by increasing the average moisture content of the contained refuse to a value of from 50 percent to 60 percent so that the activity of the degrading microorganisms would be maximized. Since the field capacity of refuse in landfill operations is between 30 and 55 percent moisture content, the provision for such a high moisture content within the fill cells obviously indicates that significant amounts of leachate will be produced, as these excess waters percolate through the contained refuse. The leachate would be contained by means of an impermeable barrier such as a compacted clay stratum and would be collected at that barrier. Collected leachate, high in organic materials and microorganisms, is then recycled into the refuse cells to promote and further accelerate the degradation processes. At the present time, no field-testing of this technique has been performed.

However, a comprehensive laboratory study of this technique was begun in 1971 at the Georgia Institute of Technology under the direction of Dr. Fred Pohland. Also, in Sonoma County, California, the U.S. Environmental Protection Agency in 1972 instituted a field study of leachate recirculation in six study landfill cells. No data are currently available from the EPA study, and no other investigations of leachate recirculation have been reported. However, the study results published by Pohland in a first annual progress report give extensive information about the process (Ref. 11-6).

In the Georgia Tech study, four cylindrical vessels were built and charged with refuse to simulate landfills; two were filled in spring 1971 and two in spring 1972. In one of the two yearly study containers, leachate was recirculated. At the initiation of the study 250 gal of tap water were added to the deposited refuse in each container; approximately 10 ft of compacted refuse weighing 2800 lb were placed in each test cylinder. The leachate produced from one test fill was collected, samples were taken, and the remaining leachate was recirculated into the refuse. Recirculation was not performed in the other test cylinder.

Leachate samples collected from the two "1971" fills were analyzed for significant parameters; the results of the tests on the samples taken are shown in Tables 11-7 and 11-8. These are based on data forwarded to EPA by Dr. Pohland in July 1972. The test results and the general conditions of the test cylinders may be summarized as follows:

1) Leachate recirculation produced more rapid stabilization of refuse than did lack of leachate recycle because of the accelerated growth of an anaerobic biological population.

2) Pollutants were removed from the refuse at a more rapid rate in the "recycle" test cell, but the eventual concentration of most organic and inorganic pollutants was lower in the recirculated leachate as compared to the "one-pass" leachate.

3) Recirculating leachate provided effective biological treatment, this treatment being more rapid and predictable than the degradation in a conventional sanitary landfill.

4) Leachate recirculation reduced and controlled the total amount of pollutants entering the environment; it should be effective in improving landfill disposal and in returning land used for fill sites to other uses in shorter periods of time than that presently required for conventional fill operations.

The authors have developed and designed a full-scale system for leachate recir-

TABLE 11-7 Concentrations of Extracted Materials in Leachates Obtained from Control Landfill

Time Since Leachate Production Began, days	0	14	24	32	39	48	81	116	125	153	173	189	197	228	249	284	312
COD, mg/l	4,320	9,150	10,380	10,260	12,000	11,700	9,200	10,100	11,700	12,200	12,300	14,400	15,600	18,100	15,600	13,300	13,800
BOD_5 mg/l	2,500	5,000	9,200	6,330	11,000	8,200	8,800	9,600	8,700	11,100	9,200	12,000	9,300	13,400	12,600	9,560	8,800
TOC, mg/l	1,230	1,910	2,622	2,622	2,802	2,835	2,864	2,259	2,418	2,680	2,696	3,049	3,409	5,000	3,590	3,000	2,930
TSS, mg/l	125	34	59	61	47	213	270	640	550	292	470	360	175	85	175	605	610
VSS, mg/l	45	20	47	52	37.6	93	160	332	314	182	268	210	104	76	141	283	286
TS, mg/l	2,442	5,819	6,323	8,300	8,736	6,789	5,530	7,250	7,358	7,620	7,875	8,320	8,130	12,500	8,780	7,716	7,167
Total alkalinity, mg/l as $CaCO_3$	558	1,610	1,640	1,920	2,280	2,110	2,420	2,650	2,120	2,350	2,100	2,482	1,760	2,480	1,580	2,430	1,930
Total acidity, mg/l as $CaCO_3$	690	1,100	1,350	1,400	1,780	2,170	1,836	1,390	2,090	2,230	2,780	2,865	3,260	3,460	2,610	2,000	2,400
pH	5.2	5.6	5.3	5.3	5.3	5.3	5.7	5.3	5.2	5.3	5.1	5.2	5.1	5.1	5.2	5.2	5.3
Total hardness, mg/l as $CaCO_3$	450	1,400	1,850	1,810	1,940	1,754	1,410	1,429	1,694	2,232	2,354	2,306	2,449	5,555	3,463	2,424	2,299
Acetic acid, mg/l	500	2,111	2,360	2,664	3,666	3,268	2,789	3,285	2,590	3,280	3,440	3,393	3,550	5,160	3,754	3,460	2,830
Propionic acid, mg/l	369	1,595	1,834	2,038	2,313	2,108	1,875	2,625	2,110	2,290	2,190	2,400	2,214	2,840	1,742	1,640	1,580
Butyric acid, mg/l	110	965	1,075	1,050	1,280	1,164	1,000	1,203	1,424	1,195	1,215	1,350	1,750	1,830	1,770	1,800	1,740
Valeric acid, mg/l	Nil	425	575	625	535	612	643	893	656	708	652	730	801	1,000	705	750	768
Phosphate, mg/l PO_4	26	3.0	5.0	7.8	2.8	2.9	3.3	4.2	3.4	2.8	1.7	1.6	1.5	1.3	1.5	0.9	1.1
Organic nitrogen, mg/l as N	56	47	61.4	62	75	48	40	177	64	6	20	12	43	107	116	76	63
Ammonia nitrogen, mg/l as N	56	150	167.6	187	185	192	148	103	130	260	214	218	264	117	52	110	103
Nitrate nitrogen, mg/l NO_3^-	13.3	32	89	84	115	15.0	—	9.5	12	—	—	—	—	—	—	—	—
Chloride, mg/l	322	385	109.8	105.1	97.9	340	—	170	240	210	208	312	308	180	300	280	295
Sulfate, mg/l $SO_4^=$	84	126	108	81	156	17	2	7	1	16	—	—	—	—	—	—	—
Calcium, mg/l Ca	125	430	470	590	750	545	430	375	420	600	578	565	545	1,250	850	550	490
Magnesium, mg/l Mg	26	71.8	67	75	68	64	52	49	53	80	85	85	75	260	210	90	65
Manganese, mg/l Mn	3	10	5	6.2	8.8	8.5	10	7.5	10	16	14	15	16	18	19	12	12
Sodium, mg/l Na	63.8	125	132	132	143	150	180	118	135	155	154	155	148	160	140	85	140
Iron, mg/l Fe	9	21	70	30	95	65	60	155	230	200	300	290	420	185	250	370	440

Source: Ref. 11-6.

TABLE 11-8 Concentrations of Extracted Materials in Leachates Obtained from Recirculating Landfill

Time Since Leachate Production Began, days	0	10	18	24	31	39	48	58	67	96	111	126	140	161	189	197	219	228	249	284	284	312
COD, mg/l	4,280	9,288	8,870	9,080	8,111	7,700	8,140	9,580	10,400	10,025	10,500	10,500	10,350	8,890	5,810	4,270	3,550	2,970	2,840	2,580	1,950	1,280
BOD$_5$, mg/l	2,750	5,200	6,900	6,800	4,300	5,400	6,202	6,400	6,380	7,200	8,700	8,500	10,100	9,405	6,650	3,500	2,860	1,400	2,500	2,420	760	760
TOC, mg/l	2,130	1,120	2,260	2,040	2,394	1,818	2,665	2,000	2,675	2,798	1,990	1,979	1,952	1,542	1,280	1,067	914	710	565	500	308	256
TSS, mg/l	93	13.6	12	36.5	70.5	25	37.0	120	301	143	222	258	385	187	232	220	131	122	145	124	67	305
VSS, mg/l	22.5	—	9	27.5	45	18.8	16.9	70	161	78	158	142	188	118	156	116	76	74	87	56	37	18
TS, mg/l	2,349	4,329	4,552	5,023	5,400	4,728	4,941	5,250	5,440	5,980	5,830	6,918	6,106	5,336	4,090	3,987	3,240	2,792	2,370	2,510	1,848	1,627
Total alkalinity, mg/l as CaCO$_3$	302	700	865	1,080	1,200	1,370	1,525	1,438	1,035	1,900	2,350	1,640	1,670	1,640	1,550	1,342	1,115	952	980	925	738	692
Total acidity, mg/l as CaCO$_3$	554	1,900	1,540	1,350	1,000	1,390	1,265	1,530	1,765	1,798	1,730	1,830	1,700	1,630	500	333	240	180	166	133	84	80
pH	5.05	4.8	5.0	5.1	5.3	5.4	5.3	5.3	5.1	5.4	5.5	5.3	5.3	5.2	6.3	6.6	6.8	6.9	7.0	7.1	7.4	7.3
Total hardness, mg/l as CaCO$_3$	370	895	880	1,010	890	1,040	1,222	1,483	1,532	1,701	1,987	1,495	2,296	1,948	1,469	1,146	978	677	539	661	513	377
Acetic acid, mg/l	1,638	556	2,000	1,843	1,475	1,583	1,795	2,146	2,438	2,742	2,438	2,470	2,380	1,877	2,925	608	734	770	670	111	234	365
Propionic acid, mg/l	960	394	1,242	1,467	1,554	1,594	1,580	1,752	1,953	2,203	1,953	1,865	2,020	1,472	1,995	714	195	111	104	57	223	110
Butyric acid, mg/l	1,300	235	1,235	1,163	1,375	1,250	1,200	1,198	1,094	1,156	1,047	1,124	937	735	665	286	194	68	65	Nil	62	44
Valeric acid, mg/l	500	735	50	833	688	670	714	800	858	857	786	842	625	556	585	276	87	65	50	Nil	35	Nil
Phosphate, mg/l PO$_4$	22	1.5	2.1	0.65	0.81	0.67	0.82	0.85	0.98	0.65	0.38	0.50	0.39	0.82	0.47	0.26	0.24	0.07	0.08	0.09	0.12	0.09
Organic nitrogen, mg/l as N	20	0	30	405	37.5	39.5	41	30	39	62	92	28	7	3	4	Nil	Nil	1	3	2	1	7
Ammonia nitrogen, mg/l as N	70	68	113.5	86.5	77.5	76.5	64	69	81	84	80	71	135	126	80	62	56	39	31	35	27	13
Nitrate nitrogen, mg/l NO$_3^-$	6.2	71.4	56.6	76.6	48	49	11.0	11.5	12.0	16.0	21.0	14.0	—	—	—	—	—	—	—	—	—	—
Chloride, mg/l	210	210	248	94.5	91	115	220	164	176	140	188	170	210	236	300	270	260	248	224	220	218	202
Sulfate, mg/l SO$_4$ =	102	138	81	51	30	12	11	Nil	12	2	1	3	—	—	—	—	—	—	—	—	—	—
Calcium, mg/l Ca	60	315	350	435	420	430	420	415	440	500	550	385	600	475	400	340	290	190	145	175	135	82
Magnesium, mg/l Mg	16.5	59	53.5	62.5	56	56	50	50	53	55	62	44	70	60	50	45	40	40	38	40	35	38
Manganese, mg/l Mn	4	30	50	65	62	62	75	75	80	80	85	60	93	80	59	50	44	19	10	19	14	8
Sodium, mg/l Na	61.5	109	81.4	91.4	85	84	95	85	88	90	98	70	84	75	61	59	50	60	55	60	55	75
Iron, mg/l Fe	4.4	19.5	19	80	43	110	25	35	40	45	110	150	150	210	90	13	5	1.4	1.9	14	4	1.2

Source: Ref. 11-6.

culation in a sanitary landfill; this system has been adopted for several sites and is considered an expeditious method for returning landfill sites to use in the shortest possible time and in using otherwise untenable sites. Difficulties arise, however, in treatment of leachate after recirculation if municipal sewage treatment systems are involved. Existing utilities may refuse to accept leachate or may affix excessive surcharges to such treatment. Long-range planning can eliminate such difficulties by providing integral treatment plants with the "recycle" landfills.

11-6. LAND DISPOSAL OF HAZARDOUS WASTES

11-6.1 General

The years since World War II have witnessed unparalleled production of new exotic organic and inorganic materials. Thousands of these substances also appear as waste materials either in their original form or as breakdown or by-products. Unfortunately, no meaningful technology has been developed to cope with this enormous problem. More basically, the question arises as to whether or not solid wastes managers have developed sufficient expertise and knowledge to properly and thoroughly define the term "hazardous wastes" as it applies to sanitary landfills. A need for such a definition exists because Congress in 1970 required planning to be initiated for the creation of a system of national disposal sites for the storage and disposal of hazardous wastes. Methodology must be developed for the storage, handling, transport, recycling, and disposal of such wastes. Current hazardous waste methodology, which is grossly inadequate, possesses potential for endangering public health or welfare and destroying the environment.

The first step in developing disposal methodology for hazardous wastes is to answer several important questions. What exactly are these wastes? Can these substances be deposited in a sanitary landfill in a technically feasible, economical, and safe way? What assurance exists that they have been permanently eliminated from the ecosystem? What compositional alterations or migrations of hazardous wastes occur in landfills? What effect will these wastes have on the air, water, and soil subsystems in the landfill ecosystem? Specifically, this section will be devoted to the development of the following items:

1. a relevant definition of hazardous wastes
2. a rating system to describe quantitatively the degree of environmental impact imposed on a landfill ecosystem by hazardous wastes
3. a rating system to delineate the capability of landfill sites to properly contain hazardous wastes following disposal

11-6.2 Definition of Hazardous Wastes

A general definition of hazardous wastes has not been developed. Most investigators in this area (for example, Lichtenstein and Schulz [1959], Rowe, Canter, and Mason [1970], and Browning [1969]) have dealt with specific substances with little regard for overall considerations. Several Battelle Memorial Institute personnel (Dawson, Shuckrow, and Swift [1970]) have combined and summarized all known hazardous-material classifications and ranking systems from the standpoint of potential for water quality impairment.

A definition of hazardous wastes particularly relevant to sanitary landfills must encompass several significant factors. Hazardous wastes in general may endanger public health or welfare and adversely affect the environment. In addition, such wastes may be dangerous to handle. Other materials may not be hazardous until they are introduced into waste management systems, at which time they may cause pollution (e.g., HCl is generated by incineration of innocuous polyvinyl chlorides). Hazardous wastes can usually be classified into four general categories:

1. Radioactive—laboratory wastes, power generation wastes, etc.
2. Toxic chemical—metals, pesticides, pharmaceuticals, warfare agents, etc.
3. Biological—antibiotics, pathogens, enzymes, warfare agents, etc.
4. Miscellaneous—flammables, explosives, irritants, corrosives, etc.

Such materials can cause acute or chronic effects on man or the environment when they require disposal. Hazardous wastes may originate in households, commercial establishments, industrial complexes, or institutional facilities. Processes and materials expected to produce hazardous wastes in the future should also be recognized as environmental hazards. Based on these observations the following definition of hazardous wastes is presented:

Hazardous waste materials are defined as those materials or combinations of materials which require special environmental management techniques because of their acute and chronic effects on:

1. The health and welfare of the public including those individuals who participate in waste management activities; or
2. Air, water, or land environments either prior to or following disposal in the ecosystem

To assist managers to develop methodology for disposal of hazardous wastes on land, rating systems will be outlined which include the interrelationships between the characteristics of the wastes, air, water, and land. Such rating systems will enable personnel to determine whether or not the placement of a hazardous waste in a specific landfill would have a deleterious effect on the total landfill ecosystem.

11-6.3 Hazardous Waste Ranking System

In order to incorporate in the proposed rating system a greater basis for comparison of potential sites, the landfill site and waste are rated or ranked independently. The sum of the waste rank and site rank given in the next section will provide a numerical indicator as to the feasibility of disposing of a particular waste at a particular landfill; however, separate waste and site ranks will allow quantitative comparisons among various wastes and among different sites.

Obviously, it would be impossible to include all possible waste parameters in the development of a hazardous waste ranking formula. Certain parameters will necessarily have a minimal effect on the hazard caused by a material, while sufficient data may not be available to directly evaluate other more important factors.

Field detection sensitivity of various wastes should play an important role in hazard ranking. Minute concentrations of some hazardous materials cannot be detected with commonly available field instruments. Catastrophic effects might result upon landfill disposal of a hazardous waste whose toxic concentration was less than its detectability limit. If such a situation occurred, the waste's environmental implications could not be determined until direct effects were observed in humans, animals, or plants. Therefore, hazardous compounds having a detectability/toxicity ratio greater than 1.0 will not be considered conducive to land disposal.

Many hazardous wastes should not be considered suitable for landfill disposal because of some specific properties such as low boiling point and flash point. Those substances which vaporize or flash may possess potential for harm to individuals either transporting the waste to the landfill or disposing of the waste material at the site. Therefore materials having boiling or flash points at or below normal air temperatures should not be considered suitable for landfill disposal.

With regard to other wastes, the Atomic Energy Commission (AEC), through the Atomic Energy Act of 1954 and subsequent amendments, is responsible for licensing and regulating the handling, storage, and disposal of radioactive waste materials. Radioactive wastes may be liquid, gaseous, or solid; furthermore, they may exhibit high or low levels of activity. Whereas domestic and industrial wastes generally lend themselves to stabilization by various chemical, physical, or biological processes, radioactive wastes do not respond sufficiently to any of these treatment methods. Only time can render radioactive wastes inactive. It is felt that because of existing regulations and the complexity and special nature of such wastes, they should not be considered acceptable for landfill disposal.

After exclusion of those wastes included in the previously outlined categories, all other compounds may be quantitatively ranked in direct proportion to their potential to cause harm to the total ecosystem including man. The five waste parameters quantitated for inclusion in the ranking system include: human toxicity, groundwater toxicity, disease transmission potential, biodegradability, and mobility. These parameters were chosen because of their importance in assessing the environmental impact of a hazardous waste on a landfill ecosystem and because data for these parameters were readily available or could be readily obtained experimentally.

11-6.3.1 Waste Rating Rationale. Obviously the development of any quantitative rating system necessitates some arbitrary weighting of parameter importance. In general, the following rationale was used in arriving at the weighted relationship between parameters in the waste rating system:

1. Those parameters which directly indicate impairment of humans, animals, or plants were assigned a first-degree priority, i.e., toxicity and pathogenicity (life state and primary disease transmission)
2. Those parameters which directly indicate persistence in the ecosystem were assigned a second degree priority, i.e., pathogenic survival and biodegradability
3. Those parameters which directly indicated mobility in landfill ecosystems were assigned a third degree priority, i.e., adsorptive capacity, and solubility

A maximum value of 40 priority ranking units (PRU) was arbitrarily assigned to first-degree parameters, whereas the maximum values assigned to second- and third-degree parameters were 24 PRU and 16 PRU respectively.

Human Toxicity. The importance of a waste's human toxicity value cannot be underestimated. Human toxicity may be defined as the ability of a substance to produce injury once it reaches a susceptible site in or on the body. Such potential hazardousness arising from the human toxicity of a waste would occur prior to, or during, landfill disposal. Obviously different waste materials would possess different levels of toxicity. To arrive at a quantitative human toxicity rank the rating compiled by Sax [1969] was used as a guide. Sax provided a single source for concise hazard analysis information regarding 10,000 materials, presenting the only quantitative rating of toxicity available in the literature. He defined toxicity ratings from 0 to 3 as follows:

1) No toxicity = 0
   This designation is given to all materials which may be categorized as follows:
   a. Materials which cause no harm under any conditions of use
   b. Materials which produce toxic effects on humans only under the most unusual conditions or overwhelming dosage
2) Slight toxicity = 1
   This designation denotes those materials which produce changes in the

human body which are readily reversible and which will disappear following termination of exposure, either with or without medical treatment.
3) Moderate toxicity = 2
   Those substances which may produce irreversible as well as reversible changes in the human body; however, these changes are not of such severity as to threaten life or produce serious permanent physical impairment.
4) Severe toxicity = 3
   Those materials which can be absorbed into the body by inhalation, ingestion, or through the skin and which can cause injury or illness of sufficient severity to threaten life or to cause permanent physical impairment or disfigurement.

A human toxicity rank (Ht) was quantitated by multiplying the Sax rating (Sr) by thirteen as below:

$$Ht = 13\, Sr$$

The human toxicity rank of a waste may be thought of as being one of the most significant factors in determining the hazardousness of the waste. The human toxicity rank will range between 0 and 39.

Groundwater Toxicity. Assuming that a toxic waste was able to enter a groundwater system, its detrimental effect on the ecosystem would be directly related to its critical concentration in water. The critical concentration must therefore be defined in terms of human toxicity, aquatic toxicity, and plant toxicity as the minimum concentration of a substance that would produce a detrimental effect; i.e., the smallest amount of waste which would cause damage or injury to humans, animals, or plants. The critical concentration is therefore the level at which damage or injury may be caused to any portion of the ecosystem. As in the 1970 Battelle Memorial Institute Study, the critical concentration for human toxicity was chosen as the maximum allowable concentration in drinking water. For aquatic toxicity a combination of T_{Lm} values and threshold values for effects on fish was used as the critical concentration. Critical concentration for plant toxicity represented the maximum amount tolerated by cultivated crops without a loss in product yield.

A survey of the literature reveals a practical range of groundwater toxicity of approximately 10^{-3} to 10^4 mg/l. To quantitate the groundwater toxicity, assigning to it a first order of priority, the following formula was developed:

$$Gt = 6(4 - \log Cc)$$

where Gt = groundwater toxicity rank
      Cc = smallest critical concentration (mg/l) for humans, aquatic life or plants

but for Cc values greater than 10^4 mg/l, $Gt = 0$ and for Cc values less than 10^{-3} mg/l, $Gt = 42$.

Therefore the range in the groundwater toxicity rank is from 0 to 42, 0 being nontoxic waste and 42 representing a very toxic material.

Disease Transmission Potential. The potential for disease transmission by various biological wastes disposed of in landfills is a relatively unexplored problem. The immediate hazard of such wastes to sanitation workers has never been fully investigated. Exposure to pathogens during the transfer of wastes to a landfill site and throughout the burial procedure potentially poses a threat to the worker and those he contacts.

Disease transmission may occur through direct contact, infection of open sores, or vectors. Although many pathogens must have a specific environment in which to live, many may survive in air, land, or groundwater. Moreover, many pathogens may be transported to man in a dormant life state.

Landfills are presently accepting potential disease transmitting wastes, including primary sludge and hospital wastes. Transmission potentials exist primarily through contamination of groundwater and its ensuing epidemic effects. Because of the grave importance of possible disease transmission by landfilled biological wastes, this term was assigned top priority with regard to quantitating its overall effects.

Information regarding concentrations of pathogenic organisms needed to produce disease is lacking in the literature. To illustrate the interrelationships of the factors which influence the probability of disease transmission when pathogenic organisms come in contact with a potentially susceptible host, the following equation was suggested by Smith (see Ref. 11-9):

$$D = NV/R$$

where D = the probability of disease
N = the number of organisms
V = the virulence of the organisms
R = the resistance of the host

When the product NV is greater than R, disease results; when less than R, it will not. Most of the terms in this equation cannot be measured quantitatively. Thus, the only feasible method of rating the disease potential of a specific waste material is to arbitrarily assign values to specific pathogenic properties.

The range of the disease transmission potential factor (Dp) was set at a minimum of 0 and a maximum of 105. To facilitate meaningful quantitating, Dp was broken down into three subgroups, each pertaining to specific disease transmission properties of the waste. Subgroup I (mode of disease contraction) represented the primary means of obtaining the disease and was subdivided into three areas:

1. direct contact—assigned a value of 40 points because of its immediate threat

248   SOLID WASTE MANAGEMENT

2. infection through open sores—given a value of 28 points since the immediate danger may not be apparent and immunization of waste handlers reduces the greatest threat
3. infection by vector—given a lower value of 16 since proper landfill maintenance should eliminate a majority of flies, mosquitoes, etc.

Subgroup I may therefore range from 0 to 40 priority ranking units (PRU).

Subgroup II (pathogen life state) was considered equally important in determining the potential of disease transmission as the mode of contraction. It was arbitrarily decided that a pathogenic organism having two life states (active and dormant) was far more dangerous than a pathogen having one life state or one which did not have the ability to survive outside its host cell. The life state persistence was therefore divided into the following areas:

1. pathogenic microorganisms with more than one life state (virus, and fungi)—assigned a value of 40 PRU
2. pathogenic microorganisms with only one life state (vegetable pathogens)—assigned a value of 20 PRU and
3. pathogenic microorganisms which cannot survive outside their host (*Treponema pallidum*)—assigned a value of 0 PRU

Subgroup II (pathogen life state) may therefore range from a minimum of 0 to a maximum of 40 PRU.

Subgroup III was considered the ability of the pathogen to survive in various environments including land, air, and water. A pathogen's ability to survive in air or water was assigned a value of 10 PRU. Air and/or water survival is mandatory for microbial persistence in, or mobility from, a landfill system. Air survival would enable possible human contact during the burial process or from a vehicle of transportation from the site. Likewise survival in water would allow for potential mobility from the site via the groundwater. A pathogen's ability to survive in the soil is of secondary importance since the soil does not afford a mechanism of mobility to the pathogen and was assigned a value of 5 PRU. Subgroup III will therefore range in value from a minimum of 0 to a maximum of 25 PRU. The total disease transmission rating (D_p) was evolved by simply adding the values of Subgroups I, II, and III.

Biological Persistence. The biological persistence of a waste deposited in a landfill may be very complex and dynamic. The chemical and biological reactions which occur between wastes must be considered individually. All reactions, and reaction by-products, should be considered from a persistence standpoint.

The method utilized in the ranking system to determine the biodegradability of a compound is based on the ratio of the Biochemical Oxygen Demand (BOD) to the Theoretical Oxygen Demand (TOD). The BOD may be considered as representing the amount of oxygen required during the biological oxidation of a waste, whereas the TOD represents the amount of oxygen required to completely oxidize a waste. Therefore, substances having a high BOD/TOD ratio would be considered highly biodegradable whereas those having low BOD/TOD ratios would be relatively nonbiodegradable. A biodegradability factor (Bd) was derived to quantitate or rank a waste's degradability as follows:

$$Bd = 16 \left(1 - \frac{BOD}{TOD}\right)$$

As a compound becomes more biodegradable the BOD/TOD ratio approaches 1 and the Bd factor approaches zero implying optimal conditions. Conversely low BOD/TOD ratio compounds will produce a Bd factor approaching 16 implying a highly hazardous state.

Waste Mobility. The mobility of a waste may be thought of as the ease with which the substance can move about in or out of a landfill. A high mobility waste will travel to the groundwater table quite easily, creating a hazardous situation. The mobility of a waste will depend on both its adsorptive capacities and solubility. The worst situation (high mobility) would consist of a waste having a high solubility and low adsorption. Conversely, the least dangerous type of waste would be one having a low solubility and high adsorption (low mobility).

The adsorptive capacity of a substance is based primarily on its charge characteristics. A waste having a net negative charge is more likely to be transported deeper into the landfill (closer to the groundwater), since it will be repelled by the net negative soil charge. To quantitate the adsorptive characteristics of a waste the following relationship was developed:

$$A = 3 + C$$

where A = adsorptive value
C = net charge of the waste determined from its molecular formula in reaction with water at a pH of 7.0

To arrive at a total waste mobility rank, the solubility of a waste must also be quantitated. A practical range of waste solubility varies from 10^{-4} mg/l to 10^6 mg/l; therefore, the following solubility value (Sv) was developed:

$$Sv = 6 - \log s$$

where s = solubility of the substance (mg/l) in water. If $s < 10^{-4}$ then $Sv = 10$ and if $s > 10^6$ then $Sv = 0$.

Therefore for solid substances the mobility rank (Ms) will be:

$$Ms = 16 - A - Sv = 7 - C + \log s$$

The range of Ms is from 0 to 16 PRU.

250 SOLID WASTE MANAGEMENT

Mobility obviously is greater for liquid wastes. The mobility rank for liquid wastes (Ml) is therefore defined as follows:

$$Ml = 16 - A = 13 - C$$

Ml will therefore range between 10 and 16 PRU.

The total ranking formula for determining the hazardousness of wastes is:

$$\text{Hazardous waste rank} = Ht + Gt + Dp + Bd + M$$

Selected Waste Ranking. Several wastes have been ranked with the above formula and compared to waste ranks developed in a study by Booz, Allen Applied Research, Inc. [1971]. The results of this comparison are as follows.

Waste Compound	PRU	Booz, Allen
Waste paper	7	—
Inert ash	18	—
Sulfur	21	—
Anthracenes	27	18
Steel wool	31	—
Benzoic acid	39	13
Ferrous sulfate	42	—
2 Ethyl hexanol-1	45	—
Propionic acid	51	18
Monoethanolamine	59	20
Furfural	62	26
Aluminum oxide	63.5	15
Malic anhydride	68	42
Naphthalene	68.5	26
Acetic acid	69	26
Acridine	72	17
Methyl bromide	72	30
DDT	74	28
Aluminum sulfite	76	37
Aniline	78	30
Copper sulfite	86	36
Phenol	88	38
Acetone cyanhydrin	91	35
Cadmium chloride	99	29
Potassium cyanide	102	40
Dieldrin	103	25
Primary sludge	104	—
Arsenic diethyl	107	31
REJECTED WASTES		
Acetaldehyde	—	28
Acetylene	—	19

Chloroform	–	23
Ethyl benzene	–	–
Lactonitrile	–	–
Methyl alcohol	–	32
Methyl isopropyl ketone	–	15
Pyridine	–	23
Toluene	–	30
Xylene	–	21

Waste rankings (PRU) can be correlated with the hazard level of wastes as follows:

Rank (PRU)	Hazard Level
0–30	Nonhazardous
31–60	Slightly hazardous
61–80	Moderately hazardous
Above 80	Hazardous

11-6.4 Landfill Site Ranking System

In order to evaluate the potential danger of depositing any waste or material in a particular landfill site, it is necessary to critically examine three general characteristics of that site: (1) the potential for precipitated surface waters to infiltrate the deposited waste material; (2) the potential for the waste material to be transported through fluid transmission away from its deposit location through underlying bottom soils to groundwater systems; and (3) other mechanisms for the removal of hazardous materials from the site and their transport to other areas. Obviously, a number of factors will have a significant bearing upon each of these three critical aspects of the disposal problem.

The thickness, permeability, and field capacity of the cover soil over the refuse, the intensity and amount of rainfall, and the slope of the surface of the cover soil are factors which define the ability of surface waters to reach the refuse. The thickness, filtering capacity, and absorptive capacity of the bottom soils are factors which determine the potential of hazardous materials escape through fluid transmission from the landfill site. Additionally, the potential for any waste material to be transported or transmitted from the disposal site must be examined and the environmental impact of such transmission from the original deposition location must be evaluated. For example, the characteristics of the groundwater beneath the refuse cache are of importance in the evaluation of any possible hazardous material escape. The factors which should be considered in the evaluation of the groundwater relate to the potential for travel of a hazardous material in the groundwater system and the potential for the transmutation (weakening or strengthening) of the hazardous material once it has entered the groundwater system. The factors which provide an insight into the potential for travel of the hazardous material in the groundwater system are the velocity of the groundwater flow system and the total travel distance for that system. Total travel distance in this instance refers to the travel of a molecule of water from a point directly beneath the landfill site through the groundwater flow system into a surface water flow system and thence to the sea. In addition to the dissipative or travel potential for the hazardous material in these systems, the changes which may occur in the hazardous material after contact with groundwater must be evaluated. The factors which define the transmutation in the hazardous material which is possible after contact with the groundwater will be the moderating or buffering capacity of the groundwater and its organic content. The buffering capacity may be related to the pH and the alkalinity of the water. The organic content can be expressed in terms of biochemical oxygen demand.

In addition to the geological and hydrological characteristics of the site, the potential there for other modes of transmission or removal of the hazardous material to other locations must also be examined. In modern sanitary landfill practice it must be assumed that daily cover will be applied to the deposited refuse and certainly in the case of a hazardous material this daily cover procedure would be followed. Moreover, special placement of a final cover layer would be accomplished. Nevertheless, the potential for escape of a volatile hazardous waste and the potential for escape by wind transportation of a lightweight hazardous waste must be taken into account. In order to adequately consider such a variation, it is pertinent to examine the prevailing direction, magnitude, and duration of winds at the particular landfill site. This information will reflect the potential danger of removal and distribution of either volatile constituents or lightweight solid constituents of a hazardous material. The danger of such an occurrence may be evaluated by determining the distribution of population around the particular site. In other words, if a site is situated such that strong prevailing winds of long duration blow across the site toward a center of very dense population, then the potential for air transport of hazardous wastes into the vicinity of possible victims is great. This danger may be evaluated quantitatively by estimating the characteristics of the winds at the site and by relating those characteristics to the number and location of persons around the site.

The above-mentioned site characteristics have been included in the site ranking formula not only because of their importance in evaluating a potential hazardous waste site but also because quantitative information regarding these characteristics can generally be obtained experimentally from soil tests, water quality tests, and local weather data.

11-6.4.1 Site Ranking Rationale. The following rationale was used to arrive at a relationship or weighting of parameters in the site rating system:

1. the parameters which would immediately affect waste transmission were assigned a first-degree priority, including infiltration potential, bottom leakage potential, and groundwater velocity
2. the parameters which would affect waste transmission after contact with water were assigned second-degree priority, i.e., filtering capacity and adsorptive capacity
3. those parameters which represented the present conditions of receiving groundwater were assigned third priority, i.e., organic content and buffering capacity
4. the parameters which represented factors outside the immediate disposal site were assigned fourth priority, i.e., the potential travel distance, prevailing wind direction, and population factor

A maximum value of 20 PRU has been arbitrarily assigned to top priority parameters. The maximum value for second, third, and fourth priority parameters are set at 15 PRU, 10 PRU, and 5 PRU respectively.

11-6.4.2 Infiltration Potential.

The potential for water to enter a waste deposit may be quantitatively expressed as the ratio of the amount of water which may enter the top surface of the cover soil divided by the amount necessary within that soil to produce a full passage of moisture from the top of the layer to the bottom and out into the contained refuse. The amount of water which could theoretically enter the site or enter the cover soil at the site (i) may be estimated as the total area under all of the rainfall intensity graphs for the site, beneath a horizontal line representing the infiltration rate of the cover soil (see Fig. 11-3). The infiltration rate of the cover soil may be expected to vary from 0.01 in./hr for bare heavy clay soils to approximately 3.5 in./hr for loose sands. The probable range of i in the United States will be from 1 to 40 in.

The amounts of water necessary for passage of moisture through the cover soil may be related to the volumetric field capacity of the layer. In other words, whereas the field capacity refers to the amount of water as a percentage of the dry unit weight of the soil required for passage of water through a unit volume, the volumetric field capacity in this instance would refer to the product of the thickness of the cover soil layer times the field capacity of the soil. Thus, $FC(H)$ was made the denominator of the infiltration potential term

where FC = field capacity of the soil expressed as a decimal
and H = thickness of cover soil layers (inches)

The FC will vary from 0.05 for a clean sand to 0.40 for a clay, whereas H will vary from approximately 30 in. to about 100 in. The infiltration potential may be finally quantitated as:

$$Ip = \frac{i}{(FC)H}$$

254 SOLID WASTE MANAGEMENT

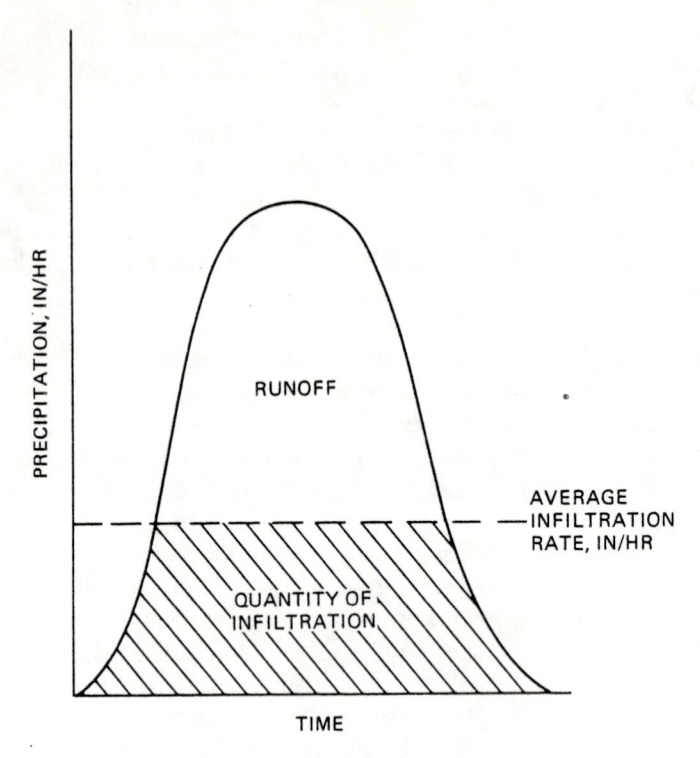

Fig. 11-3 Typical landfill site hydrograph.

having a practical range of 0.02 to 20 PRU which is the arbitrary range assigned to first-priority parameters. This infiltration potential may be thought of as one of the most significant factors in determining the site potential for waste transmission.

11-6.4.3 Bottom Leakage Potential. In addition to the problem of water entering the refuse cells and removing the contained hazardous material, consideration must be given to the action of a waste in suspension or solution in water, or in liquid form, passing through the bottom soil layer from its original location and entering the groundwater system.

The potential hazard for a waste to travel through a bottom soil from the bottom of the refuse cell through the containing soil layer and into a groundwater flow system may be evaluated in terms of the permeability of the bottom soil layer and its thickness. Since all natural geological materials possess some finite permeability it is fatuous to think in terms of an impermeable bottom in a sanitary landfill. Even in the situation where an artificial lining material has been applied to the bottom of a refuse cell, it is quite probable that the artificial liner is in truth not impermeable. For example, thin sheets of impervious polyvinyl chloride or polyethylene lining may easily be pierced and penetrated during placement or after placement by sharp-edged equipment or refuse items. Asphaltic liners likewise may crack because of distortions experienced when the bottom soils settle as a result of the applied loads of the landfill. Thus, in all cases, a certain finite permeability of the bottom confining layer must be anticipated. Therefore, in a true sense, the migration of materials from the landfill site into the substrata must always be anticipated and the only variable to consider is the time which will be required for such migration; in other words, the migration time for a hazardous substance through a bottom soil layer consisting of clay may be sufficiently long so that the substance's half-life is greatly exceeded. In such a case the virulence and hazardous nature of these substances will be diminished.

For this reason this bottom leakage factor has been quite simply expressed in the form shown below to give a measure of the time factor for migration of a hazardous material in terms of permeability and thickness of the bottom soils.

$$\text{Bottom leakage potential } (Lp) = \frac{1000 \sqrt[3]{K}}{T}$$

where K = bottom soil permeability (cm/sec)
T = bottom soil thickness (ft)

The approximate range of K for all practical situations will be about 10^{-1} to 10^{-10} cm/sec, whereas T will vary from 5 to 50 ft. The overall pragmatic range of Lp will therefore be from approximately 0.02 to 20 PRU.

11-6.4.4 Filtering Capacity. A less important characteristic of the bottom soils is their ability to remove solid particles traveling downward (through the bottom soil layer) in a fluid suspension. In general, this filtering capacity is dependent upon the sizes of the pore spaces between individual soil grains. In other words, the physical filtering ability of the bottom soil will depend upon void-space size in that soil and may therefore be related to the size of the soil particles, the degree of compaction, and the particle shape. Particle size is more important than compaction or grain shape; the physical filtering capacity is roughly proportional to the inverse of the average grain size in the soil stratum. Therefore, the filtering capacity of the bottom soil layer can be expressed as shown below.

$$\text{Filtering capacity } (Fc) = -4 \log \frac{2.5 \times 10^{-5}}{\phi}$$

where ϕ = average particle diameter (inches)

The average particle diameter of various soils will vary from about 0.25 to

2.5×10^{-5} in. Therefore the filtering capacity will vary between approximately 0 and 16.0 which is within the arbitrary range set for second-degree parameters.

11-6.4.5 Adsorptive Capacities. In addition to the removal of solid particles through physical filtering within the bottom soil layer, certain materials will be removed from suspension and solution in a migrating fluid by the physicochemical attraction of the mineral constituents within the soils. Adsorption of material both organic and inorganic in the migrating fluids will take place principally on colloidal-size particles consisting of clay minerals. The adsorptive potential of a particular soil therefore may be related directly to those properties of the clay minerals which describe the attraction of such minerals for the migrating particles. A general measure of such attraction is the cation exchange capacity of the mineral. In this rating system the greater the danger of transmission of a hazardous material from a landfill site the greater the rating factor. Therefore, the greater the ability of the bottom soil layer to adsorb migrating materials the smaller should be the adsorption factor. This ability of the soil is evaluated as an inverse quantity and a factor is obtained by dividing a numerator by cation exchange capacity in the denominator. The cation exchange capacity alone will not reflect the potential for adsorption of a material on the minerals present in the soil. Adsorption of hazardous materials will also depend upon the ability of the material in question to displace metallic ions occupying cation exchange sites. However, if the available adsorptive sites are already occupied by organic compounds and complex organic ions, very little, if any, further adsorption would be expected to occur.

Therefore the adsorptive capacity factor is comprised of the organic content as the numerator and the cation exchange capacity as the denominator as shown below.

$$\text{Adsorptive capacity } (Ac) = \frac{10 \, (Or)}{(\log CEC) + 1}$$

where Or = organic content expressed as a decimal
CEC = cation exchange capacity, meq/100 g

The range in the numerator will therefore be from 0 to 10, whereas the log CEC will range between about 0.6 and 2.2. The adsorptive capacity (Ac) will therefore vary between approximately 0 and 16 PRU. This is within the range for second-degree parameters.

11-6.4.6 Organic Content of Groundwater. Transmutations of a hazardous material following contact with groundwater must also be evaluated. Assuming that a hazardous waste has reached groundwater after disposal in a landfill, probably the most important single water parameter to be considered would be that of organic carbon content. The organic carbon content of the groundwater may be quantitated in terms of the biochemical oxygen demand (BOD). The higher the organic content (BOD) of a groundwater, the higher substrate potential, and consequently the higher the potential for pathogenic microorganism reproduction. Obviously, a high BOD groundwater system would be undesirable because of the growth potential it may afford pathogenic organisms.

Groundwater organic content was assigned a third order of priority with regard to landfill ranking factors, so its range of values is given between 0 and 10 PRU, dependent upon BOD values as follows:

$$Oc = 0.2\, BOD$$

where Oc = organic content rating (maximum value of 10)
BOD = biochemical oxygen demand of groundwater (mg/l)

11-6.4.7 Buffering Capacity of Groundwater. The buffering capacity of a groundwater is another important parameter when considering transmutation of hazardous wastes in groundwater systems. Any waste material having acidic or alkaline characteristics would be less hazardous to the groundwater ecosystem if the water it is entering possessed a high buffering capacity. In other words acidic or basic waste characteristics would be neutralized or moderated upon contact with a high-buffering-capacity water system.

Groundwater buffering capacity was assigned a third order of priority and is quantitated in relation to pH, acidity, and alkalinity. The buffering capacity ranking (Bc) is defined as follows:

$$Bc = 10 - Nme$$

where Nme = the smallest number of milliequivalents (maximum of ten) of either an acid or base required to displace the original groundwater pH below 4.5 or above 8.5

The buffering capacity ranking will therefore vary from 0 for a strong buffer to 10 for a weak buffer which is within the range assigned to third priority parameters.

11-6.4.8 Travel Distance. In addition to the interactions and effects of the wastes on the water (and the water on the wastes), the potential for dispersal of pollutants through a groundwater system should be evaluated. The potential for travel of a leached material once it enters a groundwater system may be evaluated rather arbitrarily as shown in Table 11-7, where the potential travel distance is the distance a molecule of water could travel from a point directly beneath the landfill through the groundwater and surface water systems to the sea.

TABLE 11-7

Potential Travel Distance	Rating Factor
0–500 ft	0
500–4000 ft	1
4000 ft–2 miles	2
2 miles–20 miles	3
20 miles–50 miles	4
more than 50 miles	5

11-6.4.9 Groundwater Velocity. The groundwater velocity will determine how fast a hazardous material may spread into the environment. A groundwater system having a high velocity should therefore be assigned a higher ranking since the time of waste transmission would be reduced.

The groundwater velocity, having a fourth order of priority, may be defined as:

$$V = kS$$

where V = the face velocity or quantity of water flowing through a unit cross-sectional area.
   k = the coefficient of permeability (cm/sec)
   S = the gradient, or loss of head per unit length in the direction of flow (ft/mile)

Values of k will vary between 10^{-1} and 10^{-9} cm/sec, whereas values of S will usually range between 0 and 20 ft/mile. Groundwater velocities were ranked according to the following formulation:

$$Gv = \frac{S}{\log\left(\frac{1+1}{K}\right)}$$

where Gv = groundwater velocity rank
   K = permeability (cm/sec)
   S = gradient (ft/mile)

The groundwater velocity rank will approximately range between 0 and 20 PRU, which is the range assigned to first priority parameters.

11-6.4.10 Prevailing Wind Direction. The third major site characteristic to be investigated is air. The hazardous potential of any toxin or pathogen escaping through the atmosphere from the landfill would depend upon the prevailing wind direction in relation to the distribution of population surrounding the site. Obviously the worst situation would be one in which a strong prevailing wind blew from the site to the center of a very dense population.

RECENT DEVELOPMENTS   259

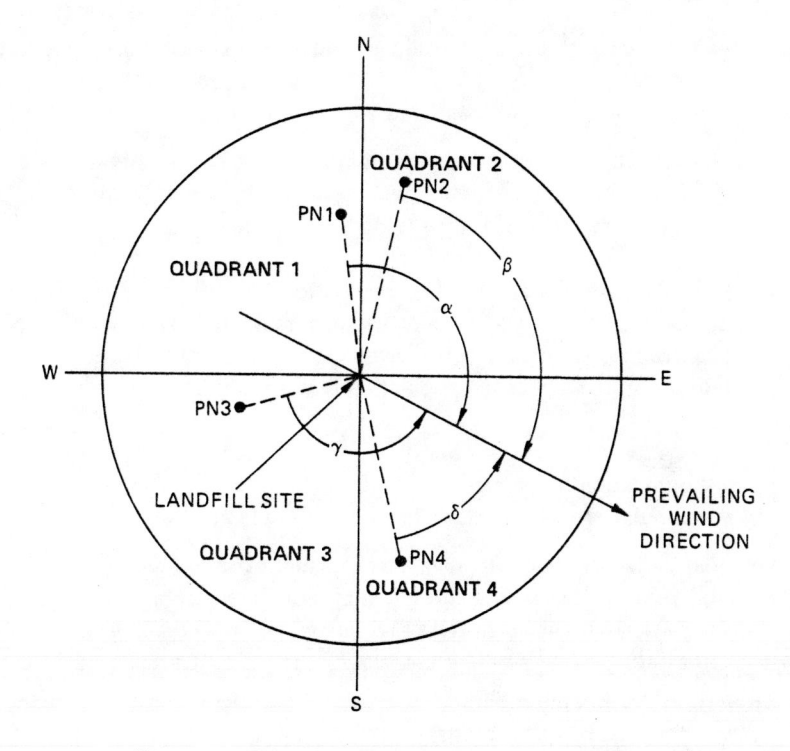

Fig. 11-4 Wind analysis diagram for landfill site.

The following procedure was therefore developed to quantitatively evaluate the potential of the prevailing wind direction. Initially, a 25-mile radius circle is constructed with the landfill site as its center (see Fig. 11-4). This circle is then divided into four quadrants by drawing two lines—one north-south and one east-west. The population of each quadrant (pi) is determined and a point representing the center of population (population node) is located in all four quadrants (PNi). A radius is then drawn from the site to each quadrant's population node. The prevailing wind direction is determined and a radius drawn in this direction from the site (center of circle). The angles from the prevailing wind direction to each site-population node radius are determined ($\alpha, \beta, \gamma, \delta$) and incorporated in the following prevailing wind potential formula:

$$Wp = \sum_{i=1}^{4} \frac{\left[\left(5 - \frac{Ai}{36}\right) \log Pi\right]}{15}$$

where Wp = prevailing wind potential rank

Ai = the angle from the prevailing wind direction to each site-population node
Pi = the population of each quadrant

Wp quantitatively interrelates the prevailing wind direction, site location, and population nodes of each quadrant. Wp has a practical range of 0 to 5 PRU, which is the arbitrary range of fourth priority parameters.

11-6.4.11 Population Factors. The population immediately surrounding the landfill site will determine how many persons could be adversely affected by escaping hazardous materials. The higher the population within a specified radius of the landfill site, the higher the population factor ranking as shown:

$$Pf = \log p$$

where Pf = population factor rank
p = population within a 25-mile radius of the landfill site

The population factor rank will range between about 0 and 7 PRU.

The total landfill site ranking formula may now be assembled by uniting the various soil, water, and air parameters as follows:

$$\text{Site Rating} = Ip + Lp + Fc + Ac + Oc + Bc + Td + Gv + Wd + Pf$$

where Ip = Infiltration potential
Lp = Bottom leakage potential
Fc = Filtering capacity factor
Ac = Adsorptive capacity factor
Oc = Organic content factor
Bc = Buffering capacity factor
Td = Potential travel distance factor
Gv = Groundwater velocity factor
Wd = Prevailing wind direction factor
Pf = Population factor

The first four parameters (Ip, Lp, Fc, and Ac) describe the soil system, the next four factors (Oc, Bc, Td, and Gv) delineate the groundwater characteristics, and the last two terms (Wd and Pf) depict air parameters. The total landfill rank may assume values from approximately 0 to 110; the lower the rank the better the landfill for hazardous waste disposal.

The data in Table 11-8 was accumulated from two existing landfill sites in Louisville, Kentucky, so that a ranking comparison could be illustrated.

Site #1 ranking parameters—Ip = 10.8, Lp = 5, Fc = 10.4, Ac = 5, Oc = 2, Bc = 7, Td = 5, Gv = 1.66, Wd = 4.05, and Pf = 6
Total landfill rating (site #1) = 56.9

TABLE 11-8

	Site #1	Site #2
Yearly rainfall	43 in	43 in
Soil type	clean sand	heavy clay
Infiltration (% of rainfall)	75	10
Field Capacity	0.05	0.40
Permeability (cm/sec)	10^{-3}	10^{-8}
Soil cover (in.)	60	24
Bottom thickness (ft)	20	15
Average particle diameter (mm)	0.25	0.002
Organic Content of soil (%)	0.5	0
Groundwater BOD (mg/l)	10	10
Cation exchange capacity (meq/100 g)	0	80
Buffering capacity (meq)	7	4
Groundwater travel distance (miles)	750	750
Gradient (ft/mile)	5	5

	Site #1	Site #2
Population within 25-mile radius	10^6	10^6
Prevailing wind direction	WNW	WNW

Site #2 ranking parameters—Ip = 0.45, Lp = 0.145, Fc = 2.0, Ac = 0, Oc = 1, Bc = 4, Td = 5, Gv = 0.625, Wd = 2.9, and Pf = 6
Total landfill rating (site #2) = 23.1

Landfill #2 having a much smaller rating than landfill #1 would obviously be more conducive to land disposal of hazardous wastes.

11-6.5 Summary

A thorough definition of hazardous wastes has been presented. In addition ranking systems have been developed to evaluate both waste compounds and landfill sites. The final numerical priority rating of a specific hazardous compound is representative of its potential threat to the environment, and therefore indicates the priority it should receive with respect to other substances. The final numerical rating of a specific landfill site will provide an indication as to the feasibility of disposing of hazardous substances at that site. Obviously, these *embryonic* systems require refinement; however, these are the first steps toward relevant and workable systems. Such systems must be developed to optimize future procedural decisions with regard to proper land disposal of hazardous wastes.

REFERENCES

11-1. "The Melt-Zit High Temperature Incinerator," Office of Solid Waste Management Programs, USPHS, Cincinnati, 1970.

11-2. Personal communication, Mr. Daniel Keller, CPU-400 Project Officer, Office of Solid Waste Management Programs, Cincinnati, Nov. 1972.

11-3. "Partial Oxidation of Solid Organic Wastes," Final Report on Research Grant EC-00263, Office of Solid Waste Management Programs, USPHS, Cincinnati, 1970.

11-4. Beckman, J. A., and Laman, J. R., "Destructive Distillation of Used Tires," *Proc.*, New Directions in Solid Waste Processing Institute, Framingham, Mass., 1970.

11-5. "Solid Waste Landfill Stabilization," an interim report by Ralph Stone & Co., to the Office of Solid Waste Management Programs, USPHS, 1969.

11-6. Pohland, F. G., "Landfill Stabilization with Leachate Recycle," Annual Progress Report Research Grant EP-00658, July 1972.

11-7. Pavoni, J. L., Hagerty, D. J., and Heer, J. E., Jr., "Evaluation of Sanitary Landfill Sites," *Public Works* (in process).

11-8. ———, "Environmental Impact Evaluation of Hazardous Waste Disposal in Land," paper presented at the 7th Amer. Water Resources Conference, Washington, 1971.

11-9. Amer. Water Works Assoc., Inc., *Water Quality and Treatment, A Handbook of Public Water Supplies*, McGraw-Hill Book Co., New York, 1971.

11-10. Booz, Allen Applied Research, Inc., *Study of Hazardous Waste Materials, Hazardous Effects, and Disposal Methods*, Report to Solid Waste Management Office, EPA, Cincinnati, Ohio, 1972.

11-11. Browning, J. E., "Hazardous Chemical Cargo: Time Bomb for Catastrophe," *Chem. Engrg.*, Vol. 76, No. 44 (1969).

11-12. Dawson, G. W., Shuckrow, A. J., and Swift, W. H., *Control of Spillage of Hazardous Polluting Substances*, Water Pollution Control Research Series, 15090 FOZ, 1970.

11-13. Lichtenstein, E. P., and Schulz, K. R., "Breakdown of Lindane and Aldrin in Soils," J. Econ. Entomol., Vol. 52, No. 118 (1959).

11-14. Rowe, D. R., Canter, L. W., and Mason, J. W., "Contamination of Oysters by Pesticides," *J. San. Engrg. Div.*, ASCE, Vol. 96, p. 1221.

11-15. Sax, N. I., *Dangerous Properties of Industrial Materials*, Van Nostrand Reinhold Co., New York, 1969.

11-16. Thompson, C. H., Ryckman, D. W., and Buzzell, J. C., Jr., *The Biochemical Treatability Index (BTI) Concept, Proc.*, 24th Purdue Indus. Waste Conference, Part I, 1969.

12
Legal aspects of solid waste management

An environmental ethic has evolved rapidly in the United States during the last several years. Although this ethic is still in process of maturing, it is becoming an integral facet of economic planning and national commitment. Environmental legislation which has developed from this ethic has exerted a tremendous impact upon private, governmental and industrial sectors of the community. New standards and limitations on the release of pollutants into the ecosphere have become an accepted part of national policy and personal value systems. These new laws, however, have placed economic restrictions on many municipal and industrial operations. The sweeping Federal legislative package for environmental protection has produced a complementary set of laws, regulations and restrictions on the state level. For example, a number of states have enacted laws analogous to the National Environmental Policy Act of 1969 within the two years after the passage of NEPA, and it is likely that eventually all states will initiate extensive programs for environmental protection and control.

264 SOLID WASTE MANAGEMENT

This chapter contains a review of existing Federal environmental legislation with special attention focussed on laws pertaining to solid waste collection and disposal and to resource recovery.

12.1. GENERAL ENVIRONMENTAL LAWS

Several significant environmentally-relevant laws passed in the late 1960's and early 1970's do not fit readily under the categories of air quality laws, water quality laws, etc. These general laws will be discussed in this section.

The most important general piece of legislation has been the Reorganization Plan No. 3 of 1970 which established the Federal Environmental Protection Agency, commonly referred to as the EPA. This reorganization plan brought together in one unit environmental divisions from the Department of the Interior, the Department of Agriculture, the Department of Health, Education and Welfare, and other departments, which previously had functioned separately and independently with little apparent coordination. The formation of EPA, a super-agency, was considered to be mandatory if federal environmental protection efforts were to be coordinated properly. The formation of EPA has been of great significance to solid wastes management, also, because of its capacity to serve as a central clearinghouse for information on solid wastes and its sponsorship of a large number of coordinated research projects on collection, disposal and recycling of solid wastes.

Another important piece of general environmental legislation is the National Environmental Policy Act of 1969 (NEPA), mentioned previously. This Act established basic policy in the Federal government for the encouragement of productive and enjoyable harmony between man and the environment. By NEPA, the Council on Environmental Quality (CEQ) was established in the Executive Office of the President and was charged with the responsibility to:

1. comment on trends in the quality and management of the environment;
2. comment on the adequacy of available natural resources;
3. review the environmental programs of the federal government; and
4. develop remedies for the deficiencies in existing federal environmental programs and activities.

Another provision of NEPA requires each Federal Agency to prepare a statement of environmental impact prior to every major action, recommendation or report on legislation that may significantly affect the quality of the human environment. These provisions of NEPA in effect have created a "watchdog" council to supervise national policy with respect to waste generation, disposal and reuse. Additionally, EPA programs for solid wastes management are subject to review when environmental impact statements are written.

The Environmental Quality Improvement Act of 1970 provided further force to NEPA by the provision of staff personnel for the Council on Environmental Quality and the authorization of fiscal appropriations for the CEQ office and staff.

Several other laws and regulations have been enacted which promote environmental protection and therefore have significance in solid wastes management:

1. The Federal Tax Laws recently have been modified to permit longer amortization periods for financing programs for pollution control facilities—the new period of sixty months is intended to promote industrial purchase of pollution control equipment;

2. The Department of Transportation Act has been amended to require that the Secretary of Transportation reject all programs or projects involving the use of publicly-owned land, unless no alternative exists;

3. The Federal Aid Highway Act now requires the Secretary of Transportation to promulgate guidelines to ensure full consideration of any adverse environmental effects of all federally supported highway projects during project planning and development;

4. The aforementioned Highway Act also requires the Secretary to promulgate highway noise level standards and air quality standards for highway construction activities;

5. Executive Orders 11507 and 11514 require Federal agency implementation of NEPA and outline the responsibilities of CEQ in comprehensive detail; and

6. Executive Order 11523 established the National Industrial Pollution Control Council, composed of 250 leaders of American industry, and charged the Council to provide the President with advice on industrial programs which affect the quality of the environment.

The federal tax laws now favor investment in solid wastes disposal and recycling facilities, and the new transportation regulations will have significant effects upon future and current collection and transportation of solid wastes. The Executive Orders mentioned above "put teeth" into NEPA and augment industrial research and development efforts in solid wastes management.

In addition to the general laws listed above, specific laws relating to certain aspects of the environment (e.g., air quality) also have importance in solid wastes management.

12.2. AIR QUALITY LAWS

Almost all Federal powers relating to the protection and enhancement of the air environment are contained in the Clean Air Act of 1970. This Act was written in three Titles, each pertaining to a different class of activities in air quality management. Title I covers air pollution control activities such as research, investigation, grants for planning and control activities, etc. Title II relates to emission standards for moving sources and control of air pollution resulting from transportation. Title III is a general statement on regulatory activities, emergency powers of regulators and similar aspects of air quality control. The Act provides for fines of as much as $25,000 per day of violation for first offenders and up to $50,000 per day of violation for repeating violators.

The Clean Air Act has tremendous significance for solid waste disposal activities, especially incineration operations. Many incinerators, particularly industrial installations, have been closed because of infractions of air pollution control regulations. Composting plants and sanitary landfills also are governed by provisions of this Act, both directly and indirectly (through such regulations imposing restrictions on the vehicles transporting collected wastes to the disposal sites).

The President has given added importance to the Clean Air Act through Executive Order 11602 by which the Environmental Protection Agency is charged with the task of compiling a list of convicted violators of the Clean Air Act who have not rectified the causes of the violation; all federal assistance is to be withheld from such violators pending efforts to remedy the offending situation. Funds would be withheld from solid wastes facilities which are in violation of air standards, according to this Order.

12.3. WATER QUALITY LAWS

The Federal Water Pollution Control Act and the 1972 Amendments thereto constitute the most comprehensive program ever enacted in the effort to ensure high quality in the Nation's waters. The goals expressed in the Amended Act include the development, by 1983, wherever possible, of waters sufficiently pure for swimming and other recreational uses and the sustenance of fish and other wildlife, and the total elimination, by 1985, of all discharges of pollutants into the Nation's waters. The specific actions (with strictly enforced deadlines and strong enforcement policies) necessitated by these goals will form the basis for a comprehensive program to improve water quality with implementation at the federal, state and local levels. This program for water quality improvement has significance in solid wastes management activities. Process water used in composting plants and municipal incinerators must be purified before being released into natural water supplies; similarly, the continued use of many sanitary landfills may be inadvisable because of the possible danger of water pollution through the generation of leachate. The significance of the newly amended Act is that strict controls and limits are now in existence to stop, even in planning stages, any possible water pollution from waste disposal or recycling activities.

Two other water-related laws with importance in solid wastes management are the Water Resources Planning Act of 1965 and the Fish and Wildlife Act. By means of the 1965 Act, the President has the power to establish river basin water and land resources commissions to coordinate plans for the development of water and land resources. Since several popular waste disposal methods use large amounts of land (sanitary landfill, windrow composting, etc.), the establishment of regulatory and planning bodies for land use is quite significant to waste disposal planning and design. Furthermore, the Fish and Wildlife Act requires that planners consult with the Fish and Wildlife Service prior to the development of any project or facility which may affect water quality; in this broad sense almost all disposal facilities would have some effect on the quality of water in bodies receiving facility wastewaters or effluents (incinerator process water, leachate from landfills, etc).

12.4. NOISE ABATEMENT LAWS

Noise pollution is somewhat more subtle than water and air pollution and only recently has public concern been focused on noise abatement. However, the Noise Pollution and Abatement Act of 1970 included provisions for the establishment within EPA of an Office of Noise Abatement and Control to study the causes and sources of noise and to develop recommendations to the President and the Congress for solutions to noise problems. Noise can be a serious problem in many solid waste management activities. The compactor-body collection truck is a source of noise familiar to many persons, especially those awakened during early-morning refuse collection. Disposal facilities also generate high levels of noise. The earth-moving and compacting equipment used at sanitary landfills and the shredders and grinders included in compost plants and recycling centers all are sources of high sound pressure levels. Table 12-1 shows noise levels recorded in the Gainesville, Florida, experimental compost plant during trial operations under HEW Demonstration Grant DO1-UI-00030, in 1967 and 1968. As is shown in the Table, noise levels at several locations in the plant exceeded limits currently considered by the federal government as indicating danger of hearing loss through prolonged exposure (90 dBA).

Considerations of noise abatement in the future will have important relevance to solid wastes collection, disposal and recycling operations because of the noise created by the powerful, automated equipment now being developed for waste transfer, compaction and processing.

12.5. SOLID WASTES LAWS

In addition to these general environmental laws and the more specific laws which pertain to other pollutants such as air and water but which indirectly affect solid waste management, there are laws at almost every level of government which pertain in some manner to solid waste alone.

TABLE 12-1 Noise Levels Within the Gainesville Plant

Location	Level (dB)	Weighting Network
Picking platform		
Middle Station	94–96	A,B,C,
Station nearest grinder	100–102	B,C
Average	96	A
Ground level at control		
station grinder	102	C
Primary grinder	100	C
Average	96	A
Digesting bins, start of		
conveyor belt	96	C
	81	B
	87	A
Outside grinder door	86	C
	81	B
	76	A
Ambient noise levels		
Picking platform	79	C
Digesting conveyor	70	C
Outside without blower	64	C

Source: USPHS, HEW

The most comprehensive of these laws and perhaps the most far reaching in its effects on solid waste management and research was the Solid Waste Disposal Act, Title II of Public Law 89-272, enacted on October 20, 1965. The purpose of this act as stated by the Congress and amended as noted below, was:

1. to promote the demonstration, construction, and application of solid waste management and resource recovery systems which preserve and enhance the quality of air, water, and land resources;
2. to provide technical and financial assistance to states and local governments and interstate agencies in the planning and development of resource recovery and solid waste disposal programs;
3. to promote a national research and development program for improved management techniques, more effective organizational arrangements, and new and improved methods of collection, separation, recovery, and recycling of solid wastes, and the environmentally safe disposal of nonrecoverable residues;
4. to provide for the promulgation of guidelines for solid waste collection, transport, separation, recovery, and disposal systems; and

5. to provide for training grants in occupations involving the design, operation, and maintenance of solid waste disposal systems.

This act was later amended by the passage of the Resources Recovery Act of 1970, Public Law 95-512 which was enacted October 27, 1970.

The original act delegated the responsibilities for enforcement with the Secretary of Health, Education and Welfare and the Secretary of Interior (for Mining and Fossil Fuel Wastes Management). At the time of the creation of the Environmental Protection Agency, the National Office of Solid Waste Management (and its responsibilities) was transferred into the newly created EPA. While both of these laws addressed themselves to the problem of solid waste disposal on a national level, specific enforcement and regulation on the local level was left to the various local governmental agencies.

The Federal laws actually regulate only two specific items with respect to waste management. In the 1970 act, P.L. 91-512, Congress gave a mandate to the Secretary of Health, Education and Welfare in regard to disposal of hazardous wastes to "submit to the Congress no later than two years after the date of enactment of the Resource Recovery Act of 1970, a comprehensive report and plan for the creation of a system of national disposal sites for the storage and disposal of hazardous wastes, including radioactive, toxic chemical, biological, and other wastes which may endanger the public health or welfare. Such report shall include: (1) a list of materials which should be subject to disposal in any such site; (2) current methods of disposal of such materials; (3) recommended methods of reduction, neutralization, recovery, or disposal of such materials; (4) an inventory of possible sites including existing land or water disposal sites operated or licensed by Federal agencies; (5) an estimate of the cost of developing and maintaining sites including consideration of means for distributing the short- and long-term costs of operating such sites among the users thereof; and (6) such other information as may be appropriate."

In addition to this provision, a committee composed of members from EPA, CEQ, and other environmental planning agencies is currently formulating a regulation which will govern the operation of sanitary landfills on federal lands. It is expected that these regulations will be used as guidelines by state and local regulatory agencies in establishing their laws.

12.6. STATE REGULATIONS

The regulation of solid waste management at the state levels can best be described as confused. In many states the responsibility for issuance and monitoring of permits for such disposal methods as incineration, landfill, etc., is split among the many agencies responsible for the control of air pollution, water pollution, solid waste disposal, mining and minerals, and zoning, to name a few.

In an attempt to bring some semblance of order to this rather chaotic condition several actions which have been taken recently are worthy of mention.

First, under grants provided by the Office of Solid Waste Management Programs, most states have formulated a management plan for the administration of solid waste disposal at the state level. Unfortunately, these plans, for the most part, are in nature only advisory to the local governments and many incorporate just the aspects of waste disposal under the control of the Divisions of Solid Waste.

A second change gradually occurring, with much more potential for reaching the heart of the problem (inter-agency squabbling, conflicting authority and regulations, and excessive and duplicated applications and permits) is the trend of the various individual states to establish a state equivalent of the federal EPA. These agencies will incorporate under a single director full authority not only for all solid waste regulations in the state, but also all other agencies responsible for the monitoring of any possible pollutions which may be caused by mismanagement of waste. A notable example of such an authority is the state of Illinois Environmental Protection Agency.

In order to assist the individual states in these areas, the Council of State Governments' Committee on Suggested State Legislation, Lexington, Kentucky, has proposed a model law entitled "State Solid Wastes Management and Resources Recovery Incentive Act". This act addresses itself to planning, regulation and enforcement, as well as operation and financing of solid wastes management within a state.

One vexing problem that none of these new proposals has discussed is that of an interstate or regional disposal authority. The matter of transporting solid waste across state lines has caused much concern in many areas; however, to date it has not been successfully solved except in a few isolated cases.

12.7. LOCAL LAWS AND REGULATIONS

On the local level the picture is even more complex. Regulations, permits and monitoring of pollution are split among numerous agencies, councils, and divisions. At this level, local politics and citizen pressures play a much more important part in decision making than they do at higher levels of government. At the local levels, the management of solid waste is generally accepted to be more of a nuisance to be tolerated at minimum cost and inconvenience to the local citizens. In these circumstances, most public hearings are held by the local politicians who are quite sensitive to the vocal groups that attend these hearings, usually to protest the action proposed.

In areas of "grass roots" politics, the disposal of solid waste has a unique sameness in all areas of the country. Everyone believes that the waste should be economically and efficiently handled, but deposited in the other end of town or in some other neighborhood.

As an example of a typical local disposal problem an actual case history is presented in Section 12.8.

12.8. A CASE HISTORY

Local politics and laws play a considerable role in the management (or non-management) of solid wastes in any community, but especially is this so in small to medium-sized cities surrounded by rural areas. To illustrate some of the potential difficulties which may be encountered in such a setting, an example of actual experience is described below. This example, is perhaps, a classic demonstration of some of these difficulties such as bureaucratic hesitancy, jealousy among regulatory agencies, lack of comprehensive and adequate land-use planning, and often mistaken but vocal public opinion.

In a Midwestern city with a population of less than 150,000 the existing solid waste disposal facility was a sanitary landfill. For over 15 years, private haulers had collected refuse in the city and deposited it in the landfill, within the city limits. The landfill in recent years had been nearing capacity and the private operator tried to anticipate the complete filling of the disposal site. The operator thus obtained an optional lease on a suitable piece of land and formally requested the county health department for permission to operate a sanitary landfill on the site; this permission was subsequently granted. However, when the operator approached the city council with an agreement to operate the approved site, the city fathers deemed the proposed operation too expensive and would not sign the agreement. Instead, after some months of delay, the city itself obtained a lease on a different site located just beyond the city boundaries. Although the state environmental conservation bureau approved the city's plans for operation of a landfill at the site, the board of supervisors of the county in which the site was located obtained a court injunction against a landfill operation by the city at that location. This move by the county government is somewhat typical; many persons working in an urban area characteristically oppose any transfer of wastes from that city into the suburban "bedroom" communities where they have their private homes. The city government subsequently appealed the decision to a higher court in an attempt to have the injunction dissolved. The higher court backed the city council and ordered the permit to be issued. In the meantime, the city council tried to circumvent any prolonged court fight by obtaining a different site. Their first attempt in this direction was also blocked when the state environmental conservation bureau refused to grant the city permission to operate a landfill at a site near a state park. The city sanitation authorities then suggested the use of a site within the city limits but near a low-income housing district. *Protests by citizen groups forced the city council to turn down this suggestion of their own personnel.*

The appeals court decision allowing the city to operate a landfill outside the city limits was granted at about this time. The county government (in County "A") then carried their appeal to the state supreme court and the operation in County "A" was temporarily abandoned, pending the supreme court's decision. The city council becoming somewhat desperate as the existing landfill site began to near completion, then accepted the offer of a private hauler to operate a landfill at a site considerably more distant from the city, in County "B". At this point the county government in County "B" enacted a local ordinance prohibiting the transport and deposition, in County "B", of refuse generated at any point outside that county. The private refuse hauler who was seeking the contract with the city then went to the courts to fight the local ban on "foreign" wastes. A lower court declared the county law to be invalid and the officials of County "B" immediately appealed that decision in a higher court. While this particular court battle was being fought, and the original struggle with County "A" was continuing, the state environmental conservation bureau officially closed the overloaded existing landfill. The city was forced to begin emergency use of a small site adjacent to the old landfill. Use of that site, which had been planned for use by an industrial developer, was granted only under the condition that any refuse deposited there would be removed as soon as a permanent site was obtained. This small site was filled to a height of more than 20 feet above the original ground level when the state supreme court finally handed down a decision in the controversy between the city and County "A". The court in its decision decided that the city was not subject to county zoning and land-use regulations. The city was granted the authority to operate a sanitary landfill at the designated site without the permission of the county government. The basis for the court's decision was a state law granting municipalities the power to purchase or acquire property "within or without the corporate limits of such city ... for the purpose of providing facilities for the disposal of garbage, refuse or ashes." However, the court did caution city officials that if the operation did not meet environmental protection standards or if the city exceeded its statutory authority, the court would intervene and enjoin the operation.

12.9 SUMMARY

The example cited in the preceding article may seem extreme, but similar situations have been created and are now existing in many other municipalities throughout the nation. While the passage of federal legislation and the implementation of national programs to encourage proper waste management have done little, to date, to improve actual waste management on the local level, it appears that the thrust for any meaningful regional legislation must come from the Congress. It is hardly realistic to expect cities and counties, or even states, to pass legislation with regard to waste management or for taxes to pay for adequate disposal that places them at an economic disadvantage with respect to their neighboring communities.

If resources recovery and recycling are ever to be successful, federal legislation will be required in the areas of transportation costs, adjustments of depreciation allowances and similar tax legislation to favor the reuse of waste material. The public must be educated to accept these materials as the equivalent of virgin materials and the federal regulatory agencies must change their requirement whenever possible to remove any social stigma from the manufacture of goods from recycled materials. Some suggested means of achieving effective solid waste management are discussed in the next chapter.

13
Effective solid waste management

13-1. GENERAL

An examination of present-day technology for solid waste collection and disposal will reveal a capability for effectively and efficiently removing such wastes from the physical environment. Applied research efforts have proven the efficacy and feasibility of recycling and salvage techniques which, when practiced in conjunction with modern collection and disposal methods, can do much to reduce depletion of non-renewable resources and to maintain valuable commodities in use. Current studies and investigations promise the development of still more effective and efficient procedures for waste disposition and salvage, as well as the potential for improvement in collection practices. Finally, the creation of new materials and new uses of existing materials designed for ease of collection and disposition after use may further alleviate much of the current solid waste dilemma. However, at this point in time it is permissible to say only that present and future technology *may* alleviate the problem; there is no guarantee that more effective techniques of collection or disposal or more suitable materials will ever be used. Before existing or future technology may be employed to lessen solid waste pollution problems, there must be generated a desire and willingness on the part of all the population to seek and use improved technology and better materials.

The situation in the field of solid waste management today is somewhat analogous to the predicament of a highly trained and educated medical doctor or dentist set down with an array of shining new equipment and apparatus in the midst of an isolated rain forest instead of an air-conditioned suburban office. Such a person has the technology and personal ability to alleviate much of the suffering and sickness of the primitive jungle dwellers. However, his most serious difficulty is, in most cases, to persuade reluctant natives to become his patients and to accept his remedies. A similar difficulty faces anyone who seeks to institute more effective solid waste management practices, with his "reluctant patients" being the general public.

Since solid wastes are generated by the entire population, by all segments of society, any solution to the problem of solid wastes must arise from and be supported by that same population. The general public, it seems, will not initiate or support any improved techniques in waste management unless and until the scope, magnitude, and severity of the solid waste crisis is demonstrated to each man and woman. Such an educational step may be the most significant single undertaking in the achievement of successful management. Of necessity, the first step in presenting the problems of collection and disposal to the populace of a city or region is to gather pertinent data on characteristics and quantities of solid wastes currently being generated in that particular city or region.

13-2. CONDUCT OF A REGIONAL SOLID WASTES SURVEY

Without a comprehensive compilation of data on solid waste generation, characteristics, and collection and disposal in a given area, it is impossible to fully appreciate the total problem of solid waste management in that area or to formulate an effective solution to it. Moreover, it will be impossible to forecast future needs for management in the area of service. Finally, without adequate data on population, waste production, and political conditions, it is not feasible to attempt the formulation of any sort of scheme or plan for systematic waste collection and disposal or recycling.

It is important to consider several categories of information for examination during a survey. Such categories include political jurisdictions and entities in the study region; information from previous planning or development studies; present waste sources including municipal, industrial, agricultural, mineral-processing, and rural (domestic) origins; present provisions for collection and transportation of wastes; distribution of present and predicted populations; land-use restrictions and zoning regulations for the study area (including designation of residential zones as "single-family dwelling," "duplex," "triplex," or "four-plex" apartment dwelling, "large multiple-unit" dwelling, etc.); community fiscal and legislative restrictions (including present expenditures for waste management, potential for tax revenues, public financing possibilities, etc.); and public awareness and appreciation of the solid waste management problem in the local community. All of these categories of information are pertinent to the development of a plan for waste management. Because there is a need for such a large quantity and diversity of data from a large number of sources, it is imperative that a professional body of specialists in information retrieval and procurement be employed to obtain any necessary data not already available. Of course, previous planning surveys or investigations may furnish significant amounts of such necessary information; for example, comprehensive demographic data may be obtained from official United States Census reports. Information on industrial production, commercial business volume, and projected economic growth or decay may be obtained from local chambers of commerce or business organizations. However, as mentioned in an earlier chapter of this book, there is a dearth of reliable information on quantities and characteristics of solid wastes generated in any given locality. At best, the information gathered during the 1968 National Solid Waste Survey can furnish a planner or engineer with a preliminary estimate of the amounts and types of solid wastes which should be anticipated from an area on the basis of the population, geographic location, and degree of urbanization of the region. Therefore, in order to obtain the information required for subsequent planning activities, in almost every case it will be necessary to conduct one or more local surveys of solid waste parameters. For guidance in the conduct of such a survey, the reader is referred to the publications of the Office of Solid Waste Management Programs, Environmental Protection Agency, and especially Refs. 13-1, 13-2, 13-3 and 13-4 listed at the end of this chapter. Actual surveys are described in Refs. 13-5 and 13-6.

13-3. USING SURVEY RESULTS—THE REGIONAL PLAN

In any attempt to utilize the information gathered in such a comprehensive solid wastes survey as described in the preceding section, several important factors should receive attention. These factors are operational "economies of size"; geographic interactions in pollutant generation; and political pitfalls associated with "compartmentalized" management.

Economies of size are realized in the operation of disposal facilities such as incinerators, composting plants and sanitary landfills because there are certain capital outlays necessary to the operation of any one of these facilities no matter what the size. Thus, for example, the cost *per ton of waste processed* for a hammermill used in conjunction with a composting plant of 100 tons/day capacity will be only three-quarters the cost for the same mill used in a plant of 75 tons/day capacity (provided, of course, that the same mill can handle the 100 tons/day amount). The economies of size related to operation of sanitary landfills have been presented graphically in Fig. 10-1 in Chapter 10. Similar economies may also be attained in the operation of incinerators and composting plants. Moreover in the operation of very large collection networks, the volume of operations may make possible the purchase and use of specialized equipment which may not be economically operable in smaller collection nets. For example, in a large urban area the volume of leaves to be collected in autumn may be quite large and may easily justify the purchase of specialized vacuum collector trucks. The purchase and use of such trucks may not be feasible in small towns or isolated communities. Consequently, more time-consuming and expensive methods of manual leaf collection may be required in such areas.

A second factor is the geographic aspects of pollutant distribution. In this context "geography" may almost be taken to mean political geography. The basic fact which waste managers must always recognize and bear in mind is that pollutants are generated and distributed with little or no dependence upon geographic and political boundaries. Air pollutants generated in one city or town may easily be carried by winds across city, state, or national boundaries into a "foreign territory." Similarly, pollutants in suspension or in solution in a stream may traverse local, state, and federal boundaries in their journey from point of origin to point of deposition. Solid wastes at first may appear to be somewhat different in character from the more mobile air and water pollutants, but a realistic examination of many localities in the United States will show the most suitable location for solid waste disposal to be situated on the other side of a township, city, or state border from the generation area for such wastes. Likewise, efficient networks for collection of solid wastes may often extend across city or state boundaries.

It should be obvious from the foregoing that the very nature of pollutant collection and disposal (since for utmost efficacy and efficiency it must traverse political boundaries and geographically intrude into various political entities) begets political difficulties and disputes. Since population distribution seldom reflects political borders in the case of township or suburb boundary limits, the distribution of solid waste generation is seldom confined within such political boundaries. Therefore, the stage is set in many localities for confrontations between planners and engineers who are seeking an effective and optimum solution to the solid waste problem by collecting and disposing waste "across county lines," and residents of the areas to be traversed by collection trucks or prospective neighbors of proposed sanitary landfills. Many solid waste managers have as a result received a "baptism of fire" in public meetings with residents of "county X" when they have suggested transporting refuse from "county Y" into or across "county X." Innumerable injunctions and stop orders have been granted on the pleas of irate citizens of the offended county. In other cases, planners and engineers have developed schemes for refuse collection and disposal which from the beginning were doomed to failure because no single political entity within the collection area possessed legal authority to impose taxation or sell bonds to pay for the proposed scheme. Political pitfalls abound in the field of solid waste management, particularly for the engineer or planner who is politically naive.

No simple solution exists to the problems outlined above, but these conditions do suggest that the best approach to management of solid wastes is the development of a regional authority vested with appropriate political powers needed for implementation of engineered plans. By the establishment of such an authority it is possible to achieve economies of size in operation, to establish ways and means of controlling the geographic distribution of pollutants, and to overcome the political pitfalls associated with piecemeal operations. Moreover, in creating such an authority it is often possible to draw together a group of professional managers—engineers, economists, planners—and remove them from the sphere of political influence. Legislative action to create such quasi-public utilities can be framed to include profit limitations, limits on bonded indebtedness, and restrictions on capital expenditures. Finally, tariff, toll, and tax imposition powers may be legislatively granted to such a regional solid waste management authority. Any operation of such an authority should of course be carried out according to modern management techniques.

The management of solid wastes collection and disposal is best accomplished if the operation of the required facilities is managed as an engineered system. Modern concepts of systems analysis should be applied to solid wastes collection, transfer, and disposal operations. Optimization techniques should be employed in the allocation of material and labor resources to various unit operations within the overall handling system. Scheduling methods such as the Critical Path Method should be used in determining focal unit operations and establishing "bottlenecks" to the flow of refuse from point of origin to final disposal destination. The use of such methods and techniques in the management of solid wastes systems will allow the realization of significant economies in resource utilization and operational costs. The dramatic impact of these management methods on the construction industry can serve as an example of what may be accomplished in the realm of solid wastes management.

13-4. PUBLIC RELATIONS ACTIVITIES

It has been indicated in an earlier chapter of this book that the most effective means of alleviating the solid waste dilemma is the interception of individual types of wastes before they are mixed with other wastes and the recycling of the separated materials into use again. In effect, the solid waste stream would be diverted at the source and transformed into reusable materials. As mentioned in that earlier discussion, such recycling and reuse of "waste" materials is not possible without the full support and active participation of the bulk of the public. If a sizable portion of the general public resists community efforts to establish a newspaper recycling program, for example, it is highly unlikely that that program will be successful. Since the homeowner can separate refuse at its point of generation before the various kinds of refuse are mixed, he is in a commanding position to create or cripple any sort of reclamation scheme. Obviously, a planner or engineer who wishes to incorporate salvage operations as part of a solid waste management plan must solicit and obtain the voluntary cooperation of the majority of the householders in the service area if the salvage operation is to succeed.

The necessity for citizen support is not limited to salvage or recycling programs. Since the individual citizen must pay for any solid waste collection and disposal, it is essential that he understand and approve of what he is asked to financially support. Also, because of the frequent necessity to conduct solid waste handling and disposition across political boundary lines it is virtually mandatory that citizens on both sides of such boundaries acquiesce in and lend support to such transport. As a consequence of the need for citizen support of solid waste management programs, it becomes necessary to educate the general public in modern technology for refuse collection and disposal. Along with this educational effort, an active program of persuasion will almost always be necessary in order to "present the case" for the proposed solid waste management program. A group of competent specialists must be retained not only because of the need for a comprehensive education and publicity campaign, but also because of the presently existing negative attitude of the average citizen toward almost all forms of solid waste collection and disposal operations.

A long history of inadequate, makeshift, or nonexistent collection and disposal operations as the normal practice in almost every community in the United States from the time of its founding until the last few years has made a deep and lasting impression on the minds of many Americans. The experience of years of burning, rat-infested dumps; odorous compost plants surrounded by clouds of flies; grimy architectural monstrosities (mistakenly termed incinerators) belching gases and particulate matter into the atmosphere; and smelly, dirty, and noisy collection trucks have generated feelings of hostility, aversion, and disgust toward refuse handling and disposal in the minds of almost all adult and adolescent Americans. Such feelings naturally predispose these same persons against improved means and methods of waste handling. Few average citizens in the United States today know that there are vast differences between a dump and a sanitary landfill, for example. The term "compost" evokes an image in the minds of many persons of an enormous pile of fly-enshrouded rotting manure. For many people the word *incinerator* recalls past incidents of glowing embers and partially burned papers floating in the sky near refuse burners. Altogether, there is an overwhelming backlog of dangerous and/or discomfiting experiences in the lives of almost every citizen who has come into contact with "traditional" methods of refuse collection and disposal.

13-5. SUMMARY

If the modern methods and techniques described in this book are ever to be employed successfully in alleviating the solid waste crisis existing today, a massive effort of reeducating the public must be undertaken. Americans during the last decade have been persuaded to recognize and take action against factors detrimental to the quality of their lives such as litter, highway accidents, and smoking. The persuasion used in the campaigns against these nuisances has been sophisticated, subtle, and massive. A similar program of publicity must be conducted on both local and national levels to present the crisis of solid waste disposal in proper perspective and to offer proposals for solutions to the problem. Above all, every man and woman in this country must be informed of (and must accept) the fact that any solution to the solid wastes crisis will involve sacrifices on their parts in the form of higher taxes, increased inconvenience, and some personal effort. Only by such personal sacrifice can any solution to this crucial national problem be attained.

REFERENCES

13-1. Muhich, A. J., Klee, A. J., and Britton, P., *Preliminary Data Analysis, 1968 National Survey of Community Solid Waste Practices*, Off. of Solid Waste Man. Prog., USPHS, HEW, Cincinnati, 1968.

13-2. Combustion Engrg., Inc., *Technical, Economic Study of Solid Waste Disposal Needs and Practices*, Off. of Solid Waste Man. Prog., USPHS, HEW, Cincinnati, 1969.

13-3. Morse, N., and Roth, E., *Systems Analysis of Regional Solid Waste Handling*, Off. of Solid Waste Man. Prog., USPHS, HEW, Washington, D.C., 1970.

13-4. Toftner, R. O., *Developing a Solid Waste Management Plan*, Off. of Solid Waste Man. Prog., USPHS, HEW, Washington, D.C., 1970.

13-5. University of Louisville, *Louisville Metropolitan Region Solid Waste Disposal Study*, Off. of Solid Waste Man. Prog., USPHS, HEW, Cincinnati, 1970.

13-6. Aerojet-General Corp. and Engr.-Science, Inc., *A Systems Study of Solid Waste Management in the Fresno Area*, Off. of Solid Waste Man. Prog., USPHS, HEW, Cincinnati, 1969.

Appendix: Recovery Act of 1970

Public Law 91-512
91st Congress, H. R. 11833
October 26, 1970

An Act

To amend the Solid Waste Disposal Act in order to provide financial assistance for the construction of solid waste disposal facilities, to improve research programs pursuant to such Act, and for other purposes.

Be it enacted by the Senate and House of Representatives of the United States of America in Congress assembled, That this Act may be cited as the "Resource Recovery Act of 1970".

Resource Recovery Act of 1970.

TITLE I—RESOURCE RECOVERY

SEC. 101. Section 202(b) of the Solid Waste Disposal Act is amended to read as follows:

79 Stat. 997.
42 USC 3251.

"(b) The purposes of this Act therefore are—
"(1) to promote the demonstration, construction, and application of solid waste management and resource recovery systems which preserve and enhance the quality of air, water, and land resources;
"(2) to provide technical and financial assistance to States and local governments and interstate agencies in the planning and development of resource recovery and solid waste disposal programs;

84 STAT. 1227
84 STAT. 1228

"(3) to promote a national research and development program for improved management techniques, more effective organizational arrangements, and new and improved methods of collection, separation, recovery, and recycling of solid wastes, and the environmentally safe disposal of nonrecoverable residues;
"(4) to provide for the promulgation of guidelines for solid waste collection, transport, separation, recovery, and disposal systems; and
"(5) to provide for training grants in occupations involving the design, operation, and maintenance of solid waste disposal systems."

SEC. 102. Section 203 of the Solid Waste Disposal Act is amended by inserting at the end thereof the following:

Definitions.
42 USC 3252.

"(7) The term 'municipality' means a city, town, borough, county, parish, district, or other public body created by or pursuant to State law with responsibility for the planning or administration of solid waste disposal, or an Indian tribe.
"(8) The term 'intermunicipal agency' means an agency established by two or more municipalities with responsibility for planning or administration of solid waste disposal.
"(9) The term 'recovered resources' means materials or energy recovered from solid wastes.
"(10) The term 'resource recovery system' means a solid waste management system which provides for collection, separation, recycling, and recovery of solid wastes, including disposal of nonrecoverable waste residues."

SEC. 103. (a) Section 204(a) of the Solid Waste Disposal Act is amended to read as follows:

Research, authority of Secretary.
42 USC 3253.

"SEC. 204. (a) The Secretary shall conduct, and encourage, cooperate with, and render financial and other assistance to appropriate public (whether Federal, State, interstate, or local) authorities, agencies, and institutions, private agencies and institutions, and individuals in the

51-913 O

283

conduct of, and promote the coordination of, research, investigations, experiments, training, demonstrations, surveys, and studies relating to—

"(1) any adverse health and welfare effects of the release into the environment of material present in solid waste, and methods to eliminate such effects;

"(2) the operation and financing of solid waste disposal programs;

"(3) the reduction of the amount of such waste and unsalvageable waste materials;

"(4) the development and application of new and improved methods of collecting and disposing of solid waste and processing and recovering materials and energy from solid wastes; and

"(5) the identification of solid waste components and potential materials and energy recoverable from such waste components."

(b) Section 204(d) of the Solid Waste Disposal Act is repealed.

Repeal.
79 Stat. 999.
42 USC 3253.
42 USC 3255, 3254.

SEC. 104. (a) The Solid Waste Disposal Act is amended by striking out section 206, by redesignating section 205 as 206, and by inserting after section 204 the following new section:

"SPECIAL STUDY AND DEMONSTRATION PROJECTS ON RECOVERY OF USEFUL ENERGY AND MATERIALS

"SEC. 205. (a) The Secretary shall carry out an investigation and study to determine—

84 STAT. 1228
84 STAT. 1229

"(1) means of recovering materials and energy from solid waste, recommended uses of such materials and energy for national or international welfare, including identification of potential markets for such recovered resources, and the impact of distribution of such resources on existing markets;

"(2) changes in current product characteristics and production and packaging practices which would reduce the amount of solid waste;

"(3) methods of collection, separation, and containerization which will encourage efficient utilization of facilities and contribute to more effective programs of reduction, reuse, or disposal of wastes;

"(4) the use of Federal procurement to develop market demand for recovered resources;

"(5) recommended incentives (including Federal grants, loans, and other assistance) and disincentives to accelerate the reclamation or recycling of materials from solid wastes, with special emphasis on motor vehicle hulks;

"(6) the effect of existing public policies, including subsidies and economic incentives and disincentives, percentage depletion allowances, capital gains treatment and other tax incentives and disincentives, upon the recycling and reuse of materials, and the likely effect of the modification or elimination of such incentives and disincentives upon the reuse, recycling, and conservation of such materials; and

"(7) the necessity and method of imposing disposal or other charges on packaging, containers, vehicles, and other manufactured goods, which charges would reflect the cost of final disposal, the value of recoverable components of the item, and any social costs associated with nonrecycling or uncontrolled disposal of such items.

Report to President and Congress.

The Secretary shall from time to time, but not less frequently than annually, report the results of such investigation and study to the President and the Congress.

October 26, 1970 -3- Pub. Law 91-512

"(b) The Secretary is also authorized to carry out demonstration projects to test and demonstrate methods and techniques developed pursuant to subsection (a).

"(c) Section 204 (b) and (c) shall be applicable to investigations, studies, and projects carried out under this section."

(b) The Solid Waste Disposal Act is amended by redesignating sections 207 through 210 as sections 213 through 216, respectively, and by inserting after section 206 (as so redesignated by subsection (a) of this section) the following new sections:

Demonstration projects.

79 Stat. 998.
42 USC 3253.

82 Stat. 1013.
42 USC 3256-3259.

"GRANTS FOR STATE, INTERSTATE, AND LOCAL PLANNING

"SEC. 207. (a) The Secretary may from time to time, upon such terms and conditions consistent with this section as he finds appropriate to carry out the purposes of this Act, make grants to State, interstate, municipal, and intermunicipal agencies, and organizations composed of public officials which are eligible for assistance under section 701(g) of the Housing Act of 1954, of not to exceed 66⅔ per centum of the cost in the case of an application with respect to an area including only one municipality, and not to exceed 75 per centum of the cost in any other case, of—

"(1) making surveys of solid waste disposal practices and problems within the jurisdictional areas of such agencies and

"(2) developing and revising solid waste disposal plans as part of regional environmental protection systems for such areas, providing for recycling or recovery of materials from wastes whenever possible and including planning for the reuse of solid waste disposal areas and studies of the effect and relationship of solid waste disposal practices on areas adjacent to waste disposal sites,

"(3) developing proposals for projects to be carried out pursuant to section 208 of this Act, or

"(4) planning programs for the removal and processing of abandoned motor vehicle hulks.

"(b) Grants pursuant to this section may be made upon application therefor which—

"(1) designates or establishes a single agency (which may be an interdepartmental agency) as the sole agency for carrying out the purposes of this section for the area involved;

"(2) indicates the manner in which provision will be made to assure full consideration of all aspects of planning essential to areawide planning for proper and effective solid waste disposal consistent with the protection of the public health and welfare, including such factors as population growth, urban and metropolitan development, land use planning, water pollution control, air pollution control, and the feasibility of regional disposal and resource recovery programs;

"(3) sets forth plans for expenditure of such grant, which plans provide reasonable assurance of carrying out the purposes of this section;

"(4) provides for submission of such reports of the activities of the agency in carrying out the purposes of this section, in such form and containing such information, as the Secretary may from time to time find necessary for carrying out the purposes of this section and for keeping such records and affording such access thereto as he may find necessary; and

"(5) provides for such fiscal-control and fund-accounting procedures as may be necessary to assure proper disbursement of and accounting for funds paid to the agency under this section.

82 Stat. 530.
40 USC 461.
Cost limitation.

84 STAT. 1229
84 STAT. 1230

"(c) The Secretary shall make a grant under this section only if he finds that there is satisfactory assurance that the planning of solid waste disposal will be coordinated, so far as practicable, with and not duplicate other related State, interstate, regional, and local planning activities, including those financed in part with funds pursuant to section 701 of the Housing Act of 1954.

82 Stat. 526.
40 USC 461.

"GRANTS FOR RESOURCE RECOVERY SYSTEMS AND IMPROVED SOLID WASTE DISPOSAL FACILITIES

"SEC. 208. (a) The Secretary is authorized to make grants pursuant to this section to any State, municipal, or interstate or intermunicipal agency for the demonstration of resource recovery systems or for the construction of new or improved solid waste disposal facilities.

"(b)(1) Any grant under this section for the demonstration of a resource recovery system may be made only if it (A) is consistent with any plans which meet the requirements of section 207(b)(2) of this Act; (B) is consistent with the guidelines recommended pursuant to section 209 of this Act; (C) is designed to provide areawide resource recovery systems consistent with the purposes of this Act, as determined by the Secretary, pursuant to regulations promulgated under subsection (d) of this section; and (D) provides an equitable system for distributing the costs associated with construction, operation, and maintenance of any resource recovery system among the users of such system.

Federal share, limitation.

"(2) The Federal share for any project to which paragraph (1) applies shall not be more than 75 percent.

"(c)(1) A grant under this section for the construction of a new or improved solid waste disposal facility may be made only if—

84 STAT. 1230
84 STAT. 1231

"(A) a State or interstate plan for solid waste disposal has been adopted which applies to the area involved, and the facility to be constructed (i) is consistent with such plan, (ii) is included in a comprehensive plan for the area involved which is satisfactory to the Secretary for the purposes of this Act, and (iii) is consistent with the guidelines recommended under section 209, and

"(B) the project advances the state of the art by applying new and improved techniques in reducing the environmental impact of solid waste disposal, in achieving recovery of energy or resources, or in recycling useful materials.

"(2) The Federal share for any project to which paragraph (1) applies shall be not more than 50 percent in the case of a project serving an area which includes only one municipality, and not more than 75 percent in any other case.

Regulations.

"(d)(1) The Secretary, within ninety days after the date of enactment of the Resource Recovery Act of 1970, shall promulgate regulations establishing a procedure for awarding grants under this section which—

"(A) provides that projects will be carried out in communities of varying sizes, under such conditions as will assist in solving the community waste problems of urban-industrial centers, metropolitan regions, and rural areas, under representative geographic and environmental conditions; and

"(B) provides deadlines for submission of, and action on, grant requests.

"(2) In taking action on applications for grants under this section, consideration shall be given by the Secretary (A) to the public benefits to be derived by the construction and the propriety of Federal aid in making such grant; (B) to the extent applicable, to the economic and commercial viability of the project (including contractual arrangements with the private sector to market any resources recovered); (C) to the potential of such project for general application to community solid waste disposal problems; and (D) to the use by the applicant of comprehensive regional or metropolitan area planning.

"(e) A grant under this section—

"(1) may be made only in the amount of the Federal share of (A) the estimated total design and construction costs, plus (B) in the case of a grant to which subsection (b)(1) applies, the first-year operation and maintenance costs;

"(2) may not be provided for land acquisition or (except as otherwise provided in paragraph (1)(B) for operating or maintenance costs;

"(3) may not be made until the applicant has made provision satisfactory to the Secretary for proper and efficient operation and maintenance of the project (subject to paragraph (1)(B)); and

"(4) may be made subject to such conditions and requirements, in addition to those provided in this section, as the Secretary may require to properly carry out his functions pursuant to this Act. For purposes of paragraph (1), the non-Federal share may be in any form, including, but not limited to, lands or interests therein needed for the project or personal property or services, the value of which shall be determined by the Secretary.

"(f)(1) Not more than 15 percent of the total of funds authorized to be appropriated under section 216(a)(3) for any fiscal year to carry out this section shall be granted under this section for projects in any one State.

Limitation.
Post, p. 1234.

"(2) The Secretary shall prescribe by regulation the manner in which this subsection shall apply to a grant under this section for a project in an area which includes all or part of more than one State.

Regulation.

84 STAT. 1231
84 STAT. 1232

"RECOMMENDED GUIDELINES

"SEC. 209. (a) The Secretary shall, in cooperation with appropriate State, Federal, interstate, regional, and local agencies, allowing for public comment by other interested parties, as soon as practicable after the enactment of the Resource Recovery Act of 1970, recommend to appropriate agencies and publish in the Federal Register guidelines for solid waste recovery, collection, separation, and disposal systems (including systems for private use), which shall be consistent with public health and welfare, and air and water quality standards and adaptable to appropriate land-use plans. Such guidelines shall apply to such systems whether on land or water and shall be revised from time to time.

Publication in Federal Register.

"(b)(1) The Secretary shall, as soon as practicable, recommend model codes, ordinances, and statutes which are designed to implement this section and the purposes of this Act.

"(2) The Secretary shall issue to appropriate Federal, interstate, regional, and local agencies information on technically feasible solid waste collection, separation, disposal, recycling, and recovery methods, including data on the cost of construction, operation, and maintenance of such methods.

"GRANTS OR CONTRACTS FOR TRAINING PROJECTS

"SEC. 210.(a) The Secretary is authorized to make grants to, and contracts with, any eligible organization. For purposes of this section the term 'eligible organization' means a State or interstate agency, a municipality, educational institution, and any other organization which is capable of effectively carrying out a project which may be funded by grant under subsection (b) of this section.

"Eligible organization."

Pub. Law 91-512 -6- October 26, 1970

"(b)(1) Subject to the provisions of paragraph (2), grants or contracts may be made to pay all or a part of the costs, as may be determined by the Secretary, of any project operated or to be operated by an eligible organization, which is designed—

"(A) to develop, expand, or carry out a program (which may combine training, education, and employment) for training persons for occupations involving the management, supervision, design, operation, or maintenance of solid waste disposal and resource recovery equipment and facilities; or

"(B) to train instructors and supervisory personnel to train or supervise persons in occupations involving the design, operation, and maintenance of solid waste disposal and resource recovery equipment and facilities.

Ante, p. 1230.

"(2) A grant or contract authorized by paragraph (1) of this subsection may be made only upon application to the Secretary at such time or times and containing such information as he may prescribe, except that no such application shall be approved unless it provides for the same procedures and reports (and access to such reports and to other records) as is required by section 207(b)(4) and (5) with respect to applications made under such section.

Study.

"(c) The Secretary shall make a complete investigation and study to determine—

"(1) the need for additional trained State and local personnel to carry out plans assisted under this Act and other solid waste and resource recovery programs;

"(2) means of using existing training programs to train such personnel; and

84 STAT. 1232
84 STAT. 1233

"(3) the extent and nature of obstacles to employment and occupational advancement in the solid waste disposal and resource recovery field which may limit either available manpower or the advancement of personnel in such field.

Report to President and Congress.

He shall report the results of such investigation and study, including his recommendations to the President and the Congress not later than one year after enactment of this Act.

"APPLICABILITY OF SOLID WASTE DISPOSAL GUIDELINES TO EXECUTIVE AGENCIES

"SEC. 211. (a)(1) If—

80 Stat. 379.

"(A) an Executive agency (as defined in section 105 of title 5, United States Code) has jurisdiction over any real property or facility the operation or administration of which involves such agency in solid waste disposal activities, or

"(B) such an agency enters into a contract with any person for the operation by such person of any Federal property or facility, and the performance of such contract involves such person in solid waste disposal activities,

Compliance.

then such agency shall insure compliance with the guidelines recommended under section 209 and the purposes of this Act in the operation or administration of such property or facility, or the performance of such contract, as the case may be.

"(2) Each Executive agency which conducts any activity—

"(A) which generates solid waste, and

"(B) which, if conducted by a person other than such agency, would require a permit or license from such agency in order to dispose of such solid waste,

shall insure compliance with such guidelines and the purposes of this Act in conducting such activity.

"(3) Each Executive agency which permits the use of Federal property for purposes of disposal of solid waste shall insure compliance

288

October 26, 1970   -7-   Pub. Law 91-512

with such guidelines and the purposes of this Act in the disposal of such waste.

"(4) The President shall prescribe regulations to carry out this subsection.   Presidential regulations.

"(b) Each Executive agency which issues any license or permit for disposal of solid waste shall, prior to the issuance of such license or permit, consult with the Secretary to insure compliance with guidelines recommended under section 209 and the purposes of this Act.

"NATIONAL DISPOSAL SITES STUDY

"SEC. 212. The Secretary shall submit to the Congress no later than two years after the date of enactment of the Resource Recovery Act of 1970, a comprehensive report and plan for the creation of a system of national disposal sites for the storage and disposal of hazardous wastes, including radioactive, toxic chemical, biological, and other wastes which may endanger public health or welfare. Such report shall include: (1) a list of materials which should be subject to disposal in any such site; (2) current methods of disposal of such materials; (3) recommended methods of reduction, neutralization, recovery, or disposal of such materials; (4) an inventory of possible sites including existing land or water disposal sites operated or licensed by Federal agencies; (5) an estimate of the cost of developing and maintaining sites including consideration of means for distributing the short- and long-term costs of operating such sites among the users thereof; and (6) such other information as may be appropriate."   Report to Congress.

(c) Section 215 of the Solid Waste Disposal Act (as so redesignated by subsection (b) of this section) is amended by striking out the heading thereof and inserting in lieu thereof "GENERAL PROVISIONS"; by inserting "(a)" before "Payments"; and by adding at the end thereof the following: Ante, p. 1229.
84 STAT. 1233
84 STAT. 1234

"(b) No grant may be made under this Act to any private profit-making organization."   Grants, prohibition.

SEC. 105. Section 216 of the Solid Waste Disposal Act (as so redesignated by section 104 of this Act) is amended to read as follows: Appropriation.

"SEC. 216. (a)(1) There are authorized to be appropriated to the Secretary of Health, Education, and Welfare for carrying out the provisions of this Act (including, but not limited to, section 208), not to exceed $41,500,000 for the fiscal year ending June 30, 1971.

"(2) There are authorized to be appropriated to the Secretary of Health, Education, and Welfare to carry out the provisions of this Act, other than section 208, not to exceed $72,000,000 for the fiscal year ending June 30, 1972, and not to exceed $76,000,000 for the fiscal year ending June 30, 1973.

"(3) There are authorized to be appropriated to the Secretary of Health, Education, and Welfare to carry out section 208 of this Act not to exceed $80,000,000 for the fiscal year ending June 30, 1972, and not to exceed $140,000,000 for the fiscal year ending June 30, 1973.

"(b) There are authorized to be appropriated to the Secretary of the Interior to carry out this Act not to exceed $8,750,000 for the fiscal year ending June 30, 1971, not to exceed $20,000,000 for the fiscal year ending June 30, 1972, and not to exceed $22,500,000 for the fiscal year ending June 30, 1973. Prior to expending any funds authorized to be appropriated by this subsection, the Secretary of the Interior shall consult with the Secretary of Health, Education, and Welfare to assure that the expenditure of such funds will be consistent with the purposes of this Act.

Pub. Law 91-512   -8-   October 26, 1970

Program evaluation.

"(c) Such portion as the Secretary may determine, but not more than 1 per centum, of any appropriation for grants, contracts, or other payments under any provision of this Act for any fiscal year beginning after June 30, 1970, shall be available for evaluation (directly, or by grants or contracts) of any program authorized by this Act.

Funds, availability.

"(d) Sums appropriated under this section shall remain available until expended."

TITLE II—NATIONAL MATERIALS POLICY

Citation of title.

SEC. 201. This title may be cited as the "National Materials Policy Act of 1970".

SEC. 202. It is the purpose of this title to enhance environmental quality and conserve materials by developing a national materials policy to utilize present resources and technology more efficiently, to anticipate the future materials requirements of the Nation and the world, and to make recommendations on the supply, use, recovery, and disposal of materials.

National Commission on Materials Policy. Establishment. Membership.

SEC. 203. (a) There is hereby created the National Commission on Materials Policy (hereafter referred to as the "Commission") which shall be composed of seven members chosen from Government service and the private sector for their outstanding qualifications and demonstrated competence with regard to matters related to materials policy, to be appointed by the President with the advice and consent of the Senate, one of whom he shall designate as Chairman.

Travel expenses, etc.
84 STAT. 1234
84 STAT. 1235
Study.

(b) The members of the Commission shall serve without compensation, but shall be reimbursed for travel, subsistence, and other necessary expenses incurred by them in carrying out the duties of the Commission.

SEC. 204. The Commission shall make a full and complete investigation and study for the purpose of developing a national materials policy which shall include, without being limited to, a determination of—

  (1) national and international materials requirements, priorities, and objectives, both current and future, including economic projections;
  (2) the relationship of materials policy to (A) national and international population size and (B) the enhancement of environmental quality;
  (3) recommended means for the extraction, development, and use of materials which are susceptible to recycling, reuse, or self-destruction, in order to enhance environmental quality and conserve materials;
  (4) means of exploiting existing scientific knowledge in the supply, use, recovery, and disposal of materials and encouraging further research and education in this field;
  (5) means to enhance coordination and cooperation among Federal departments and agencies in materials usage so that such usage might best serve the national materials policy;
  (6) the feasibility and desirability of establishing computer inventories of national and international materials requirements, supplies, and alternatives; and
  (7) which Federal agency or agencies shall be assigned continuing responsibility for the implementation of the national materials policy.
(b) In order to carry out the purposes of this title, the Commission is authorized—

October 26, 1970   -9-   Pub. Law 91-512

84 STAT. 1235

(1) to request the cooperation and assistance of such other Federal departments and agencies as may be appropriate;   Agency cooperation.

(2) to appoint and fix the compensation of such staff personnel as may be necessary, without regard to the provisions of title 5, United States Code, governing appointments in the competitve service, and without regard to the provisions of chapter 51 and subchapter III of such title relating to classification and General Schedule pay rates; and   Personnel. 80 Stat. 443, 467.

(3) to obtain the services of experts and consultants, in accordance with the provisions of section 3109 of title 5, United States Code, at rates for individuals not to exceed $100 per diem.   5 USC 5101, 5331. Experts and consultants.

(c) The Commission shall submit to the President and to the Congress a report with respect to its findings and recommendations no later than June 30, 1973, and shall terminate not later than ninety days after submission of such report.   80 Stat. 416. Report to President and Congress. Termination.

(d) Upon request by the Commission, each Federal department and agency is authorized and directed to furnish, to the greatest extent practicable, such information and assistance as the Commission may request.   Agency assistance.

SEC. 205. When used in this title, the term "materials" means natural resources intended to be utilized by industry for the production of goods, with the exclusion of food.   "Materials."

SEC. 206. There is hereby authorized to be appropriated the sum of $2,000,000 to carry out the provisions of this title.   Appropriation.

Approved October 26, 1970.

LEGISLATIVE HISTORY:

HOUSE REPORTS: No. 91-1155 (Comm. on Interstate and Foreign Commerce)
              and No. 91-1579 (Comm. of Conference).
SENATE REPORT No. 91-1034 accompanying S. 2005 (Comm. on Public Works).
CONGRESSIONAL RECORD, Vol. 116 (1970):
     June 23, considered and passed House.
     July 31, Aug. 3, considered and passed Senate, amended in lieu of
         S. 2005.
     Oct. 7, Senate agreed to conference report.
     Oct. 13, House agreed to conference report.

Index

Accessory Operations
  classification of, 30
  for composting, 39
  for incineration, 38-39
  for sanitary landfill operations, 31-32, 37-38
Adsorptive capacity. *See* landfill site ranking system
Aerobic Decomposition, in sanitary landfills, 203
Aerobic Degradation of Wastes, in landfills, advantages of, 112
Agricultural Wastes, amounts (generation), 5
  categories of, 5. *See also* crop residues; animal wastes
  classification of, 52
  collection of, 52
  disposal of, 53
  recycling of, 6, 53
Air Pollutant Control Methods, for incinerators, 169-174
Air Pollution, from incineration 159-166
Air Quality Laws, influence on solid wastes, 265-266
Air Quality Standards, as applied to incinerators, 169
Algae
  degradation of solid wastes, 104
  in the soil, 101, 104
Altoona, Pennsylvania, composting plant, 42-43, 119, 125
Aluminum
  problems in recycling, 69
  recyling from solid wastes, 65
  Reynolds recycling system, 67-68
Aluminum Association, integrated recycling plant, 91-92
Anaerobic Decomposition, in sanitary landfills 203
Animal wastes, 5
Area Method
  advantages, 197-198
  limitations, 198
  of sanitary landfill, 194
  vs. trench method, 197
Atomic Energy Commission, 244

Automobile Shredders, 64
Auxiliary Fuel, for incinerators, 140-141, 155-156

Bacteria
  degradation of solid wastes, 103
  in the soil, 101-103
    aerobic versus anaerobic, 103
    conditions for growth, 102-103
    most common species, 103
    populations/life spans, 102-103
Baling Operations
  for sanitary landfill, 32
  Los Angeles County, 32
Basic data
  availability of, 4
  for sanitary landfill, 180
  for selection of collection equipment, 4
  necessity for, 3
  on amounts of solid wastes, 3
  on characteristics of solid wastes, 3
  preliminary uses for, 3
Batch feeding, of incinerators, 150
Battelle Memorial Institute, 243, 246
  study of recycling 94
Bearing Capacity, of completed landfills, 217-218
Biochemical Oxygen Demand, as a persistence index, 248-249
Black Clawson Company
  Franklin, Ohio, plant, 84
  Franklin plant (separation), 79
  Hydrapulping, 97
  Hydrasposal pulper, 38
  Hydrasposal System, 82-84
  materials recovery system, commodities recycled, 84
Bottom Leakage Potential. *See* landfill site ranking system
Buffering Capacity Factor. *See* Landfill site ranking system
Bulky Wastes, incineration of 133, 174-175

Carbon Cycle, in solid waste degradation 106-107

INDEX

Carbon Dioxide
  "fixing" by microbes, 108
  production by microbes, 107
Cell Construction. *See* Sanitary Landfilling
Chilton County Alabama. *See* Collection Systems
Chlorinated Polyethylene (CPE), 72
Clean Air Act (1970), influence on solid wastes, 265-266
Collection, frequency of, 10
Collection Systems
  agricultural wastes, 52
  Chilton County, Alabama, 48
  collection agencies, 44
  compactor trucks, 47
  design of, 44-46
  detachable containers, 48
  equipment innovations, 47-48
  equipment, selection of, 46-47
  innovative methods, 47-48
  miscellaneous transport systems, 50
  municipal system, advantages and disadvantages, 45
  private contract, 45
  private haulers, advantages and disadvantages, 45
  "refuse trains", 48
  separate vs. don't separate, 10
  transfer stations, 48-50
Combustion, in incinerators, 144
Combustion Power Company, 229
Commercial Wastes, collection of, 10
Comminution. *See* Size Reduction
Compaction Systems, 32-37
Compactor trucks
  advantages of, 47
  capacity, 31, 47
  European designs, 31
Components of solid wastes, 4
Compost, definition of product, 116
Composting
  addition of sewage solids, 117, 119, 127-128
  advantages, 130-131
  "backyard" methods, 24
  basic operations, 117
  capital costs, 125
  comminution apparatus, 118
  decomposition time, 116
  definition of process, 116
  digestion methods, 118
  disadvantages, 129-130

Fairfield system, 119-120, 125
  final curing, 125
  financing, 129
  granulation, 125
  historical development, 116
  IDC system, 119, 121-122, 125
  labor requirements, 125
  market development, 126
  mechanical systems, 119
    comparison of systems, 124
    detention time, 119, 122-123
  Metro Waste system, 122
  microbial control parameters, 113
  microbiology of, 113
  moisture requirements, 113, 117, 119, 127
  nutrient requirements, 113, 117, 127
  of animal wastes, 53
  operating and labor costs, 125
  oxygen enrichment, 123-124
  oxygen requirements, 114
  plugged-flow digester, 121
  potential for process, 126-127
  process costs, 126
  product upgrading, 125
  reduction in detention time, 123-124
  refuse preparation for, 117-118
  sale of end-product, 129
  size reduction power requirements, 118
  survival of pathogens, 122, 130
  systems in the United States, 117
  temperature requirements, 113, 116-117, 119
  windrow methods, 118-119
Construction/Demolition Wastes, 6
Costs of Solid Waste Management, 15-16
Council of State Governments, Committee on Suggested State Legislation, 270
Council on Environmental Quality (CEQ), 264
Cover Soil, for sanitary landfills, 192
CPU-400, 88
  advantages, 231
  basic configuration, 230
  components, 231
  factors favoring, 229
  fluidized-bed combustor, 231
  initial costs, 232
  process economics, 231-233
  production of electricity, 232
  waste heat recovery, 231
Critical Path Method, in solid waste management, 278
Crop Residues, 5
Cyclones, 170

INDEX

Deamination, production of ammonia, 109
Degradation of Solid Wastes
  acceleration of, 237-242
  by algae, 104
  by bacteria, 103
  by fungi, 104
  by non-rhizospheric organisms, 105
  herbicides and insecticides, 106
  humus as an end-product, 106
  in composting by microbes, 114, 116
  in sanitary landfills, 111-113, 216-217
    by aerobic species, 112
    optimal conditions, 112
    primary organisms, 112
    products of microbial activity, 112
Denitrification, by soil anaerobes, 109
Density of refuse, in sanitary landfills, 195
Department of Transportation Act, influence on solid wastes, 265
Destructive Distillation. *See* Pyrolysis
Disease Transmission Potential
  mode of disease contraction, 247
  pathogen life state, 248
  survival ability, 248
  See Hazardous Wastes
Disease Vectors, in composting operations, 130
Disposal of Solid Wastes
  composting, 115
  sanitary landfill, 178
Domestic Wastes, daily generation, 5
Dumping Area, for incinerators, 142

Economic Comparison, sanitary landfill vs. incineration, 182
Economies of Size, in solid waste management, 276-277
Effluent Control, from incinerators, 158-159
Electrostatic precipitators, 172
Electrostatic precipitators, for incinerators, 171-173
Emission Control, from incinerators, 166-174
Energy Recovery, 87-92
  CPU-400, 88
  Melt-Zit incinerator, 88
  Norfolk, Virginia incinerator, 88-89
  pyrolysis systems, 89-91
  St. Louis experiment, 87-88
    advantages of system, 87-88
    components of system, 87
  waste heat recovery, 88-89
  See Pyrolysis

Environmental Impact Statements (EIS), 264
Environmental Protection Agency (EPA), 264
Environmental Quality Improvement Act (1970), influence on solid wastes, 265
Equipment, for sanitary landfills, 209-217
Exhaust gases
  from incinerators, 152-154, 165-166
  temperature control of, 153-154

Fabric Filters, 172
Fairfield Composting System, 119-120, 125
Federal Aid Highway Act, influence on solid wastes, 265
Federal Tax Laws, influence on solid wastes, 265
Federal Water Pollution Control Act, influence on solid wastes, 266
Fencing, for sanitary landfills, 208
Filtering Capacity. *See* Landfill site ranking system
Firestone Tire and Rubber Company, 235
Fish and Wildlife Act, influence on solid wastes, 266-267
Flail Mills, description of, 42
Flyash, from incinerators, 159-160
Forced Aeration, in sanitary landfills, 238
Fungi
  degradation of solid wastes, 104
  importance to agriculture, 104
  in the soil, 101, 103
Furnaces
  furnace curing and warm-up, 156
  in incinerators, 144-146
  multicell rectangular, 144-146
  "mutual assistance" furnace, 146
  rectangular design, 144
  rotary kiln furnace, 146
  vertical circular furnace, 146

Gainesville, Florida
  composting plant, 122
  noise levels, 267-268
Garbage, definition of, 4
Garbage grinders, 24
Garchey System. *See* Hydraulic transport systems
Garrett Research Corporation. *See* Pyrolysis
Gases
  automatic burn-off of landfill gases, 206
  from sanitary landfills
    control of movement, 204, 206
    hazards to structures on fills, 219-220
    migration, production, 203-204

296　INDEX

Geographic Aspects, of solid waste management, 277
Geologic Aspects, of landfill sites, 190–193
Georgia Institute of Technology, 239
Glasphalt, 75–76
Glass
　glass-composite containers, 77–78
　recyclable (soluble) glass, 76–77
　recycling potential, 73–74
Glass Containers, 74
Grates, for incinerators, 148–151
Gravel Trenches, control of gas from landfills, 204–205
Groundwater Velocity Factor. *See* Landfill site ranking system
Gulf Oil Company, compactor truck, 48

Hammermills, description of, 40–41
Hand sorting, 78
Hazardous Biological Wastes, 243
Hazardous Wastes
　adsorption vs. mobility, 249
　biodegradability, 245, 248–249
　biological persistence, 248
　categories, 242
　definition, 243
　description, 242–243
　detectability/toxicity ratio, 244
　disease transmission potential, 245, 247–248
　disposal of, 223, 242–251
　example rankings, 250
　groundwater toxicity, 245–246
　human toxicity, 244–246
　low flash point wastes, 244
　mobility, 245, 249
　national disposal sites, 269
　ranking parameters, 244–245
　ranking system for, 244–251
　ranking vs. hazard level, 251
　Sax toxicity rating, 245
　solubility vs mobility, 249
　summary, 261
　waste ranking, 250
　waste ranking rationale, 245
High Temperature Incineration
　advantages, 224, 226–227
　description, 224–229
　disadvantages, 226–227
　economics, 227–228
　residue, 228
Heil-Gondard Hammermill, 37–38, 41
Horner-Shifrin. *See* St. Louis Experiment

Humus, contribution to soil systems, description of, 106
Hydrapulper. *See* Black Clawson Company
Hydrasposal Pulper. *See* Black Clawson Company
Hydraulic Transport Systems, 51–52
Hydrogen Chloride, from incinerators, 170
Hydrogen Sulfide, produced by soil microorganisms, 111

Incineration
　advantages, 133–134
　air pollutants from, 159–166
　air pollution control regulations, 137
　air requirements, 152
　apartment house units, 25–26
　auxiliary fuels, 155–156
　average costs, 138
　basic principles, 132
　batch vs. continuous feeding, 150
　capital costs, 137–138
　combustion stages, 144, 151
　definition of, 132
　disadvantages, 133
　effluent control methods, 169–174
　effluents, 159–160
　electrical power requirements, 140
　emissions criteria, 166–169
　energy recovery, 87–89
　exhaust gases, 152–153
　fluctuation in wastes, 137
　influence of heat content, 136
　influence of refuse characteristics, 134
　influence of water content, 136
　instrumentation and control, 158
　need for weight records, 141
　odor control, 172–174
　of hospital wastes, 175
　operating costs, 137–138
　pathological wastes, 175
　physical plant design and location, 140
　primary combustion, 152
　products of, 132
　quantities of refuse, 136
　relative costs, 138
　removal of residue, 151
　residue recycling (USBM), 85–87
　secondary combustion, 152
　special wastes, 174
　temperature control, 153
　time-temperature, turbulence, 151
　waste heat recovery, 156
　wastewater characteristics, 161

INDEX

Incineration (*continued*)
  water requirements, 140
  water vapor plumes, 165–166, 172–173
  within sanitary landfills, 238
Incinerators
  auxiliary fuel, 140–141
  average downtime, 137
  basic design, 134, 138–140
  batch-fed units, 150
  circular vertical furnaces, 146
  combustion chamber design, 155
  components of, 140–142
  dumping or tipping areas, 142
  exhaust system design, 153
  external features of plant, 139–140
  facilities for operating personnel, 139
  for bulky wastes, 174
  furnaces and appurtenances, 144–146
  grate systems, 148–151
  location vs. collection area, 139
  materials handling in, 142–143
  particulate collectors, 160–166
  physical factors of site, 139
  process waters from, 160
  rectangular furnaces, 144
  refractory materials, 154–155
  removal of siftings, 151
  rotary kiln furnaces, 146
  scales for, 141
  site selection criteria, 138–139
  storage pits for, 143
  suspended-wall construction, 155
  transfer cranes for, 143
  utilities for, 140
  zoning regulations on, 137
Industrial Wastes
  amounts of, 5
  characteristics of, 5
  collection of, 10
Infiltration Potential. *See* Landfill site ranking system
Innovations in Disposal
  CPU-400, 223, 229–233
  geo-ecological studies, 223, 242–261
  landfill stabilization, 223, 237–242
  Melt-Zit high temperature incinerator, 223
  pyrolysis techniques, 223, 233–237
  organic wastes, 233–234
  used tires, 234–237
Institutional Wastes, 7
Instrumentation, for process control in incinerators, 158–159

International Disposal Corporation (IDC), composting system, 119, 122, 125, 131

Landfill compactors, 31
Landfill Site Ranking
  adsorptive capacity, 256
  bottom leakage potential, 254–255
  buffering capacity, 257
  cover soil influence, 251
  dispersion factors, 252
  example rankings, 260–261
  filtering capacity, 255–256
  geological/hydrological factors, 252
  groundwater influence, 256–258
  groundwater velocity, 258
  infiltration factors, 251
  infiltration potential, 253
  migration factors, 251–252
  organic content of groundwater, 256–257
  population distribution influence, 260
  population factor, 260
  ranking rational, 252–253
  subsoil/bedrock influence, 254–255
  travel distance, 258–259
  wind direction, 258–260
  wind influence, 258–259
Lantz Converter. *See* Pyrolysis
Laws on Solid Wastes, 267–273
  Resources Recovery Act (1970), 269
  Solid Waste Disposal Act (1965), 268
  state regulations, 269–270
Leachate
  characteristics of, 201–202
  collection and control systems, 200–201
  from sanitary landfills, 189
  movement out of fills, 191–192
  pollution of groundwater, 200–201
  recirculation of, 203, 239–241
  treatment of collected leachate, 242
Legal Aspects of Management, 263–273
Legislation Affecting Solid Wastes
  air quality laws, 265–266
  noise abatement laws, 267
  solid wastes laws, 267–268
  water quality laws, 266–267
Litter, 95–96
Local Laws and Regulations, influence on solid wastes, 270–271

Materials Recovery, pyrolysis, 90–91
Materials Recovery Systems, 82–87
  See CPU-400, Black Clawson Company
  U.S. Bureau of Mines system, 82, 84–87

298   INDEX

Melt-Zit Incinerator, 88
Melt-Zit High Temperature Incinerator
  advantages, 226-227
  coke use, 225
  disadvantages, 226-227
  economics, 227-228
  evaluation, 226-227
  operation, 224-225
  pilot plant, 224-225
  rates of burning, 227
  recycling, 228
  residue, 228
  water use, 229
Metals, recycling ferrous metals, 62-64
Methane
  in sanitary landfills, 203-204
  production by microbes, 107
Metro Waste System
  composting plant, 119, 122-125
Microbes
  in compost, 105, 113-114, 116
  in sanitary landfills, 105, 111-113
Microbial Populations, in the soil, 101
Microbiology of Waste Disposal, 100
Microorganisms
  amino acid breakdowns, 108
  breakdown of cellulose, 106
  controlled degradation (composting), 113
  degradation of nitrogen compounds, 108
  fixation of nitrogen, 109
  fungi and actinomycetes in compost, 113
  influence of soil constituents on, 102
  interaction with soil particles, 102
  in the soil, 101
  mineralization in the soil, 106
  oxidizing sulfur to sulfates, 110
  production of CH_4, 107
  production of CO_2, 106
  reduction of nitrates to ammonia, 109
  role in solid waste disposal, 100
  soil parameters affecting, 105
  transformation of carbon compounds, 106
  transformation of CO_2, 107
  transformation of H_2S, 111
  types active in degradation, 101
Mil-Pac Compactor, 36
Munich Incinerator, 89
Municipal Solid Wastes
  collected density, 135
  composition of, 135
  heat content, 135-136
  quantities generated, 136
  ultimate analysis of, 135-136

Municipal Wastes, chemical and physical characteristics, 6
"Mutual Assistance" Furnace. See Furnaces

National Environmental Policy Act (NEPA), 263-265
1968 National Survey
  adequacy of disposal, 16-17
  amounts of wastes collected, 15
  collection cost, 15
  collection systems, 7, 9-10
  community ownership, 7-8
  conical burners, 18
  disposal cost, 15
  equipment for collection other than compactor trucks, ownership, 13
  equipment for collection, ownership, 12. See also compactor trucks
  incinerators, general cost, 18-20
  incinerators, performance of, 19-20
  labor, collection, job distribution, 11
  labor, employment data, collection, 10
  land disposal, 17
  population sample, 7
  quantities collected, 14
  summary, 22
  transfer stations, 18
Nitrification, 109
Nitrogen Cycle, in solid waste degradation, 108, 110
Nitrogen Oxides
  from incineration, 163, 167
  in incinerators, 153
Noise Abatement Laws, influence on solid wastes, 267
Noise Pollution and Abatement Act (1970), 267
Norfolk, Virginia Incinerator, 88-89

Office of Solid Waste Management Programs (OSWMP), 4, 238, 269
Organic Content Factor. See Landfill Site Ranking System
Overfire Air, for incineration 152
Owens Illinois Company
  glass-composite containers, 77-78
  glass recycling program, 77

Packaging
  composition of, 95
  control of packaging wastes, 96-97
  recycling, 95-97
Paper
  as a replaceable fuel, 62

INDEX

economics of recycling, 59
newsprint recycling, 61
non-recoverable waste, 60
reasons for low recovery rates, 59–60
recycling mills, 60–61
recycling rates, 59
virgin-fiber vs recycled-fiber, 60
virgin yields in managed forests, 59
Particulate Collectors
  types of collectors, 160
  variation in characteristics, 161–162
Particulates
  as a function of underfire air, 162–163
  electrical resistivity of, 162
  from incinerators, 133, 159–160
  properties of, 162
Pathogenic Microorganisms, transmission from landfills, 247
Pepsi Cola Company, recycling experiment, 97
Pipe Vents, control of gas from landfills, 205–206
Plastics
  as an energy source, 72
  compatibilizers for recycling, 72
  conditions for recycling, 71–72
  methods of recycling, 71
  obstacles to recycling, 73
  recycling potential, 71–73
  returnable milk bottles, 72–73
Pneumatic Transport System, 50–52
Pohland, Dr. Fred, 239
Political Pitfalls, in solid waste management, 277–278
Power Generation from Wastes. *See* CPU-400
Primary Combustion, during incineration, 152
Proteolysis, 108
Protozoa
  degradation of solid wastes (balancing bacteria), 105
  in the soil, 101, 104
Pulverator. *See* Rasp Mills
Pyrolysis
  compared to incineration, 89
  description of process, 89–90, 233–234
  economic analysis, 90
  gases produced, 89
  Lantz Converter, 90–91
  of organic wastes, 223, 233–234
  of used tires, 223, 234
  products of combustion, 234–236
  products of process, 89–90
  technical feasibility, 234, 236

  test results, 234
  trends in products, 236

Radioactive Wastes, 243–244
Ralph Stone Company, 238
Ramp Method
  advantages of, 198
  of sanitary landfilling, 198
Rasp Mills
  description of, 41–42
Reclamation of Strip Mines, 130
Reclamation Systems Compactor
  description of, 36
  final density achieved, 37
  operating time and costs, 37
Recycling
  advantages 55
  bans of non-returnable goods, economic effects, 58
  beneficial actions for, 94–95
  compost marketing, 126
  consumer education, 97
  effect of purchasing policies, 93–94
  energy recovery systems, 87–92
    pyrolysis 89–91
    waste heat utilization, 87–89
  factors in reuse of solid wastes, 55–56
  glass in solid wastes, 73–78
    advantages of glass, 73–74
    advantages of glass-composites, 77–78
    current recovery operations, 74
    filler in asphalt (glasphalt), 75–77
    new glass containers, 76–77
    problems with glasphalt, 75–76
    separation techniques, 74–75
  incentives for reuse of packaging, 96
  integrated recovery systems, 91–92
  lack of simple solution, 56
  limitations of, 56
  major problem areas, 92–95
  materials recovery systems, 82–87
    USBM 85–87
  metals in solid wastes, 62–69
    advantages of aluminum recovery, 68–69
    aluminum cans, 65–68
    automobile shredders, 64
    copper precipitation, 64
    criteria for aluminum recycling, 66
    ferrous metal recovery, 62–64
    home separation, 66
    limitations on aluminum reuse, 66–67
    metallurgical problems, 64–65
    problems in aluminum recovery, 69

INDEX

Recycling (*continued*)
    Reynolds aluminum program, 67–68
    tin can recovery, 62–64
    transportation costs, 64
  necessity for markets, 58, 92–94
  need for separation of wastes, 78
  new technology, 82, 97
  packaging wastes, 61–62, 95–97
  paper in solid wastes, 58–62
    decreasing rate, 59
    economic factors, 59
    influence of packaging, 61
    limitations, 58–61
    Madison, Wisconsin experiment, 61
    percentage for reuse, 58–59
    reasons for declining rate, 59
    recycling mill requirements, 61
    regulatory limitations, 60
    saving trees, 58
    socioeconomic problems, 60
    use as fuel, 62
  plastics in solid wastes, 70–73
    obstacles to recycling, 73
    returnable bottles, advantages, 73
    three principal methods, 71
    uses for recycled plastics, 72
  potential in solid wastes, 55–56
  problems of tax equalization, 92–93
  reasons for, 56
  resource recovery systems, 82–97
  returnable vs non-returnable packages, 96
  separation techniques, 78–82
  transportation rates problems, 93
Refractory Walls
  deterioration zones, 155
  in incinerators, 154–155
Refuse, definition of, 4
Refuse Trains. *See* Collection Systems
Reorganization Plan No. 3 (1970), 264
Residue, from incineration, 133, 151, 159
Resources Recovery Act (1970)
  provisions, 269
  site selection for hazardous wastes, 269
Resource Recovery Systems, 82–97
Reynolds Metals Company
  aluminum recovery system, 67–68
  can reclamation, 65–67
  homeowner separation program, 66
Rhizosphere. *See* Section 7–8
Ringelmann Number, emission control by, 167
Rubbish, definition of, 4

St. Louis, energy recovery experiment, 87–88
Salvage. *See* Recycling

Sanitary landfill, 178–181
Sanitary landfilling
  accessory equipment, 209–210, 212–216
  advantages, 179–180
  aerobic decomposition, 203
  basic steps, 179
  cell construction, 195–196
  cell dimensions, 196
  construction on completed fills, 218–220, 237
  contaminants from, 185–186
  control of gas movement, 204–205
  cover soil requirements, 192, 195–196, 218
  definition of, 179
  design factors, 193–194
  design philosophy, 199
  development costs, 181
  difficulties in using sites, 237
  drainage provisions, 200
  economics, 180–183
  equipment costs, 214–216
    capital costs, 214–215
    direct operating costs, 215
    indirect operating costs, 215–216
    useful life, 214
  equipment difficulties, 213–214
  equipment hazards, 211–212
  equipment requirements, 209–213, 217
  fire protection, 208
  forced aeration in fills, 238
  gas production, 203–205
  geological aspects, 190
  inspection of operation, 220
  landfill incineration, 238
  leachate collection, 89
  leachate generation, 89
  leachate treatment, 89
  liners for landfills, 201
  methane production, 203–204
  methods of filling, 194–198
  method vs topography, 196–197
  operating costs, 182
  physical factors in siting, 186–188
  preliminary planning, 179
  prevention of pollution from, 198–207
  problems with recreational use, 219
  rapid decomposition, 199
  recirculating leachate, 238–242
  refuse "cache" technique, 199
  required basic data, 180
  rock/soil permeability, 190–191
  segregating wastes, 220
  site improvements, 207
  site operation plans, 208–209
  site ranking system, 251–261

INDEX

site selection, 183–193
  basic considerations, 184
  criteria, 185
  physical factors, 186
small sites, 213
soil investigations for, 193
stabilization techniques, 223, 237–242
surface water pollution, 199–200
use of completed sites, 216–220
  agriculture, 218–219
  green space, 218
  recreational areas, 219
  structures, 219–220
utilities for, 207
wet weather operations, 208
Scales
  for incinerators, 141
  maintenance of, 142
Secondary Combustion, during incineration, 152
Separation
  at a central location, 79–82
  basis for mechanical devices, 78
  binary sorters, 79
  by multiple-output sensors, 79
  color separation of glass, 74–75
  conflict with collection methods, 79
  for composting operations, 117–118
  in the home, 78–79
  methods for solid wastes, 78–79
  "signature" methods, 79–80
    impact deceleration, 80–82
    infrared spectroscopy, 80–81
Settlement
  in completed landfills, 217–220
  of structures on landfills, 220
Settling Chambers, in incinerators, 170
Sewage Sludge, incineration of, 175
Sewage Solids
  addition to compost, 117, 119, 127–128
  characteristics of, 127
Shredding, of bulky wastes, 133
Siftings, in incinerators, 151
Site Selection, for sanitary landfills, 183–193
Size reduction
  during composting, 39–43, 118
  effects on disease vectors, 38
  equipment, 30, 40–43
  for sanitary landfill, 37–38
    advantages, 37–38
    increase in density, 38
    wet pulping, 38
  Madison, Wisconsin Experiments, 37–38
  prior to incineration, 39
  advantages, 39
  bulky waste reduction, 39
Soil
  "crumb" structure from fungi, 104
  influence on microorganisms, 105
  physical phases vs microbial populations, 102
  structure produced by algae, 104
Soil Requirements, for sanitary landfills, 190–193
Soil Studies, for sanitary landfills, 193
Solid Waste
  amounts generated, 15
  as supplementary fuel, 87–88
  classification of, 4
  collected density of, 5
  management of, 1
Solid Waste Disposal Act of 1965, 1, 268–269
Solid Waste Disposal, incineration, 132
Solid Waste Management
  educational programs, 280
  legal aspects, 263
  legal case history, 271–272
  politics vs. management, 277–278
  public relations activities, 279–280
  regional authority, 278
  regional planning
    economies of size, 276–277
    geographic interactions, 277
    political pitfalls, 277–278
    using survey results, 277–278
  regional survey
    categories of information, 275
    conduct of survey, 276
    need for survey, 275
    1968 Solid Waste Survey, 276
    public relations value, 276
  state-of-the-art, 275
  systems engineering, 278
  technology vs. management, 274–275
Solid Waste, the third pollution, 1
Sonoma County, California, recirculated leachate study, 239
Sortex Separators, 74–75
Spalling and Sagging, refractory material, 155
Special compactors, for sanitary landfilling, 209–211
State Regulatory Agencies, influence on solid wastes, 269–270
Stationary compactors, 27–28
Storage Pits, for incinerators, 143
Structures, on completed landfills, 219–220
Sulfur Cycle, in degradation of solid wastes, 111
Sulfur Oxides, from incineration, 163, 167

INDEX

Sundyberg, Sweden Collection. *See* Pneumatic Transport Systems
Supplementary Fuel. *See* Energy Recovery

Temperature Control
  in incinerators, 153
  methods of, 154
Tezuka Kosan Compressor
  capital costs, 35
  components of system, 34
  density achieved, 34
  installation and operating costs, 35-36
  operating time, 34
  stability of bales, 35
Theoretical Oxygen Demand, as a persistence index, 248-249
Tollemache Hammermill, description of, 41
"Total" Recycling Systems, 91-92
Toxic Chemical Wastes, 243
Transfer Stations
  advantages, 48-49
  criteria for location and design, 49-50
  description of, 48-49
  economic analysis, 50-51
  economic data, 50
  in Los Angeles County, 50
  in Orange County, 50
  justification of, 49
  reasons for, 32
Travel Distance Factor. *See* Landfill Site Ranking System
Trench Method
  of sanitary landfill, 194-196
    advantages, 197
    limitations, 196

Underfire Air, for incineration, 152
United States Bureau of Mines (USBM)
  "back-end" recycling system, 82, 84, 87
    recovery rates, 87
  Coal Research Center, 235
Used Tires, 234-237

Viruses, in the soil, 101, 105
Visual Effluent Efficiency, 168
Volume Reduction
  at the site
    apartment house incinerators, 25-26
    home compaction units, 25
  at the source
    economic reasons for, 23
    health effects of, 24
    home composting, 24
    home garbage grinders, 24
    incentives for, 23-24
    methods of, 24
    stationary compactors, 27
    wet pulverization, 26-27
  by incineration, 133
  centralized, 28
    advantages, 29
    purpose and means, 29
    prior to incineration, 38-39

Waste biodegradability Rating. *See* Hazardous Wastes
Waste Groundwater Toxicity Rating. *See* Hazardous Wastes
Waste Heat Recovery
  advantages, 156, 158
  dissipation of excess heat, 157
  expenses, 157-158
  generation of steam, 156
  in incinerators, 156-157
  United States Practice, 157
Waste Human Toxicity Rating. *See* Hazardous Wastes
Waste Mobility Rating. *See* Hazardous Wastes
Wastepaper Mills, 60
Water Pollution
  groundwater, sanitary landfills, 200-201
  impermeable barriers for landfills, 200-201
  observation wells for landfills, 203
  surface water, sanitary landfills, 199-200
Water Quality Laws, influence on solid wastes, 266-267
Water Requirements, for incinerators, 140, 160
Water Resources Planning Act (1965), influence on solid wastes, 266-267
Water-Tube Walls, in incinerators, 154
Water Vapor Plumes
  detrimental effects of, 166
  from incinerators, 165-166
Wet Pulping, 26-27
Wet Scrubbers
  advantages, 171
  costs, 171
  disadvantages, 171
  efficiency, 171
  water requirements, 171
Wetted Baffle Sprays, costs efficiency, 170
Wilson, Dr. David, (separation technology), 80
Wind Direction Factor. *See* Landfill Site Ranking System